Nuclear Structure

NATO ADVANCED STUDY INSTITUTES SERIES

A series of edited volumes comprising multifaceted studies of contemporary scientific issues by some of the best scientific minds in the world, assembled in cooperation with NATO Scientific Affairs Division.

Series B. Physics

Recent Volumes in this Series

This series is published by an international board of publishers in conjunction with NATO Scientific Affairs Division

A Life Sciences	Plenum Publishing Corporation
B Physics	London and New York
C Mathematical and Physical Sciences	D. Reidel Publishing Company Dordrecht, Boston, and London
D Behavioral and Social Sciences	Sijthoff & Noordhoff International Publishers
E Applied Sciences	Alphen aan den Rijn, The Netherlands, and Germantown, U.S.A.

Nuclear Structure

Edited by
K. Abrahams
Netherlands Energy Research Foundation
Petten, The Netherlands

K. Allaart
Free University
Amsterdam, The Netherlands

and
A. E. L. Dieperink
Nuclear Accelerator Institute
Groningen, The Netherlands

PLENUM PRESS • NEW YORK AND LONDON
Published in cooperation with NATO Scientific Affairs Division

Library of Congress Cataloging in Publication Data

Netherlands Physical Society International Summer School on Nuclear Structure
  (1980: Dronten, The Netherlands)
  Nuclear structure.

  (NATO advanced study institutes series. Series B, Physics; v. 67)
  "Proceedings of the Netherlands Physical Society 1980 International Summer
School on Nuclear Structure, held August 12-23, 1980, in Dronten, The Netherlands"
— T.p. verso.
  Includes bibliographical references and index.
  Contents: Application of the shell model to nuclear spectroscopy/J. B. McGrory —
Effective interactions/J. P. Elliott — An introduction to the interacting Boson model/
F. Iachello — [etc.]

  1. Nuclear structure — Congresses. I. Abrahams, K. II. Allaart, K. III. Dieperink,
A. E. L. IV. Nederlandse Natuurkundige Vereniging. V. Title. VI. Series.
QC793.3.S8N47 1980   539.7′4   81-7291
                                                                          AACR2

ISBN 978-1-4684-3952-6   ISBN 978-1-4684-3950-2 (eBook)
DOI 10.1007/978-1-4684-3950-2

Proceedings of the Netherlands Physical Society 1980
International Summer School on Nuclear Structure, held
August 12—23, 1980, in Dronten, The Netherlands

© 1981 Plenum Press, New York
Softcover reprint of the hardcover 1st edition 1981

A Division of Plenum Publishing Corporation
233 Spring Street, New York, N.Y. 10013

FOREWORD

After the success of the previous summer schools organized by
the Nuclear Physics Division of the Netherlands' Physical Society
in 1975 and 1977, we thought it worthwhile to continue this
tradition. The immediate very positive reactions received from all
invited speakers encouraged us to proceed with the orgaization.
Although the number of students had to be restricted to about one
hundred, the international character of the School was evident from
about thirty nationalities which were represented.

The material contained in this book covers the talks given by
all speakers invited to lecture on the subject of nuclear structure
research. These proceedings should therefore serve as an excellent
introduction to many topics of current interest in this exciting
field. We hope that the lectures and discussions as well as the
many informal contacts made during the various social activities
will greatly stimulate interest in nuclear structure investigations
among all the participants.

The organization of the summer school has been made possible
by substantial support given by the Scientific Affairs Division
of the North Atlantic Treaty Organization, the Netherlands' Ministry
of Education and Science and the Netherlands' Physical Society.

The invaluable help of the "Bureau Congressen" of the Ministry
of Education and Science and the friendly assistance of the manage-
ment of the College of Agriculture in Dronten contributed greatly
to the pleasant atmosphere during the summer school.

P. W. M. Glaudemans

PREFACE

This book presents a review of nuclear structure research as presented during the international summer school, Dronten, the Netherlands, 12-23 August 1980. It includes not only intoductions to the shell model, to effective interactions, and to the interacting boson model, but also a detailed treatment of the interactions of nuclei with electrons, muons, protons, pions and heavy ions. In addition applications are discussed in the field of exotic nuclei, fission, astrophysics and proton induced x-rays.

To speed the publication, the corrected and sometimes revised manuscripts are published without consulting all authors and is the responsibility of the editors.

The assistance of the secretaries of the Physics Department of the Netherlands Energy Research Foundation is gratefully acknowledged.

K. Abrahams
K. Allaart
A.E.L. Dieperink

CONTENTS

APPLICATION OF THE SHELL MODEL TO NUCLEAR SPECTROSCOPY

J. B. McGrory

Oak Ridge National Laboratory*

Oak Ridge, Tennessee 37830

INTRODUCTION

The intent in these lectures is to present a cursory overview of existing nuclear shell model technology, to give an indication of how well nuclear phenomena can be described by the nuclear shell model, and what are possible future directions of the field. The nuclear shell model has been with us for a long time, and it is a widely used model. It is also very popular in many circles to deride the model for its lack of elegance and sophistication. Thus, I believe it is appropriate to spend a minute at the beginning of these lectures to recall some of the reasons for our interest in the nucleus. The combination of features that make the nucleus unique are: 1) the microscopic dimensions of the nucleus are such that one must use quantum mechanics; 2) at the level we study, it is a many-body system with $1 \sim 300$ particles; and 3) of the four "fundamental" interactions (weak, strong, electromagnetic, and gravitational), only the gravitational force is generally ignorable. Beta-decay, particle transfer reactions, and electromagnetic transitions are direct manifestations of the first three of these forces, respectively. The nucleus is the only known quantum-mechanical many-body system where one can conceive of treating the constituent coordinates microscopically. There now exist volumes of nuclear data on a vast array of nuclear observables. Taken as a body, these data can be interpreted to suggest that the nucleus

*Research sponsored by the Division of Basic Energy Sciences, U. S. Department of Energy, under contract W-7405-eng-26 with the Union Carbide Corporation.

sometimes behaves collectively, i.e. it rotates or vibrates, and
sometimes exhibits properties of a system of independent particles.
A longstanding goal of nuclear physics is to formulate one model
which can give a consistent picture of all nuclear observables, and
which is derivable from more-or-less well-known properties of the
constituent protons and neutrons and of their mutual interactions.
The starting point of virtually all such models is the nuclear shell
model. The model is conceptually very simple, but very tedious to
implement. The model became generally accepted roughly 30 years
ago, but only in the last decade or so has shell model technology
evolved to the point that truly systematic studies of nuclear
structure could be undertaken. Perhaps fortuitously at the same
time that shell model technology evolved to a really useful level,
the theory of effective interactions made considerable progress.
One can, with some degree of confidence, construct the input to
shell model calculations from well-known properties of interactions
between free nucleons. These two developments mean that one can
really investigate the question how do observed properties of nuclei
develop from the known properties of nucleon-nucleon interactions.

I will not include in this written version of the lectures the
material on shell model methods, on existing shell model codes, and
on the extensive calculations of nuclear properties in the sd shell.
These topics are already adequately covered in existing publications.[1,2] The points from those sections to note particularly here
are that one can treat the entire sd shell with an $(sd)^{A-16}$ model
with no truncation, that a large body of nuclear data on these nuclei
can be correlated with the nuclear shell model, and that shell model
dimensions escalate so rapidly with increasing numbers of shell model
orbits that there is scant hope that any other major shell can be
treated as completely as the sd shell. Thus, in these printed notes
I will discuss some truncation methods which might be useful to extend the domain of applicability of the shell model.

APPROXIMATIONS TO LARGE-SPACE SHELL MODEL CALCULATIONS

There is now a clear case of the usefulness of the shell model
where it is applicable. There are also relatively few areas of the
periodic chart where the model can be applied in any completeness.
The question then is: are there systematic ways to truncate the
shell model spaces to viable sizes? I would now like to discuss some
of the possible ways to truncate the shell model. This is a curious
field in which many have nibbled, but few have moved much past the
appetizer. There are many unpublished studies I have heard of but
relatively few detailed studies published.

A first approach would be to truncate the shell model matrix as
calculated in the complete space. A first method is to calculate
all diagonal elements and choose to use only the states whose

diagonal energies are lowest algebraically, using perturbation theory
arguments that high levels won't mix with low levels. We tried some
cases in the sd shell for five and six particles, using an unrenor-
malized interaction. The lowest eigenvalues converged rather slowly, and usually on the order of 50% of the space must be included to
obtain reasonable results. Some examples of this are shown in
Brussaard and Glaudemans.[14] A modification of this is to go a step
beyond a simple diagonal truncation scheme. Consider a matrix of a
given Hamiltonian in a given space where the basis space has been .
reordered according to diagonal energies. A simple diagonal trunca-
tion scheme involves treating the submatrix $|a_{ij}|$ i < N, j < N, where
N is the number of eigenstates included in the truncated space. An
extension of this could involve including in the matrix the lowest
N complete rows (columns) plus all diagonal matrix elements. If N
is significantly less than the dimension in the complete space, then
there are a lot of 0 matrix elements. The time to calculate the
matrix is thus reduced, and one might use sparse matrix techniques
to obtain eigenvalues and eigenvectors rapidly. In this form, one
is including the effects of the omitted space to first order. I've
seen no results using such an approach, although Pieter Glaudemans
tells me they have had some encouraging results with this approach
in the Ni region.

Another general approach, on which I'd like to spend some time,
involves ideas which might be called cluster coupling. A general
description is to divide the shell model space into two smaller subspaces or clusters, each of which can be handled more or less exactly. One then forms a truncated "complete" space by including only
low-lying eigenstates in these model spaces. I'll make the ideas
more explicit below. These ideas are discussed in some detail by
Arima and Hamamoto.[4] As an example, consider ^{20}Ne, a system of two
neutrons and two protons. This can be considered in one way as one
n-p system coupled to a second n-p system, i.e. ^{18}F coupled to ^{18}F.
One could diagonalize ^{18}F, and then couple the low-lying states of
the resulting spectrum together, i.e.

$$|^{20}\text{Ne}\rangle = |^{18}\text{F},J_i,T_i\rangle \times |^{18}\text{F},J_j,T_j\rangle. \tag{1}$$

Physically, this is an appealing representation. The n-p interaction
is believed to be responsible for deformation, or rotational behavior. We already know that ^{20}Ne is a "rotational" nucleus, as indeed are many light sd shell nuclei. There are, however, problems
with such a basis. It is overcomplete, non-orthogonal, and non-
normalized. These are not insurmountable problems. One can form
an orthonormal basis. In a complete space, it is much more work
than using the conventional shell model codes from the start. There
may be some virtue if one can select only a few low-lying states of
^{18}F and construct a four-particle basis from these. Wong and Zuker
have reported[5] an investigation along these lines. They proceeded

as follows: Select a few low-lying eigenstates from ^{18}F, for which
the creation operator is $z_i^\Gamma(^{18}F)$, where

$$z_i^\Gamma(^{18}F) = \sum_{\alpha,\beta} C_{\alpha,\beta}^{\Gamma,i} \frac{(A_\alpha \times A_\beta)^\Gamma}{\sqrt{1 + \delta_{\alpha\beta}}} \tag{2}$$

Construct a basis from coupling two of these states

$$z_{ij}^{\Gamma_1\Gamma_2\Gamma} = \left[z_i^{\Gamma_1} \times z_j^{\Gamma_2} \right]^\Gamma \tag{3}$$

Now expand $z_{ij}^{\Gamma_1\Gamma_2\Gamma}$ in terms of four-particle states in the original
conventional shell model basis, i.e.,

$$z_{ij}^{\Gamma_1\Gamma_2\Gamma} = \sum_\alpha C_{ij\alpha}^{\Gamma_1\Gamma_2\Gamma} \, z^\Gamma \left(\begin{matrix} n_1 & n_2 & n_3 \\ \rho_1 & \rho_2 & \rho_3 \end{matrix} ; \alpha \right). \tag{4}$$

The expansion coefficients $C_{i,j\alpha}^{\Gamma_1\Gamma_2\Gamma}$ are found in terms of matrix elements of the form

$$< z_j^{\Gamma_2} \| z_i^{\Gamma_1} \| z \left(\begin{matrix} n_1 & n_2 & n_3 \\ \rho_1 & \rho_2 & \rho_3 \end{matrix} ; \alpha \right) >. \tag{5}$$

These matrix elements are evaluated easily in terms of the general
expression (3.10) of Ref. 1. One then diagonalizes the given effective interaction in this truncated space, taking appropriate care
of normalization and orthogonality problems. Such an approach has
been taken to calculate ^{20}Ne, ^{22}Na, ^{24}Mg in a variety of coupling
schemes; i.e., ^{24}Mg = ^{20}Ne × ^{20}Ne or ^{18}F × ^{22}Na. The results suggest that one can reproduce full space calculations in small spaces
chosen in this way, and there seems to be considerable freedom as to
how the clusters are chosen. This is shown in Table 1 for the ground
state of ^{24}Mg, a J=0, T=0 state. In the full (sd)^8 space, there are
325 basis states with J = 0, T = 0. Table 1 shows the principal
components of the ground state for two truncation schemes, one where
^{24}Mg = ^{20}Ne × ^{20}Ne, and one where ^{24}Mg = ^{18}F × ^{22}Na. In both cases,
there are one or two clusters that dominate the state. Reasonable
spectra for a number of sd shell nuclei are obtained in this fashion.
Thus, a useful truncation scheme seems suggested by these results.
However, the problems with non-orthogonality, overcompleteness, etc.
are not trivial, and with increasing dimension size, these problems
limit the usefulness of the scheme. As a measure of the problem with

Table 1. Large components in ground state wave functions in ^{24}Mg.
J_i, T_i, x_i labels the x_i^{th} eigenstate J_i, T_i. α is the
amplitude of the given term.

^{24}Mg $= ^{20}$Ne $\times ^{20}$Ne			^{24}Mg $= ^{18}$F $\times ^{22}$Na		
$J_1 T_1 x_1$	$J_2 T_2 x_2$	α	$J_1 T_1 x_1 (^{18}F)$	$J_2 T_2 x_2 (^{22}Na)$	α
0 0 1	0 0 1	−0.82	0 1 1	0 1 1	−0.82
2 0 1	2 0 1	0.08	1 0 1	1 0 1	0.29
4 0 1	4 0 1	−0.48	1 0 1	1 0 2	0.08
6 0 1	6 0 1	0.09	1 0 2	1 0 1	−0.07
			2 1 1	2 1 1	0.22
			2 1 2	2 1 1	0.11
			3 0 1	3 0 1	0.11
			3 0 2	3 0 1	−0.03

non-orthogonality, the overlap of the state (^{20}Ne J=0$_1^+$, T=0) × (^{20}Ne
J=0$_1^+$, T=0) with the state (^{20}Ne J=4$_1^+$, T=0) × (^{20}Ne J=4$_1^+$, T=0) is 0.77.
These calculations do suggest that one could capitalize on a cluster
structure, but there is a lack of uniqueness in the prescription.

      The problem in the cluster description suggested here is that
one is coupling states involving equivalent particles. If one di-
vides the shell model space into two subspaces with no single-particle
orbit in common, life is much less complicated. Suppose $Z^{\Gamma_1}(n_1)$ and
$Z^{\Gamma_2}(n_2)$ are state creation operators for states in spaces 1 and 2,
where these spaces are inequivalent (no orbit is common). It is easy
to show that the state operator formed by coupling these two operators together in a direct product, i.e.

$$Z^{\Gamma_1 \Gamma_2 \Gamma}(n_1 + n_2) = \left[Z^{\Gamma_1}(n_1) \times Z^{\Gamma_2}(n_2)\right]^{\Gamma} \tag{6}$$

will create a normalized and anti-symmetrized state. Further, if one
includes the complete set of states in the two inequivalent spaces,
then the direct product space will be a complete space. I would like
now to discuss some investigations of truncation schemes involving
inequivalent spaces.

      First, let me briefly discuss the structure of the Hamiltonian.
In Ref. 1, it is shown that in a second-quantized, tensor operator
representation, the Hamiltonian can be written

$$H = H_1 + H_{12} = \sum_{r<s} \langle \rho_r | H_1 | \rho_s \rangle \; A \underset{0}{\overset{\rho_r \quad \rho_s}{\triangle}} B$$

$$- \sum_{\substack{\Delta, r\leq s; t\leq \mu; \\ r\leq t; s\leq \mu (\text{if } r=t)}} \hat{\Delta} \; \langle W^{\Delta}_{rstu} \rangle \quad A \overset{\rho_s \quad \rho_t}{\underset{0}{\pentagon}} B \quad \rho_r \; A \; \Delta \; \Delta \; B \; \rho_u \tag{7}$$

Here $A^{\rho 1}_{\mu 1} = a^{\dagger}_{\rho,\mu}$, the usual creation operator, and $B^{\rho}_{\mu} = (-)^{\rho+\mu} a_{\rho-\mu}$.
W^{Δ}_{rstu} is a matrix element of the residual interaction between normal-
ized and anti-symmetrized states. The triangles are symbols for
angular momentum coupling, as discussed in Ref. 1. $\Delta = \sqrt{2\Delta+1}$.
r,s,t, and u run over all orbits in the model space. In this form
of the two-body term, the operator "scatters" two particles from the
orbits t and u into the orbits r and s, acting first with the de-
struction operators and then with two creation operators. It is a
straightforward operation to restructure the two-body term so that
the operator is written as the product of two "particle-hole" creation operators, taking appropriate care of angular momentum coupling
and anti-commutation properties of the operators.

$$H_{12} = \sum_{\substack{r\leq s \\ \Delta}} [\Delta] \frac{\xi^2_{rs}}{\hat{r}} W^{\Delta}_{rsrs} \quad A \underset{0}{\overset{r \quad s}{\triangle}} B$$

$$+ \sum_{\substack{rstu \\ \Delta\Omega}} (-)^{1+s+t-\Delta-\Omega} \xi_{rs}\xi_{tu} \hat{\Delta} W^{\Delta}_{rstu} U(rstu;\Delta\Omega) \quad B \overset{t \quad s}{\underset{0}{\pentagon}} A \quad A^r \; \Omega \; \Omega \; B^u$$

$$= \sum_{r<s} G_1(r,s) \quad A \underset{0}{\overset{r \quad s}{\triangle}} B \; \delta_{\Gamma_r \Gamma_s}$$

$$+ \sum_{\substack{rstu \\ \Omega}} G_2(rstu;\Omega) \quad B \overset{t \quad s}{\underset{0}{\pentagon}} A \quad A^r \; \Omega \; \Omega \; B^u \tag{8}$$

In (8) $[\Delta] = (2\Delta+1)$, and $U(rstu;\Delta\Omega)$ is a normalized Racah coefficient, $G_1(r,s)$ and $G_2(rstu;\Omega)$ are simply symbols for the sums over Δ. In developing these expressions, no assumption is made as to the equivalence or inequivalence of orbits. Note that in writing these expressions, we have assumed an ordering of orbits. Suppose we divide the set of orbits up into two sets, where one set consists of the first n_1 orbits and the second set consists of the last n_2 orbits, so $n_1 + n_2$ is the total number of orbits. Then it is possible to break the Hamiltonian up into three pieces

$$H = H(n_1) + H(n_2) + H(n_1,n_2) \tag{9}$$

where $H(n_1)$ and $H(n_2)$ involve only orbits in the separate spaces.

If we ignore $H(n_1,n_2)$ for the moment, we can consider $H(n_1)$ and $H(n_2)$ as our "cluster generators". Assume we can diagonalize each term separately, i.e.

$$H(n_1)|\psi_{\alpha_1}^{\Gamma_1}(n_1)> = E_{\alpha_1}|\psi_{\alpha_1}^{\Gamma_1}(n_1)>$$

$$H(n_2)|\psi_{\alpha_2}^{\Gamma_2}(n_2)> = E_{\alpha_2}|\psi_{\alpha_2}^{\Gamma_2}(n_2)>. \tag{10}$$

Then from the complete sets of states $|\psi_{\alpha_1}^{\Gamma_1}(n_1)>$ and $|\psi_{\alpha_2}^{\Gamma_2}(n_2)>$, we can generate a complete set of states for the total space $(n_1 + n_2)$, i.e.

$$|\psi_{\alpha_1\alpha_2}^{\Gamma_1\Gamma_2\Gamma}(n_1 + n_2)> = \left[|\psi_\alpha^{\Gamma_1}(n_1)> \times |\psi_\alpha^{\Gamma_2}(n_2)> \right]^\Gamma. \tag{11}$$

If then one diagonalizes $H(n_1,n_2)$ in this direct product space, exact solutions of the eigenvalue problem for H in the model space $(n_1 + n_2)$ are obtained. In such an approach, the mechanical steps are first to obtain eigenvalues and eigenvectors of $H(n_1)$ and $H(n_2)$ using conventional codes such as the Oak Ridge-Rochester or Glasgow codes. Then one must construct and diagonalize the matrix

$$<\psi_{\alpha_1\alpha_2}^{\Gamma_1\Gamma_2\Gamma} |H(n_1;n_2)|\psi_{\alpha_1'\alpha_2'}^{\Gamma_1'\Gamma_2'\Gamma} >. \tag{12}$$

If we transformed $H(n_1,_2n)$ as we transformed the full H in (9), there would be one difference. The orbits and t are necessarily inequivalent, so the operator A^s and B^t "anti-commute," and the one-body term in b) is not present. Then (12) becomes

$$\sum_{\Omega} \sum_{r \le t} \sum_{s \le u} G_2(rstu;\Omega) \frac{(-)^{\Gamma_1+\Gamma_2'-\Gamma-\Omega}}{\hat{\Omega}} W(\Gamma_1\Gamma_2\Gamma_1'\Gamma_2';\Gamma\Omega)$$

$$\langle \psi_{\alpha_1}^{\Gamma_1}(n_1) \| (A^r \times B^t)^{\Omega} \| \psi_{\alpha_1'}^{\Gamma_1'}(n_1) \rangle \langle \psi_{\alpha_2}^{\Gamma_2}(n_2) \| (A^s \times B^u)^{\Omega} \| \psi_{\alpha_2'}^{\Gamma_2'}(n_2) \rangle . \tag{13}$$

Given the set of states $\psi_{\alpha_1}^{\Gamma_1}(n_1)$ and $\psi_{\alpha_2}^{\Gamma_2}(n_2)$, the major part of the work to calculate the coupling matrix element $H(n_1,n_2)$ is in the calculation of the reduced matrix elements of operators of the form $(A^s \times B^t)^{\Omega}$ between the eigenstates. Each matrix element involves only one of the two subspaces. They play the role here of the well-known single shell matrix elements of the Oak Ridge-Rochester shell model program. The eigenstates $\psi_{\alpha_1}^{\Gamma_1}(n_1)$ are expanded in terms of basis states in the subspace $|\Gamma_1, n_1, x_1\rangle$

$$|\psi_{\alpha_1}^{\Gamma_1}(n_1)\rangle = \sum C_{\alpha_1}^{\Gamma_1} |\Gamma_1, n_1, x_1\rangle . \tag{14}$$

The basis states $|\Gamma_1, n_1, x_1\rangle$ in the space n_1 are, of course, independent of the interaction $H(n_1,n_2)$, so one can calculate the matrix elements of the density operator, $(A^s \times B^t)^{\Omega}$ and store them as "mixed configuration" single shell matrix elements. Given a set of eigenvectors of interest, it is a trivial matrix transformation to transform to the physical space. Given the physical density matrices, the calculation of the interaction matrix is straightforward. Since in the cluster procedure one has reduced the problem in the final stage to essentially a two-shell problem, there is a particular economy that the sums over r,s,t,u in (13) must only be done once, i.e. (13) can be written as

$$\sum_{\Omega} \frac{(-)^{\Gamma+\Omega}}{\hat{\Omega}} W(\Gamma_1\Gamma_2\Gamma_1'\Gamma_2';\Gamma\Omega) \, F(\Gamma_1\Gamma_2\Gamma_1'\Gamma_2';\Omega) . \tag{15}$$

For the matrix for a given total angular momentum Γ, one merely re-sums the $F(\quad)$ coefficients with different Racah coefficients.

This has all been simply formalism for carrying out shell model calculations in a direct product space of two inequivalent spaces. The major obvious virtue of doing "cluster" calculations in inequivalent spaces is that the bases are orthonormal and complete, and the calculations are straightforward, if tedious. The obvious question is "Having given all this formalism, how is it useful?" As a simple example, consider again the $(sd)^4$ model of ^{20}Ne. This is a system of two neutrons and two protons. Previously, I discussed treating ^{20}Ne as coupling ^{18}F to ^{18}F, and the difficulties of over-completeness were pointed out. Now instead, consider ^{20}Ne as the coupling of two protons and two neutrons. Then, in obvious notation, the Hamiltonian can be decomposed as $H = H_{pp} + H_{nn} + H_{pn}$. As a first step, diagonalize H_{pp} and H_{nn} in the space of states $(sd)^2$, T=1. There are 14 such states (0^3, 1^2, 2^5, 3^2, 4^2). Next, form the set of states $[\psi(^{18}O,J_i) \times \psi(^{18}Ne,J_j)]$.^{J} (Obviously, in our model, the states ($^{18}O,J_i$) and ($^{18}Ne,J_i$) are identical.) There are 196 states if we ignore the total angular momentum. Some dimensions for fixed J in the coupled space are 0^{46}, 1^{97}, 2^{143}, 3^{124}, 4^{110}. Note that in directly coupling neutron and proton states, we have not formed states of good isotopic spin. Thus, the dimensions of the $(sd)^4$, T=0 space are 0^{21}, 1^{31}, 2^{56}, 3^{45}, 4^{44}. By ignoring isospin in the construction of the basis in this particular case, we have matrices which are roughly three times as large as in the T=0 space. Obviously, the method is not a good one for doing complete calculations. The approach is useful only if one can pick out a subset of ^{18}O and ^{18}Ne eigenstates which generate a direct product space which has a large overlap with the low-lying eigenstates of ^{20}Ne. The most obvious first guess is to pick the lowest eigenstates to generate the basis. If H_{p-n} is not too strong, then the mixing of high-lying states will be suppressed by the energy denominator. ^{20}Ne is an interesting first laboratory in which to study the approach since the calculations in the complete $(sd)^4$ space can be done economically in both the conventional approach and in the weak-coupling approach. As an experiment, I calculated the shell model structure of ^{20}Ne in a series of calculations which differed in the number of states selected from the two-neutron (and two-proton) spaces. The criteria selected essentially was to start with the lowest three states, and then increase the number of states until all two-particle states are included. Some results are shown in Fig. 1. The fifth (rightmost) column shows the spectrum of the lowest J=0^+, 2^+, and 4^+ states of the two-particle system (^{18}O, ^{18}Ne), using the Chung-Wildenthal interaction.[6] The fourth column shows the results of extreme weak coupling, i.e. the n-p interaction is ignored, and the spectrum results from simple coupling in ^{18}O states to an ^{18}Ne state to a good J, and adding the excitation energies of the two states. The third column is the spectrum obtained by diagonalizing H_{np} in the direct product space formed by the lowest three states of the n=2, T=1 space, i.e. the lowest J=0^+, 2^+, and 4^+ states. The second column shows the results for the space generated from six states (0^2, 2^2, 4^2), while the first column shows the results of an exact diagonalization. The number identified as B.E. below each spectrum is roughly the contribution to the binding energy of the system due to the two-body interactions (i.e., the single-particle energy is

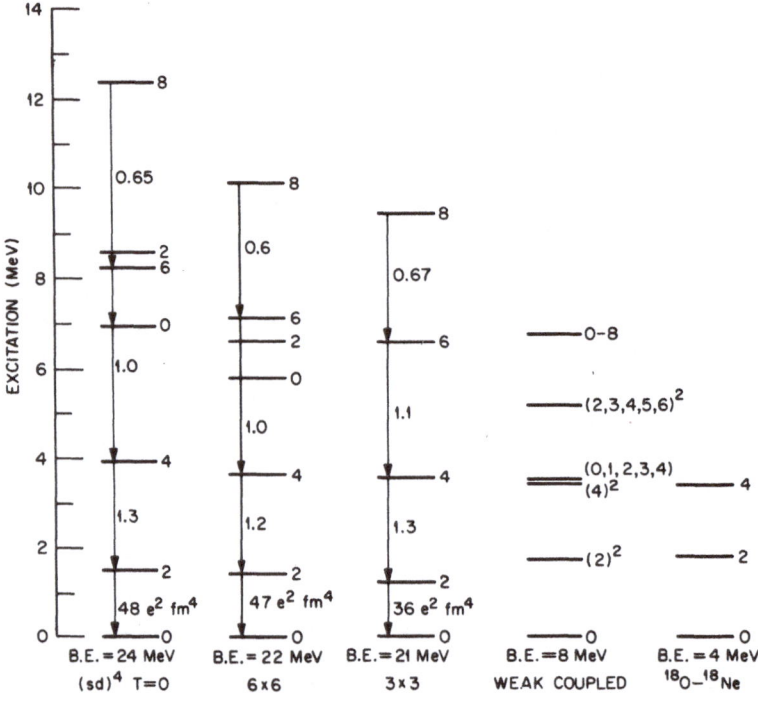

Fig. 1.   Spectrum of ^{20}Ne ground band and J=0_2^+, 2_2^+ states in various
          models. See text for details of figure.

subtracted out approximately by subtracting 4x the $d_{5/2}$ binding
energy (-4.15 MeV) in ^{17}O). Shown in the spectra are the ground
state band and the first excited J=0^+ and 2^+ states. Also shown in
the figure are the absolute B(E2)-values in e^2fm^4 for the ^{20}Ne
$2_1^+ \rightarrow 0^+$ transition with an effective charge 0.5e, and the relative
B(E2)-values for transitions between excited states in the ground
state band. ^{20}Ne is a typical sd shell rotor. The ground state band
is rotational up to the 6^+ state. The 8^+ state is depressed and the
E2 rate is significantly suppressed relative to the rigid rotor value.
The lowest excited states outside this ground state band are the
second J=0^+ and J=2^+ states. The pure weak-coupling model (column 4)
is clearly wrong. There are many extra low-lying states. The effect
of n-p correlations is to pick out one J=0^+, 2^+, 4^+, 6^+, and 8^+
states and depress them in energy with respect to the other states.
Most of these correlations already exist in the (3 × 3) model. The
n-p interaction accounts for ~ 16 MeV of the 24 MeV two-body binding.
13 MeV of the 16 MeV are introduced by correlations among the lowest
three states. Absent in the 3 × 3 spectrum are the first excited

$J=0^+$ and $J=2^+$ states. These are present in the (6×6) space. The (6×6) space essentially spans the entire "important" space for the low-lying spectrum. From the figure, it is seen that the E2 observables are also well described in these small spaces. In the 3×3 space, the E2 matrix element for the 2-0 transition (i.e. $\sqrt{B(E2)}$) is 87% of the exact value. The results certainly suggest that the (3×3) space would provide an excellent starting point for a perturbation theory calculation.

The wave functions of the ground state band states of ^{20}Ne in this weak-coupling picture are listed in Table 2. Only the wave functions in the 6×6 space and the complete spaces are given. The 3×3 wave functions are insignificantly different from the 6×6 wave functions. In this table, all amplitudes greater than 0.2 are listed. As the energy levels and B(E2) values suggest, the structure of the band is essentially given in the 3×3 space. Note also the strong mixing of the $J=2^+$ state in this rotational band. Although the state $(2_1^+ \times 2_1^+)0$ is 4 MeV above the $(0_1^+ \times 0_1^+)0$ state in the absence of n-p correlations, the n-p interaction admixes the two states almost equally into the ground state. It is known that these ^{20}Ne states are very similar to the states of the SU(3) representation $|(\lambda\mu)K,J> = |(8,0)0,J>$. Thus, these wave functions are very close to what SU(3) wave functions would look like in this cluster representation.

I have made similar studies of ^{24}Mg and ^{28}Si. A number of calculated results for energy levels and E2 observables for selected low-lying states in these nuclei in the complete $(sd)^8$ and $(sd)^{12}$ spaces are available in the literature.[3] For these nuclei, the full-space calculations are prohibitively expensive in the weak-coupling basis. Therefore, I have made the calculations in two spaces using 6 and 14 states respectively from the ^{20}O(^{20}Mg) and ^{22}O(^{22}Si) spectra. The (6×6) space includes the $0_{1,2}^+$, $2_{1,2}^+$, $4_{1,2}^+$. The 14×14 space includes the lowest fourteen states in the relevant spectrum. In Fig. 2, the results for ^{24}Mg in the $(sd)^8$, (6×6), and (14×14) models are shown. The distinctive qualitative features of the ^{24}Mg spectra are: 1) a ground state rotational band, 2) an excited "K=2" rotational band where the 2^+ band head is roughly degenerate with the $J=4^+$ member of the ground state, and c) a band crossing at the $J=8^+$ state so that the 8^+ member of the ground state band is no longer the YRAST state. In Fig. 2, the ground state band of the complete diagonalization is essentially reproduced in the (6×6) space. The (6×6) space calculation also reproduces the band crossing. The bulk of the ground state n-p correlation energy is again in the 6×6 space. The excited K=2 band is not reproduced in the 6×6 space. Only with the inclusion of the lowest fourteen states is the agreement indicated in Fig. 2 achieved. The structure of the band is reproduced in the 14×14 calculation. The excitation energy of the entire band is off by roughly 2 MeV. The structure of the E2 observables in the ground

Table 2. ^{20}Ne Wave Functions. $(J_i \times J_j')$ labels coupling of ith
eigenstate of ^{18}O with given J to jth eigenstate of
^{18}Ne with J'. All amplitudes greater than 0.2 are
shown. States $J_i \times J_j'$ and $J_j' \times J_i$ appear symmetri-
cally, so coefficient is only shown once, but $\alpha_{J_1 J_2}$,
the amplitude, is labeled with an asterisk to remind
that the $\alpha_{J_1 J_2}$ appears twice. 6 × 6 indicates calculation where lowest two J=0$^+$, 2$^+$, and 4$^+$ states are included in space. <u>Full</u> labels the complete calculation.

J	$J_1 \times J_2$		6 × 6 $\alpha_{J_1 J_2}$	Full $\alpha_{J_1 J_2}$
0	0_1	0_1	0.67	0.63
	2_1	2_1	0.59	0.55
	4_1	4_1	0.24	0.23
2	0_1	2_1	0.51*	0.49*
	2_1	2_1	0.36*	0.34*
	2_1	4_1	−0.26*	−0.25*
4	0_1	4_1	0.40*	0.40*
	0_1	4_2	0.21*	0.22*
	2_1	2_1	−0.50*	−0.47*
	2_1	4_1	0.30*	0.29*
6	2_1	4_1	−0.62	−0.62
	2_1	4_2	−0.21	−0.20
	4_1	4_1	0.24	0.22
8	4_1	4_1	0.97	0.97

state band of ^{24}Mg are well reproduced by the (6 × 6) calculation.
This includes the "band-crossing" of the J=8$^+$ states. The relative
B(E2)-values for the K=2 band are not very well reproduced in the
calculations. There <u>is</u> a high degree of collectivity in the band,
and this is reproduced in the truncated calculation. The wave functions for the ground state band of ^{24}Mg in the weak-coupling model
are shown in Table 3. We see again that the structure is dominated
by very few "clusters". The structure of the states is little

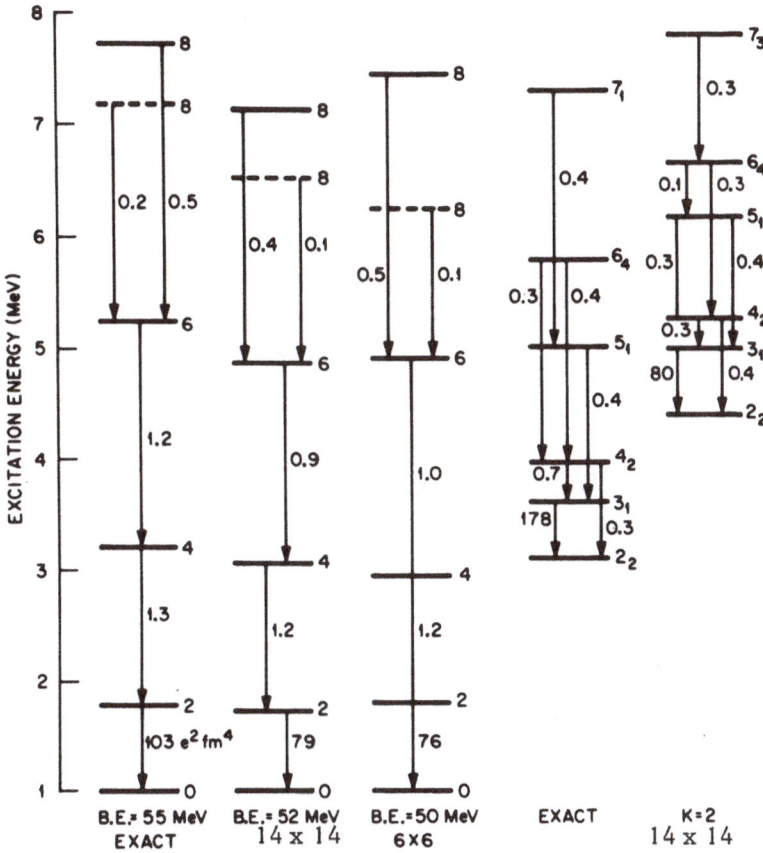

Fig. 2.   Spectrum of ground state band and "K=2" band in ^{24}Mg in
           various models. See text for details of figure.

altered if the space is expanded to the lowest fourteen states.
Again, the low-lying states of the ground state rotational band are
dominated by states with $J = 2^+$. As with ^{20}Ne, these results all
summarize to say that this weak-coupling approach, at a minimum, provides an excellent first-order approximation.

    The wave functions for the K=2 band are shown in Table 4. There
is little collectivity in the (6 × 6) calculation for this excited
band. There is considerable collectivity in the excited band of the
(14 × 14) space. The wave functions do not have as simple a
structure here as in the ground state band. I have not yet extended
the calculations to larger spaces.

Table 3. Wave functions of ^{24}Mg ground state band states. See caption to Table 2 for explanation. Here ^{20}O states are coupled to ^{20}Mg states.

J	$J_1 \times J_2$	6 × 6 $\alpha_{J_1 J_2}$	14 × 14 $\alpha_{J_1 J_2}$
0_1	0_1 0_1	0.70	0.63
	2_1 2_1	0.65	0.68
2_1	0_1 2_1	-0.56^*	-0.54^*
	2_1 2_1	0.44	0.44
4_1	0_1 4_1	-0.29^*	-0.23^*
	0_1 4_2	-0.26^*	-0.24^*
	2_1 2_1	-0.71_*	-0.70
	2_1 4_1	0.20^*	
6_1	2_1 4_1	-0.56^*	-0.53^*
	2_1 4_2	-0.38^*	-0.36^*
8_2^+	4_1 4_1	-0.30_*	
	4_1 4_2	-0.63^*	-0.57^*
	4_2 4_2	-0.31	-0.28_*
	4_1 5_1		0.22^*

The final sd shell nucleus I have studied in this cluster approximation is ^{28}Si. The spectra of ^{28}Si as calculated in various models are shown in Fig. 3. The only results I have for the full $(sd)^{12}$ calculation are the energy levels for states with $J \leq 4$, T = 0, and a few B(E2) values. There is no obvious qualitative description of the spectrum, i.e. it is neither rotational nor vibrational. I was somewhat surprised to see how well the truncated calculations did. In the full $(sd)^{12}$, T=0 spectrum, the dimensions of the $J=0^+$, 2^+, and 4^+ states are 839, 3276, and 3793. Thus, the dimensions in the (6 × 6) space, ~ 36, and in the (14 × 14) space, ~ 200) are significantly smaller. Since the total binding energy of the ground state in the $(sd)^{12}$, T=0 space is 138 MeV, the deviation of 1-2 MeV in the excitation energy is relatively small.

Of the lowest nine states in the exact calculation, there are plausible analogs for eight of them in the (14 × 14) calculation. The wave functions of the ^{28}Si states are summarized in Table 5. The simple structure persists. The states I list here are connected by strong B(E2)-values, so in a sense they are members of the ground

Table 4. Wave functions of "K=2" band in ^{24}Mg. See captions to
          Tables 2 and 3 for explanation of headings.

J	$J_1 \times J_2$		6×6	14×14
			$\alpha_{J_1 J_2}$	$\alpha_{J_1 J_2}$
2_2	0_1	2_2	-0.57^*	0.27^*
	0_1	2_3		-0.33^*
	2_1	2_2		-0.28^*
	2_1	3_1		0.24
3_1	2_1	2_1	0.68	
	2_1	2_2	0.22	0.29^*
	2_1	4_1	-0.36^*	
	4_1	4_1	0.23	
	0_1	3_1		0.33^*
	2_1	2_3		0.35^*
4_2	0_1	4_1	0.46^*	0.20
	0_1	4_2	-0.29^*	
	2_1	4_1	0.34^*	
	2_1	4_2	-0.21^*	
	0_1	4_3		0.25^*
	2_1	2_1		-0.27^*
	2_1	2_3		-0.27^*
	2_1	3_1		-0.20^*
	2_2	4_2		-0.22
5_1	2_1	4_1	0.58^*	-0.48
	2_1	4_2	-0.30^*	
	4_1	4_2	-0.25^*	
	0_1	5_1		0.21
	2_1	4_2		0.24

state band. These states shift somewhat between the 6 × 6 and the
(14 × 14) spaces. In particular, there is evidence for "band
crossing" at the 6$^+$ state. The ground state band J = 6$^+$ and 8$^+$ are
the third J=6$^+$ and 8$^+$ states in the calculation. I do not have the
results of the exact complete space calculations for these states.
It would be interesting to look for these states experimentally and
"theoretically".

These represent all the calculations I have made for the sd
shell nuclei. Some study of the odd-even nuclei is in progress. I
consider the results rather encouraging. In formulating this

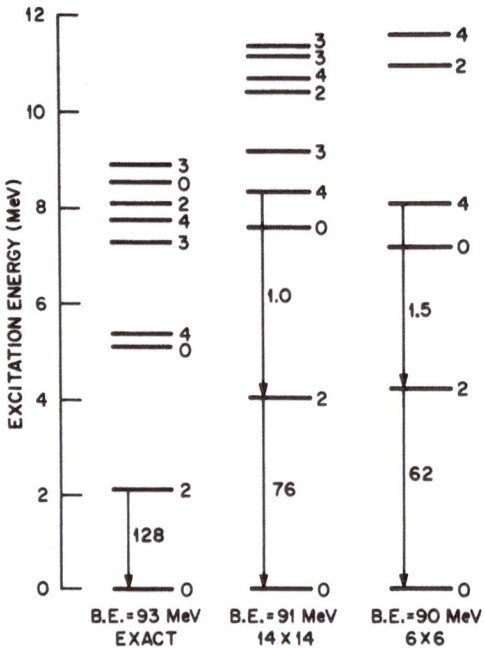

Fig. 3.   Spectrum of low-lying states in ^{28}Si in various models.
          See text for details of figure.

cluster approach in this way, one is treating the proton-proton and
neutron-neutron correlations exactly, and then treating the proton-
neutron interaction as a perturbation. This is in contradiction to
the fact that the p-n interaction is stronger than the p-p and n-n
interactions in the sd shell, as evidenced by the spectrum of ^{18}F,
where the three T=0 states lie well below the first T=1 state. One
can get some insight into what is going on from a consideration of
the surface delta interaction (SDI).[7] We know that the SDI can be
parameterized to be a fairly "effective" interaction in the sd shell.
If we assume that the $H_{n,p}$ is the SDI, the matrix elements of $H_{n,p}$
in our neutron-proton direct product space; i.e., Eq. (13) takes the
following simple form

Table 5. Wave functions of ^{28}Si as the coupling of ^{22}O states to
          ^{22}Si states. See the caption to Table 2 for explanation
          of the table headings.

J	$J_1 \times J_2$		6×6 $\alpha_{J_1J_2}$	14×14 $\alpha_{J_1J_2}$
0	0_1	0_1	0.86	0.83
	2_1	2_1	0.41	0.41
	4_2	4_2	0.25	0.24
2_1^+	0_1	2_1	0.61^*	0.60^*
	2_1	2_1	0.39	0.34
4_1^+	0_1	4_2	0.46^*	0.54^*
	2_1	2_1	0.66	0.51
6_1^+ $(6_3)^+$	2_1	4_1	0.45^*	0.31^*
	2_1	4_2	0.51^*	-0.58^*
8_3	3_1	5_1		-0.38^*
	4_1	4_1		0.22^*
	4_1	4_2		-0.41^*
	4_2	4_2		0.53

$$\langle \psi^{J_1}(p) \times \psi^{J_2}(n)J \,|\, SDI \,|\, \psi^{J_1'}(p) \times \psi^{J_2'}(n)J \rangle$$

$$= \sum_K (-)^{J_1+J_2'-J}\, W(J_1 J_2 J_1' J_2'; J) \tag{16}$$

$$\times\ \langle \psi^{J_1}(p)\|Y^K\|\psi^{J_1'}(p)\rangle \langle \psi^{J_2}(n)\|Y^K\|\psi^{J_2'}(n)\rangle.$$

Here Y^K is the usual spherical harmonic. The matrix element is then
the sum of products of matrix elements of spherical harmonics in the
separate spaces. Thus, suppose the set of states in the proton (or
neutron) space could be divided into two sets of states $\psi^\alpha(p)$ and
$\psi^\beta(p)$ with the following properties

$$\langle \psi^\alpha(p)\|Y^K\|\psi^\beta(p)\rangle = 0 \qquad \text{for all K.} \tag{17}$$

Then one could carry out an exact diagonalization of the SDI in two
separate steps, where either $\psi^\alpha(p)$ or $\psi^\beta(p)$ are included in the

direct product space. This suggests that, to the extent that the real residual interaction is like a surface delta interaction, if there is a set of identical particle states which are essentially closed among themselves with respect to matrix elements of the Y^K's, then this set of states would form a good basis for a set of truncated states. I have looked at this possibility for the space of neutron states outside ^{56}Ni. This is the $(f_{5/2}, p_{3/2}, p_{1/2})^n$ space which, dimensionally, is identical to the sd shell model of the oxygen isotopes. I have calculated the matrix elements of Y^2 between most low-lying states in ^{58}Ni, ^{60}Ni, and ^{62}Ni. The results for ^{60}Ni are shown in Table 6. Ideally, one would like such a table as is shown there to be diagonal, i.e., one given state is connected only to itself. If this were true for all k, there would be no n-p interaction. As is seen from inspection, Table 6, only those states which lie close to a given state are strongly connected by the Y^2 operator. This suggests a truncation scheme based simply on energies in the correlated n-n and p-p spaces should be pretty good. This is our finding in the sd shell. The implication of this is that no matter how strong the n-p interaction is, if there is a concentration of collectivity of the multipole operators in the separate spaces, a weak-coupling approach is reasonable.

One final note about this cluster scheme. There is no a priori guarantee that the coupled neutron-proton states have good isotopic spin. If we include the complete space, and use an isospin invariant interaction, diagonalization will produce states with good spin. However, as a trivial example, if I directly couple one proton to one neutron state, the resulting state is, the principle, a linear combination of $T_{min} = \dfrac{|n_1 - n_2|}{2} \rightarrow T_{max} = \dfrac{n_1 + n_2}{2}$ where n_1 and n_2 are the number of neutrons and protons in the system. Suppose the coupled space includes at least $(T_{max} - T_{min})$ states. Each such state can be written as

$$|J_1^p(n_1) \times J_2^n(n_2)J\rangle = \sum_{i=T_{min}}^{T_{max}} \alpha_i |J, T_i\rangle. \tag{18}$$

If for each i, $\alpha_i \neq 0$ for at least one state in the coupled space, then the diagonalization of an isospin invariant H_{int} will produce states of good T. In the calculations I've done, calculated properties appear to vary smoothly with increasing dimension. This could indicate that the isospin problem is really minor.

J_f ↓ \ J_i →	0	2	4	2	0	2	3	4	4	6	2	0	3	2	1	1	5	3	4
0	100																		
2		16	84	101	14														
4			0	38															
2		9	55	67	34	23													
0				13	116	118													
2		2	22	107	30	65	13	21	17										
3			12	54	3		17	25											
4			37	30	14	8	3	6	93										
4				34	31	4	1	68											
6						63	104	47	68	118									
2		9	22	100	25	35	33	63	8			17	19	41	27	2	14	12	45
0				12	21		38	41			31								
3			30	54	8	1	22	70	85		80	23	18	78	15	68			
2				14	4	40	11	5			20	19	4	45					5
1				31	63	50	40	8	68			14	45						
1						47	104	4	40				7						
5								85					27				8		
3								68	40								16		
4																			

Table 6.   Matrix elements $\langle J_f \| r^2 Y^2 \| J_i \rangle$ in ^{60}Ni between lowest energy levels normalized to 100 for $2_1 \to 0_1$ transitions

APPLICATION OF NEUTRON-PROTON CLUSTERING IN THE Fe and Pb REGIONS

It is possible for this neutron-proton cluster model to be useful in regions where the n-p interaction is strong, but obviously
the approach would be more useful in a region where the n-p interaction is not strong. This is more likely to be the case where there
is a neutron excess such that the valence neutrons and protons are
in different single-particle orbits. The consequent loss of overlap
of the single-particle wave functions would then diminish the
strength of the interaction. The general approach I've outlined has
been applied in several regions where this is the case, in particular
in studies of the Fe isotopes and in the lead region. I will spend
the remainder of the talk discussing these calculations. The calculations in the iron isotopes have the virtue that modestly large
shell model calculations account rather well for the observed data,
so that it is perhaps the heaviest region where the cluster coupling
schemes can be compared to more complete models. In the Fe isotopes,
there are interesting examples of "band crossing" which are accounted
for by the calculations,[8] so the structure of these wave functions
is of interest. The lead region is a well-known weak-coupling region
and is a region where one might expect these cluster coupling approaches to be particularly useful. It is a region where some applications of the Interacting Boson Model have been made, and a comparison of the two approaches there is useful.

The Fe isotopes were among the first nuclei studied with the
Oak Ridge-Rochester shell model codes. The original interest arose
because there were data available on a relatively large number of
isotopes so that systematic studies were possible, and it is for
such studies that the shell model is most useful. In recent years,
the availability of good quality heavy-ion beams has made it possible to study higher spin states, so there has been renewed interest.
The shell model most often used[8] is a modest one. A ^{48}Ca core is
assumed. The valence protons are restricted to the $f_{7/2}$ shell. The
active neutron orbits are the $1p_{3/2}$, $0f_{5/2}$, and $1p_{1/2}$ orbits. I
will concentrate here on the three isotopes ^{56}Fe, ^{58}Fe, and ^{60}Fe
which involve, respectively, 2, 4, and 6 active neutrons. All
Pauli-allowed configurations of these neutrons distributed in the
three orbits are included in the model space. The largest matrices
are on the order of 400 × 400. The residual Hamiltonian is taken
completely from the literature with no parametric adjustment. The
points of interest on which I want to focus here are: 1) the existence of ground state rotational bands in ^{56,58,60}Fe up to J = 8$^+$,
2) the existence of a "band-crossing effect" in ^{56,58}Fe which is
suggested by experiment, and 3) a prolate-oblate shape transition
underlying this "band crossing" or backbending. These features are
in the complete shell model and, as I'll now demonstrate, they are
accurately reproduced in the cluster approach with wave functions
dominated by relatively simple structures.

Given the exact shell model calculations, I have repeated the Fe calculations in the n-p cluster approach; i.e., I coupled the ^{54}Fe states to low-lying eigenstates of ^{58}Ni, ^{60}Ni, and ^{62}Ni. In the case of ^{56}Fe, the calculations can be done in the complete space of all ^{58}Ni eigenstates. For ^{58}Fe and ^{60}Fe, I gradually increased the number of Ni eigenstates until good agreement with the exact results were obtained. In Fig. 4 the results for ^{56}Fe are shown. In this figure I have shown the eigenspectrum of ^{54}Fe and the calculated eigenenergies of ^{58}Ni for which the associated eigenstates are the basis space building blocks. The first column in the figure represents the exact diagonalization. The second column are the results when the lowest J=0, 2, and 4 states of ^{58}Ni are coupled to the states of ^{54}Fe. The only states shown are the ground state band and the two "band-crossing" J=6$^+$ levels. The numbers on the figures are $B(E2)_{J_i \rightarrow J_f}$-values in units of e^2fm^4. The quadrupole moment of the J=2$^+$ state in efm is also shown. I would emphasize that the exact values are in good agreement with experiment. The structure of the shell model calculation is quite accurately mirrored in the truncated calculation. The collective B(E2) strength is contained almost completely in this space, and the "band crossing" is accurately reproduced.

Results of similar calculations for ^{58}Fe are shown in Fig. 5. Here three cluster calculations are shown. The first calculation shows the complete shell model results, the third column shows the results of coupling in the lowest J=0$^+$, 2$^+$, and 4$^+$ states of ^{60}Ni, and the middle column shows the results where 11 states (0^2, 2^2, 4^2, 6^2, 1^1, 3^1, 5^1) are included. Beyond this space, little significant happens as more states are added. The complete spectrum is characterized by sets of "doublets" for J = 4, 6, 8, and 10. There is a band-crossing here, but at the J=8$^+$ instead of at the J=6$^+$ state. There is no clearcut collective J=10$^+$ state. On close inspection, much of this structure occurs already in the spectrum where only the first J=0$^+$, 2$^+$, and 4$^+$ states of ^{60}Ni are coupled in, but the 4$^+$, 6$^+$, and 8$^+$ doublets are inverted.

The final example here is for ^{60}Fe in Fig. 6. In this case, the simplest coupling of the lowest J = 0, 2, 4 is not adequate. The result with 23 states is in quite good agreement with the exact calculations. There is not too much data on the decay properties of the higher spin states in ^{60}Fe. There is again, in the theoretical calculations, a doubling of the higher spin states, and the E2 strength is split over all members of the doublet so that there is not a clearcut band crossing, but considerable band mixing.

The wave functions for YRAST levels with J $\leq$ 8 generated in the "best" cluster calculations in Figs. 4-6 are tabulated in Table 7.

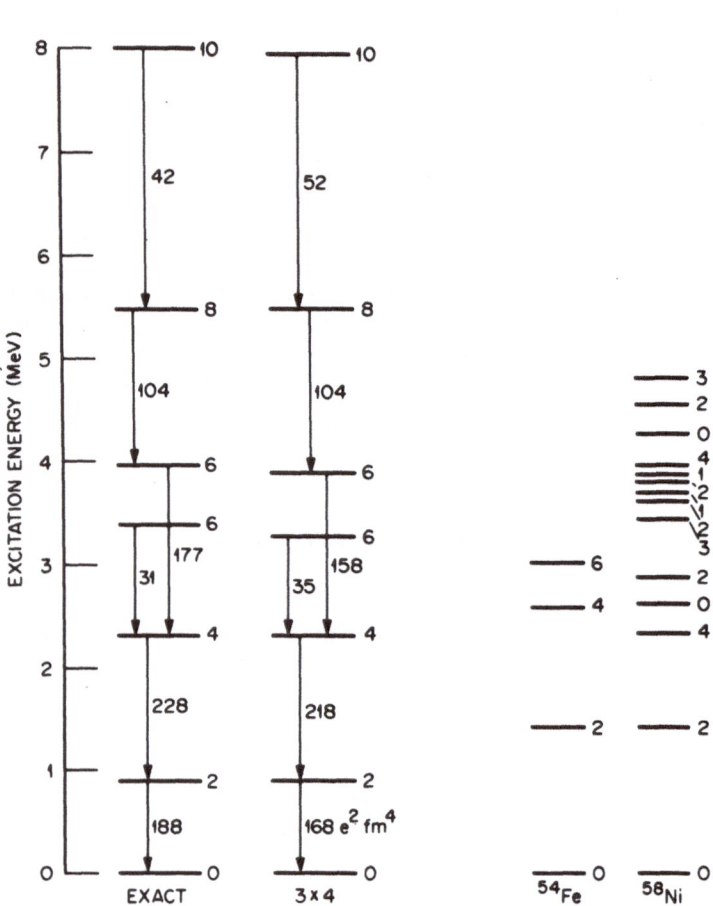

Fig. 4. Weak cluster coupling spectrum of ^{56}Fe. See text for details of the figure.

    Several general comments can be made about the wave functions. First, the wave functions in general are relatively simple in the cluster representation. 80-90% of the wave functions are in 2-6 basic components. For the $J=0^+$ and $J=2^+$ states, the wave functions have a very similar structure in ^{56}Fe, ^{58}Fe, and ^{60}Fe; i.e., there is an intrinsic state whose character changes little from one mass to to the next. The structure of the $J=4^+$ states varies more significantly from one mass to the next. In ^{56}Fe the lowest 4^+ state of

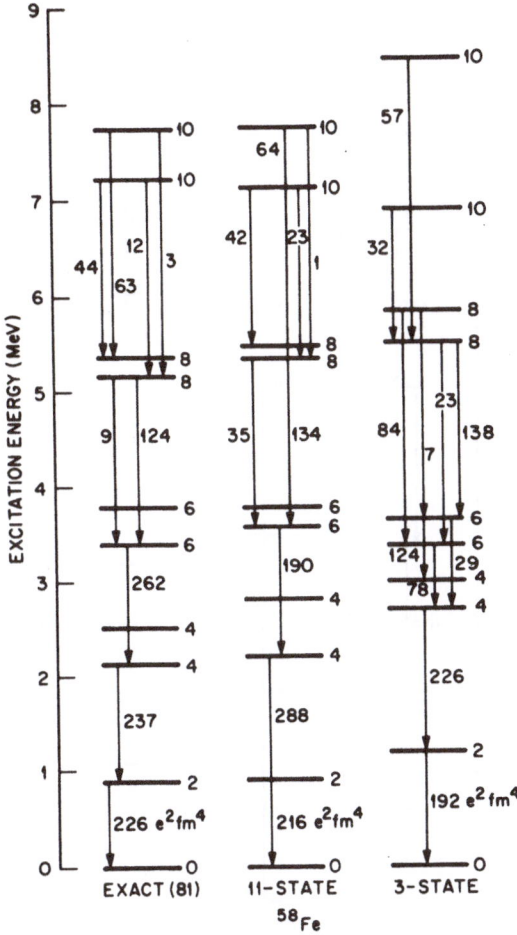

Fig. 5. Weak cluster coupling spectrum of ^{58}Fe. See text for details of the figure.

^{54}Fe and ^{58}Ni are important elements. For the "large" components
shown in the table, 44% involve the lowest J=4$^+$ states. In ^{58}Fe
this number is 19%, and in ^{60}Fe this number is 13%. These lowest 4$^+$
states are g-bosons in the IBM language. (See the lectures in this
school by Iachello.) On the other hand, the 4_2^+, 4_3^+, and 2_2^+ states
increase in importance. These states have significant overlap with
states formed by coupling two-particle states with J = 2$^+$. Thus,
as there are more particles and increased collectivity, states formed
from J=2$^+$ pairs do seem to increase in importance, consistent with
the assumptions of the IBM.

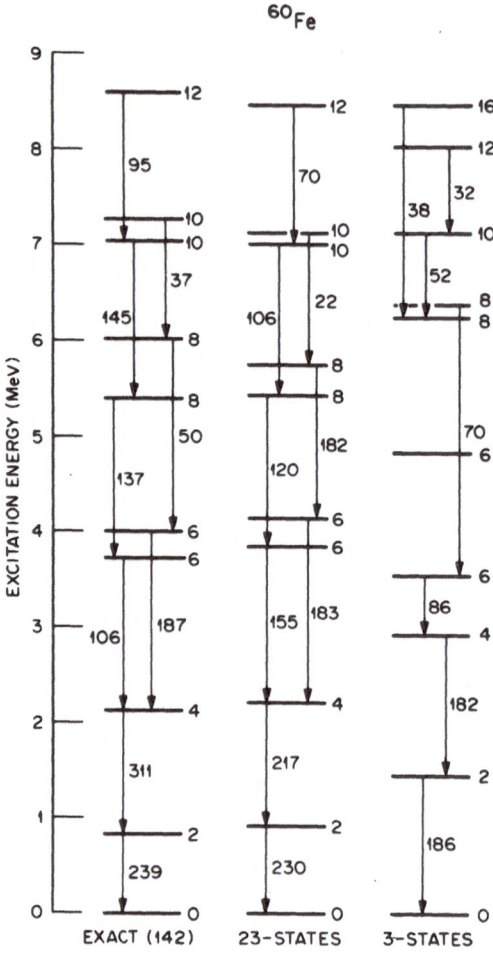

Fig. 6. Weak cluster coupling spectrum of ^{60}Fe. See text for de-
         tails of the figure.

The J=6$^+$ states in ^{56}Fe are the "band-crossers" in the current
jargon. As we see from the wave functions, this band-crossing has
a relatively simple explanation in this cluster picture. The largest
components in the YRAST J = 4$^+$ (2$_1^+$ × 2$^+$) and J = 6$^+$ (0$_1^+$ × 6$^+$) are
not connected by the E2 operator. The other J=6$^+$ states are more
collective combinations of J=2$^+$ and 4$^+$ neutron states. The 0$_1$ × 6
component does not dominate the ^{58}Fe YRAST J=6$^+$ state, and there is
no resultant inhibition of the transition. There is not as clear a
picture of the source of the inhibition for the 8$^+$ state in ^{58}Fe.

Table 7.　Cluster wave functions for ^{56}Fe, ^{58}Fe, and ^{60}Fe. All amplitudes greater than 0.2 in the wave functions are listed. For ^{56}Fe, J = 6$^+$ and ^{58}Fe J = 8$^+$, the first and second states of that spin are shown.

J_i(Ni) × J_j(Fe)	^{56}Fe	^{58}Fe	^{60}Fe
J = 0			
0_1 × 0	0.83	0.81	0.75
2_1 × 2	0.52	0.51	0.50
4_1 × 4		0.20	0.22
J = 2			
0_1 × 2	0.60	0.60	0.51
2_1 × 0	0.64	0.57	0.50
2_2 × 0			0.25
2_1 × 2	0.39	0.33	0.27
2_2 × 2			0.22
4_2 × 2			0.23
J = 4			
0_1 × 4	0.43	0.38	0.36
2_1 × 2	0.71	0.59	0.47
2_2 × 2			0.24
2_1 × 4	0.21		
4_1 × 0	0.39	0.22	
4_2 × 0		0.49	0.39
4_3 × 0			0.29
4_1 × 2	0.25		
4_2 × 2			
4_3 × 2			
J = 6			
0_1 × 6	0.78(0.06)	0.47	0.65
2_1 × 4	0.21(0.59)	0.37	0.39
2_1 × 6	0.50(0.40)	0.21	0.31
4_1 × 2	0.61	0.36	
4_2 × 2		0.36	0.32
4_1 × 4	0.30	0.21	
4_1 × 6	0.22	0.42	
6_1 × 0			
J = 8$^+$			
2_1 × 6	0.38	0.70	
2_2 × 6		0.49(0.22)	0.22
2_3 × 6			0.23
3_1 × 6		0.31(0.35)	
4_1 × 4	0.66	(0.50)	
4_2 × 4		(0.27)	0.34
4_3 × 4			0.30
4_1 × 6	0.59	0.28(0.22)	
6_1 × 2		(0.58)	
6_2 × 2			0.22

The final region I would like to discuss is the region near the ^{208}Pb core. There is good reason to use weak-coupling schemes in this mass region. There is a larger neutron excess here, so the low-lying neutron and proton single-particle orbits have relatively small overlaps. There are a number of relatively high-j single-particle orbits, so it is possible to generate states with rather large spins with only a few particles. There is experimental evidence of the onset of deformation as the number of neutron (proton) holes and particles increases. There have been some applications of the Interacting Boson Model to nuclei in this region, so it would be of interest to treat the nuclei here in some good shell model calculations to serve as a check on the IBM. For illustrative purposes, I'd like to consider the isotopes near ^{208}Pb which can be treated as proton particles coupled to neutron holes, for example ^{208,206}Po and ^{208}Rn. The protons and neutrons are both filling the same set of single-particle orbits, i.e. the $0h_{9/2}$, $1f_{7/2}$, $0i_{13/2}$, $2p_{3/2}$, $1f_{5/2}$, and $2p_{1/2}$, but the protons are filling the shell from the bottom up, while the neutrons are actually hole states of orbits at the top of the shell. Thus, the protons first fill the $g_{9/2}$ level, while the neutron holes are in $j = 1/2$ and $3/2$ orbits. One criterion for strong n-p interactions is not met here (i.e., $n_p, \ell_p \approx n_n, \ell_n$) so the coupling should be weak. Blomqvist and others in the Stockholm group[9] have already treated high-spin states in the mass region in a weak-coupling approximation with great success.

For a specific example, consider the Po and Rn isotopes. If a ^{208}Pb core is assumed, there are two or four proton particles and two or four neutron holes in these isotopes. The goal is to couple low-lying states in ^{208}Po or ^{212}Rn to low-lying states in ^{206}Pb or ^{204}Pb. The first question is how well can the identical particle nuclei be treated in the shell model? There are published results[10] on shell model calculations of these nuclei. The models assume a ^{208}Pb core, they include all the active proton-particles or neutron hole orbits, and the realistic interactions of Kuo-Herling are used. These are based on the Hamada-Johnston potential and the proton-proton interactions include the Coulomb interaction. In general, the structure of the low-lying levels of these nuclei are well described by the shell model. The calculations were performed in the complete six-shell space for ^{204}Pb. This case, the matrices are on the order of 500 × 500. If one were to try to perform the same calculation for 6 and 8 holes, the calculations would, at a minimum, be economically undesirable. Therefore, even before the neutron-proton weak-coupling is applied, one would like to truncate in the separate neutron and proton spaces. It would be desirable to find a reasonable truncation scheme which would give a good representation of the maximum number of low-lying states.

The states of the even-even ^{206,204,202}Pb nuclei have been studied in weak-coupling models[11,12] similar in spirit to the calculations of Wong and Zuker of ^{20}Ne as ^{18}F × ^{18}F. Their results are not uniformly good, and the problems of overcompleteness remain tedious. For the nuclei involving identical particles outside the ^{208}Pb core a simple seniority truncation is useful. I have calculated the spectra of the nuclei ^{212}Rn and ^{204}Pb in a complete shell model space. To the extent there are data available, the calculated spectra are in excellent agreement with experiment. I have repeated the calculations where only those states are included for which the sum of seniorities of all single shells is ≤ 2. The complete shell results for ^{212}Rn are compared with experiment in Fig. 7. In the same figure, the complete shell results are shown for ^{204}Pb, together with the results in the seniority-2 truncation. For the lowest 2-3 eigenstates of each spin, the truncated calculations are almost identical to the full-space calculations. The seniority-limited space is easily extended to many neutrons or protons. The cluster coupling calculations are thus easily implemented in this restricted space.

There is already evidence the weak-cluster coupling approach is useful in the Pb region. The Stockholm group have analyzed[9] the

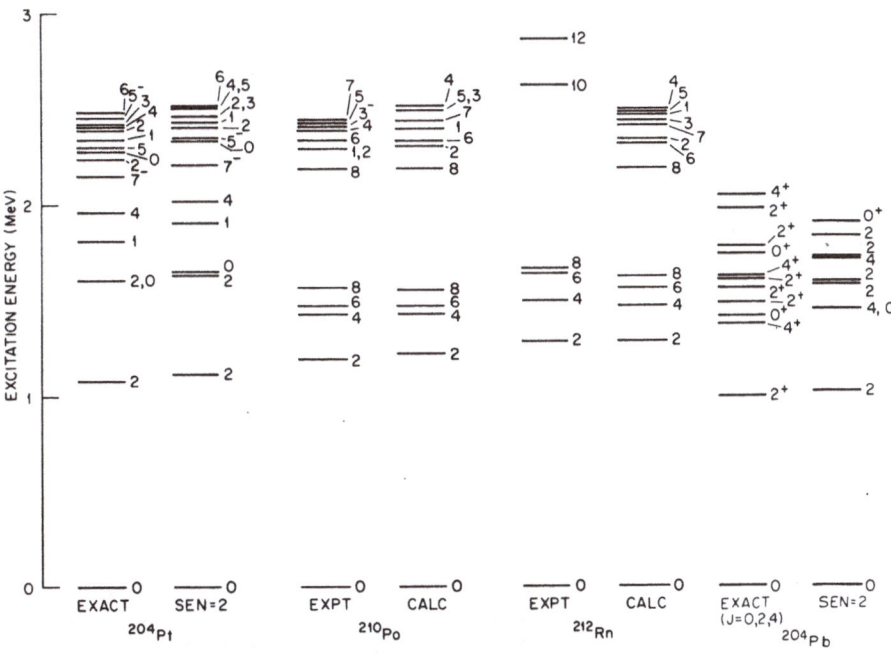

Fig. 7.   Spectra of ^{204}Pt, ^{210}Po, ^{212}Rn, and ^{204}Pb. For ^{204}Pt and ^{204}Pb spectra calculated in complete space shell model calculations are shown, as are calculations in bases which include only basis states with total seniority-2. The results for ^{210}Pt and ^{212}Rn are compared to experiment.

high-spin spectra of states in this mass region with such an approach
with considerable success. As I'll show here, an approach quite
analogous to the Stockholm approach is successful for the low-spin
states. ^{208}Po can be viewed simply as ^{210}Po coupled to ^{206}Pb.
^{206}Po is pictured as ^{210}Po coupled to ^{204}Pb. The lowest states in
^{208,206}Po then are the states where the ground state of ^{210}Po is
coupled to the low-lying ^{206,204}Pb states, and the ground state of
^{206,204}Pb is coupled to the low-lying spectrum of ^{210}Po. Such
spectra are compared with the observed low-lying spectra of ^{208,206}Po
in Fig. 8. Virtually all the lowest 10 states are reasonably accounted for by this "non-calculation". The major discrepancy is the
position of the first excited J=2$^+$ state which comes too high in the
constructed spectrum. The two observed spectra are quite similar,
the major difference being the existence of an extra low-lying J=0$^+$
state in ^{208}Po. This is directly traced to the vanishing of the
$(p_{1/2}^2)^2$ configuration in ^{204}Pb due to the Pauli principle.

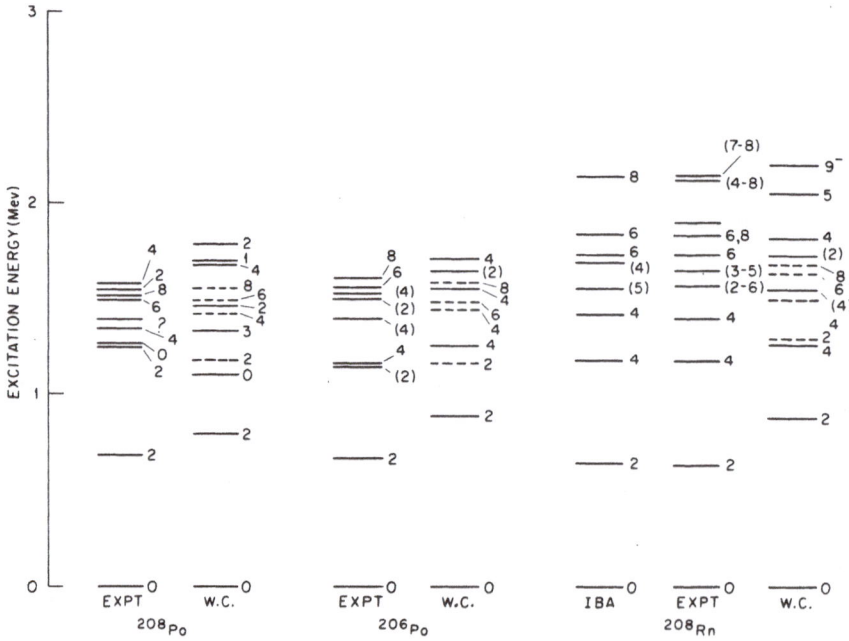

Fig. 8. Calculated and observed spectra of ^{208}Po, ^{206}Po, and ^{208}Rn.
The calculated spectra of ^{208}Po and ^{206}Po result from summing
observed energies of states in ^{210}Po and ^{206,204}Pb. In the
calculated spectra solid lines result from coupling Pb
states to the ground state of ^{210}Po, while dashed lines are
^{210}Po states coupled to the ground state of ^{206,204}Pb. The
^{208}Rn spectrum similarly shows results of coupling ^{212}Rn to
^{204}Pb.

There is recent experimental data[13] on ^{208}Rn from on-line measurements of decay gammas produced in the reaction $\text{Ir}_{\text{natural}}$(^{20}Ne (~ 100 MeV), xn) ^{206,208}Fr → ^{206,208}Rn. ^{208,206}Rn are describable in our weak-cluster terms as ^{212}Rn + ^{204,202}Pb. Shell model calculations and the experimental spectrum imply that the lowest states of ^{212}Rn are ($g_{9/2}^4$, J, sen=2). As shown in Fig. 7, the observed lowest five states of ^{210}Po and ^{212}Rn are identical, except for a slight shift upward in the excitation energies of the J≠0 states. Thus, one would expect the spectra of ^{206}Po and ^{208}Rn to be extremely similar. The two experimental spectra are compared in Fig. 8. There are excellent analogs for each state in ^{206}Po in the ^{208}Rn spectrum, with the exception that the higher-spin states in ^{208}Rn are at somewhat higher energies. This reflects the similar behavior between ^{210}Po and ^{212}Rn. The weak-cluster construction of the ^{208}Rn spectrum which results from coupling the ground state of ^{204}Pb to ^{212}Rn states and vice-versa is shown in Fig. 8. The existence of a low-lying second J=2$^+$ state in ^{206}Po and in the constructed spectrum both suggest such a state remains unobserved in the observed spectrum of ^{208}Rn. The excitation energy of the first J=2$^+$ state is again too high in the constructed spectrum. Presumably, the inclusion of the weak n-p correlations would account for this shift. I have made some calculations including these correlations, where I used a surface-delta interaction for the residual n-p interaction, since the matrix elements of such a force were readily available. The results are unsatisfactory. I am convinced it is a force problem and not a breakdown of the cluster coupling picture. The SDI assumes all radial matrix elements are equal, and so the "loss-of-overlap" due to the fact that neutrons and protons are in quite different orbits is missing. I am repeating the calculations with a finite-range force.

Assuming the existence of the missing J=2$^+$ state in ^{208}Rn, the weak cluster coupled model describes the spectrum of ^{208}Rn quite well. A calculation of the ^{208}Rn spectrum has been made in terms of IBM-1, the Interacting Boson Model which ignores distinctions between neutrons and protons. The results are shown in Fig. 8. The agreement is excellent, and there is no "missing" J=2$^+$ state in the calculation. There are quite significant differences between the two calculations. I've shown above that all the low-lying states of ^{212}Rn and ^{204}Pb are seniority-two states. Only the ground states of ^{212}Rn and the first excited J=2$^+$ states in ^{212}Rn and ^{204}Pb can be correlated with the building blocks of the IBM. In the weak-cluster model, there are no states with J > 4 which can overlap IBM states, and only one J=4$^+$ state can be an IBM state. Thus, the weak-cluster-coupled calculation suggests the IBM may be fitting low-lying states of ^{208}Rn which it should not.

The Pt isotopes have been studied in terms of the IBM also. There is no experimental information on the four-proton hole system

in ^{204}Pt. I have made complete space shell model calculations of
^{204}Pt, and I have calculated the spectrum of ^{204}Pt in the seniority-2
truncation scheme described above. A similar situation exists here
as in ^{204}Pb. The low-lying states are well described in a
seniority-two calculation. Thus, there are no obvious two-boson
states at low energy, but there are s'-, d'-, g'-, etc. bosons.

This latter discussion is offered as a caveat to experimental-
ists rather than as a criticism of the IBM. The proponents of the
IBM have warned of applying the model to nuclei close to closed
shells. There may be "two-boson" states at higher excitation energy
in the neutron (proton) clusters, and as protons (neutrons) are
added, these states may come down in energy. But no code should be
used as a black box, and some care must be taken in using the IBM.

In final summary, the shell model technology has matured considerably in the past decade. It is unlikely that increased technical capability will significantly increase the domain of applicability of "complete-shell" shell model calculations. This is mainly
true because the Hilbert space increases much more rapidly with the
addition of new j-shells than one can conceive of technologically
doing. Even a one-small-step increase in Hilbert space will require
a 1000-fold increase in technology. In addition to this, the sensitivity of super-large calculations to small changes in two-body
matrix elements is a source for considerable concern.

Some variants of the cluster-coupling calculations I've discussed must be pursued. I strongly believe an intelligent development of new fields of nuclear science must involve the continued in-
volvement of shell model theory, both as a check on new models and
as a means of separating the flowers from the weeds in the increasingly complicated experimental studies of nuclear properties.

REFERENCES

1. J. B. French, E. C. Halbert, J. B. McGrory, and S. S. M. Wong,
    Advances in Nuclear Physics, 3, 193 (1969).
2. J. B. McGrory and B. H. Wildenthal, Annual Reviews of Nuclear and
    Particle Science 30, 383 (1980).
3. B. H. Wildenthal, Nucleonika 33, 459 (1978).
4. A. Arima and I. Hamamoto, Ann. Rev. of Nucl. Science 21, 55
    (1971).
5. S. S. M. Wong and A. P. Zuker, Phys. Lett. 36B, 437 (1971).
6. W. Chung, Empirical Renormalizations of Shell Model Hamiltonians
    and Magnetic Dipole Moments of sd Shell Nuclei, Ph.D. Thesis,
    Michigan State University, East Lansing, 135 pp.
7. P. W. M. Glaudemans, P. J. Brussaard, and B. H. Wildenthal, Nucl.
    Phys. A102, 593 (1967).
8. J. B. McGrory and S. Raman, Phys. Rev. C20, 830 (1979).

9. C. G. Lindén, I. Bergström, J. Blomqvist, and C. Roulet, Z. Phys. A284, 217 (1978).
10. J. B. McGrory and T. T. S. Kuo, Nucl. Phys. A247, 283 (1975).
11. C. M. Ko, T. T. S. Kuo, and J. B. McGrory, Phys. Rev. C8, 2379 (1973).
12. C. Pomar and R. J. Liotta, Phys. Lett. 92B, 229 (1980).
13. B. C. Ritchie, Experimental and Theoretical Investigations of the Decays of ^{206}Fr and ^{208}Fr, University of South Carolina, Columbia, 1979.
14. P. J. Brussaard and P. W. M. Glaudemans, Shell-model applications in nuclear spectroscopy, North-Holland, 1977.

EFFECTIVE INTERACTIONS

J.P. Elliott

School of Mathematical & Physical Sciences
University of Sussex
Brighton, England

INTRODUCTION

Many different nucleon-nucleon interactions have been used in nuclear structure calculations and the aim of these lectures is to describe and compare some of the more important interactions and to relate them where possible to the real nucleon-nucleon interaction. In practice different interactions have been used depending on whether one is fitting to total binding energies and densities with a Hartree Fock (HF) calculation or fitting to spectra and spectroscopic data in a shell model calculation. Both types of calculation will be discussed after two preliminary sections concerned with notation and with the philosophy underlying the use of model spaces and effective interactions.

There are a number of good reviews on effective interactions including Ellis and Osnes[1] who concentrate on the first principles derivation of effective aperators using many-body theory, Schiffer and True[2] who are concerned with the deduction of an effective interaction from shell-model calculations and by Kuo[3] who covers both of these aspects. A review by Svenne[4] discusses the effective interactions used in self-consistent field calculations.

NOTATION AND THE TWO-NUCLEAN STATES

To achieve antisymmetry the states of two nucleons are limited to the following four channels:

$$T = 0, \quad S = 1, \quad \text{even } \ell, \quad J = \ell \pm 1, \ell \qquad \text{(triplet-even)}$$
$$T = 1, \quad S = 0, \quad \text{even } \ell, \quad J = \ell \qquad \text{(singlet-even)}$$
$$T = 0, \quad S = 0, \quad \text{odd } \ell, \quad J = \ell \qquad \text{(singlet-odd)}$$
$$T = 1, \quad S = 1, \quad \text{odd } \ell, \quad J = \ell \pm 1, \ell \qquad \text{(triplet-odd)}$$

which are commonly denoted by the expressions in brackets describing the spin multiplicity and the parity. Invariance requirements prevent coupling between channels for the real interaction and the same constraints are usually applied to effective interactions although they need not be.[5] The J-dependence of the interaction within the triplet channels is determined by the spin-dependence. If it is scalar (central force) there is no J-dependence while if it contains a vector spin operator such as $\sigma_1 + \sigma_2$ (vector or spin-orbit force) or a second rank tensor $\sigma_1 \times \sigma_2$ (tensor force) the J-dependence is given by a 6-j symbol $(-1)^J \begin{Bmatrix} J & S & \ell \\ k & \ell & S \end{Bmatrix}$ where k = 0, 1 or 2 is the degree of the spin tensor. It is usually a matter of taste whether one writes the interaction as a single expression involving the spin and isospin operators or specifies the interaction separately in each channel. For example the traditional expression $(W + M P_{12}^r + H P_{12}^r P_{12}^s + B P_{12}^s) V(r)$ in terms of space and spin exchange operators P_{12}^r and P_{12}^s with four constants, W,M,H and B is precisely equivalent to a specification that the interaction has the same radial shape V(r) in all channels with strengths (W+M+H+B), (W+M-H-B), (W-M+H-B) and (W-M-H+B) in the order as listed above. Identical interactions can also appear in different forms following use of the two identities $P^r P^s \equiv -P^t$, where P^t exchanges isospin and $P^s \equiv \frac{1}{2}(1 + \sigma_1 . \sigma_2)$.

THE NEED FOR AN EFFECTIVE INTERACTION

In atomic structure the inter-electron force is sufficiently weak (Z^{-1}) compared to the central electron-nucleus attraction that the energy levels of many electron atoms can be calculated quite well in first order perturbation theory starting from a self-consistent central field. In nuclear structure it is well established experimentally through the success of the shell model that a central field also dominates the behaviour of nuclei although it is also clear from experiment that there is a very much greater mixing of configurations in nuclei than in atoms. A further obstacle to the use of a central field approximation in nuclei is that the nucleon-nucleon phase shifts show the existence of a very strong repulsion at short distance. So strong is this repulsion that any perturbation based on independent particle motion would be very slow to converge. In spite of these objections the simplicity of the central field approximation is so appealing and the empirical evidence for its features are so convincing that we suppose there to be a correspondence between the shell model wave functions in some fairly pure configuration and the low lying states of nuclei.

In more precise terms we consider a simple "model space" and the projections $\phi = P\psi$ of the real wave functions ψ onto this space. Although the operator P has no inverse we can write $\psi = \Omega \phi$ where Ω is a complicated many-body operator which builds into ψ all the short range correlations and configuration mixing which is absent from ϕ. It is then formally possible to recast the original Schrodinger equation in the full space to become a Schrodinger equation in the model space but with an effective interaction replacing the original interaction. Every physical operator is also replaced by a corresponding effective operator acting in the model space. To deduce the effective operators (including the interaction) from the original operators is of course almost equivalent to the solution of the original problem and can be attempted only with the use of a perturbation expansion based on some single particle hamiltonian. By appropriately rearranging the series (the reaction matrix method) one can ensure that the effective interaction has finite matrix elements in independent particle wave functions even though the corresponding matrix elements of the real interaction may be infinite.

In practice the problem of calculating an effective interaction from the real interaction raises many important theoretical questions which have received much attention but there are still many uncertainties and heavy computation is necessary. I shall, in these lectures, spend most time on the more empirical approaches in which a simple form is chosen for the effective interaction in a given model space and the parameters are then deduced from fitting many-body data.

The model space is usually defined in terms of possible configurations for occupied states in a single particle hamiltonian. Thus for nuclear matter the model space would be a Fermi gas with only one degree of freedom, the density. For a simple nucleus which is known to have the characteristics of a closed shell the model space is again a single wave function, the Slater determinant of occupied states. If the single particle hamiltonian were assumed to be a harmonic oscillator this model space would again have only one degree of freedom, the length parameter $b = (\hbar / m\omega)^{\frac{1}{2}}$ for the oscillator but a less restrictive model space would retain the Slater determinant but allow each single particle wave function to have its radial dependence freely chosen. This defines the Hartree-Fock problem. More ambiguity arises when we move away from closed shells. For example in ^{18}O, with two neutrons beyond the closed shell ^{16}O, we might restrict the two valence neutrons to the next lowest orbit, the $0d_{5/2}$, but a more realistic model space would allow mixing with $1s_{\frac{1}{2}}$ which lies very close and also with the other member $0d_{3/2}$ of the spin-orbit doublet. It is clear that the effective interaction will depend on the choice of model space and that the larger the model space the more will the effective interaction resemble the real interaction. The effective interaction may also differ from nucleus to nucleus and may change with excitation energy in a given nucleus. Although the real interaction is predominantly two-body the effective interaction will contain three-body and even many-body terms. Also whereas the real interaction is relative, depending only on the distance between the interacting nucleons the effective interaction will depend on the environment (e.g. the density) in which the interacting particles find themselves and hence can be a function of the centre of mass of the interacting pair.

We see the study of effective interactions as a two-way process. On the one hand the many-body theorists can try to deduce from the real interaction the effective interaction appropriate to a particular model space in a particular nucleus. At the same time work continues with simple empirical effective interactions designed to fit many-body data in specified model spaces. One hopes eventually that these two approaches will agree in their effective interactions.

Before beginning the discussion of particular effective interactions we give a greatly oversimplified illustration. Let E_α and Ψ_α denote zero-order energies and wave functions in some central field perturbation treatment of a hamiltonian H so that to second order the energy is given by

$$E = <\psi_o|H|\psi_o> - \sum_{\alpha \neq 0} |<\psi_o|H|\psi_\alpha>|^2 / (E_\alpha - E_o)$$

$$= <\psi_o|H|\psi_o> - <\psi_o| \sum_{\alpha \neq 0} \{H|\psi_\alpha>(E_\alpha - E_o)^{-1}<\psi_\alpha|H\}|\psi_o>$$

$$= <\psi_o|H_{eff}|\psi_c>$$

where

$$H_{eff} = H - \sum_{\alpha \neq 0} H|\psi_\alpha>(E_\alpha - E_o)^{-1}<\psi_\alpha|H$$

defines the effective hamiltonian which when used in the zero-order wave function gives the energy to second order. This example shows how H_{eff} depends on the energies E_α of the central field. Hence if the density of the system is increased, so that the E_α also increase, the effective interaction becomes less attractive. We shall see that many of the popular effective interactions contain a repulsive density dependence whose origin could lie partly in this effect and partly in the fact that the intermediate state sum over α excludes the "occupied" ground state. This "Pauli effect" is greater at high density.

EFFECTIVE INTERACTIONS IN HARTREE-FOCK CALCULATIONS

It is important to distinguish between two types of calculation (a) for global nuclear properties like total binding energy and size and (b) for spectroscopic properties like excitation energies and transition probabilities which involve only a few nucleons, the valence particles (or holes) beyond the nearest closed shell. For the first of these it is sensible to restrict attention to the closed shells while in the second the closed shells play no part. (The study of excited, particle-hole, states in a closed shell nucleus would come in the second category.) The effective interactions in these two types of problem will surely differ but one would hope to see some similarity and to have some understanding of the differences. Calculations of the second type are probably of interest to more people since they relate to much more experimental data but in this section I give a brief account of calculations of the first type.

To find an effective interaction which reproduces the energy and size of a <u>single</u> closed shell nucleus would be a very limited objective but in practice simple effective interactions have been found which give a satisfactory account of these quantitites for the <u>full range</u> of closed shells from ^{4}He to ^{208}Pb and including nuclear matter. There is of course no direct experimental data for nuclear matter but analyses[6] of the semi-empirical mass formula with a liquid drop model suggest a binding energy per nucleon of 15.68MeV and a Fermi momentum of $k_F = 1.30$ fm^{-1}. Since these are extrapolated numbers they are necessarily less reliable than the corresponding numbers for finite nuclei and in particular the consistency of this value of k_F with estimates of the incompressibility modulus has recently been questioned.[7] (A value of k_F in nuclear matter defines the density through the relation $\rho = 2k_F^3/3\pi^2$ and sometimes authors quote the mean inter-nucleon distance given by:

$$r_0 = (9\pi/8)^{1/3}/k_F.)$$

Skyrme Interactions

The most widely used interactions in HF calculations belong to a family originated by Skyrme[8] more than twenty years ago which was used by Vautherin and Brink[9] and more recently by Beiner et al.[10] It is essentially a short range approximation containing the δ function and its first and second derivatives. (Any function may formally be expanded as an infinite series in derivatives of the δ function.) The usual form of the Skyrme interaction contains six parameters t_0, t_1, t_2, t_3, x_0 and W, and is written

$$V = t_0(1 + x_0 P^s)\delta(\underline{r}) + \tfrac{1}{2}t_1\{k'^2\delta(\underline{r}) + \delta(\underline{r})k^2\} +$$

$$+ t_2\underline{k}'\cdot\delta(\underline{r})\underline{k} + iW\underline{k}'\cdot\delta(\underline{r})(\underline{\sigma}_1 + \underline{\sigma}_{2\wedge}\underline{k}) + \tfrac{1}{6}t_3\,\rho(\underline{R})\,\delta(\underline{r})$$

where $\underline{r} = \underline{r}_1 - \underline{r}_2$, $\underline{R} = \tfrac{1}{2}(\underline{r}_1 + \underline{r}_2)$, $\underline{k} = -\tfrac{1}{2}i(\underline{\nabla}_1 - \underline{\nabla}_2)$ is the relative momentum with $\underline{k}'$ its adjoint (acting to the left) and ρ is the density. The first and second terms act only in relative s-states with x_0 governing the ratio of singlet-even to triplet-even strength and t_1 introduces a departure from zero range. The term in t_2 allows for a central force in p-states with shortest possible range and W governs the two-body spin-orbit force. The final term with strength t_3 is again a δ- force, acting only in s-states, but now with a linear dependence on density. This term was originally written by Skyrme[8] as a three-body δ-force without density dependence but the two forms are equivalent in a Hartree-Fock calculation and the form given above is the more likely to occur naturally in any derivation of the effective interaction. Beiner[10] finds that the sizes and binding energies of seven closed shells from ^{16}O to ^{208}Pb can be fitted within a few per cent with the same set of values for the six parameters of the Skyrme interaction. However the set is not unique and in fact there is a linear continuum of solutions. Table 1 shows three sets of parameter values all of which fit the data equally well as would any other set obtained by linear interpolation.

Table 1. Some Skyrme force parameters. Units are MeV fm^3 for t_0, MeV fm^5 for t_1, t_2, W and MeV fm^6 for t_3.

	t_0	t_1	t_2	t_3	x_0	W
S III	−1129	395	−95	14000	0.45	120
S IV	−1206	765	35	5000	0.05	150
S V	−1248	971	107	0	−0.17	150

It is encouraging to find that these fundamental quantities of energy and size can be fitted so simply with a universal interaction but in view of the success of the semi-empirical mass formula this is perhaps not surprising. In some ways the ease with which the fit was obtained is disappointing because, for example, the strength t_3 of density dependence is not well determined. However, Beiner[10] remarks on two quantities which do appear to be sensitive to t_3. One is the single-particle and hole energies in the closed shells for which he finds a large value of t_3 is required to produce the levels near the Fermi surface while a moderate t_3 is better for the deep levels. The parameter set labelled SIII in table 1 is an adequate compromise. The set SV with $t_3 = 0$ would place all hole levels much too deep. The other quantity sensitive to t_3 is the incompressibility which in nuclear matter is defined as the curvature of the energy v density curve at the saturation minimum,

$$K = k_F^2 \ d^2(E/A)/d \ k_F^2$$

It should be stressed that the parameter sets in table 1 were not fitted to the nuclear matter data but they all give values for E/A and k_F close to the accepted ones, see table 2. Nevertheless the value of K, shown also in table 2, changes considerably and is always greater than the currently proposed "experimental" value of K=220 MeV deduced by Blaizot et al[11] from the excitation energies of monopole vibrational resonances in finite nuclei. As one would expect, the value of K decreases with t_3 but even SV with $t_3 = 0$ gives K= 306 MeV.(Negative t_3 are not considered, presumably because they are physically unreasonable and would lead to collapse.) Beiner shows also that the radii of the closed shell nuclei are given better by the interactions with small t although the differences are small. This is consistent with the values of k_F in table 2 which show a greater density for SV than for SIII.

A number of other variants of the Skyrme force have been proposed. Kohler[12] argues that a density dependence of $\rho^{1/3}$ gives a better value for K = 263 MeV and suggests a spin-dependence for the density dependent term. Both suggestions are reasonable but they

Table 2.   Some nuclear matter parameters for various effective interactions. The experimental values are necessarily model-dependent.

	SIII	SIV	SV	Ska	D1	B1	Exp
E/A(MeV)	−15.87	−15.98	−16.06	−16.00	−16.30	−15.70	−15.68
k_F(fm^{-1})	1.29	1.31	1.32	1.32	1.36	1.45	1.30
K(MeV)	356	325	306	263	228	193	210 ± 30

introduce two new parameters'. This variant is called SKa and in
comparison with other Skyrme forces has a rather large value of
k_F = 1.32 fm^{-1}. Further refinements include[13,14] a term which depends
on both momentum and density whereas previously these features
appeared separately, governed by t_1 and t_3 . This development is
equivalent to the introduction of a finite range for the density
dependent term which is clearly a reasonable improvement.

Finite Range Interactions

    The reason why the Skyrme forces have been so popular is that
the δ-function radial shape makes HF calculations very much simpler
because the total energy can be written in terms of the density and
it is not necessary to calculate the large number of two-body matrix
elements which otherwise make HF calculations very time-consuming.
Nevertheless, several finite range effective interactions have been
proposed and used for HF calculations.

    Brink and Boeker[16] used a sum of two gaussians $\exp(-r^2/\mu^2)$,
one being attractive with long range μ = 1.4 fm and the other repulsive
with short range μ = 0.7fm. The latter was also strongly repulsive
in odd states. This interaction, often referred to as B1, had no
density dependence and was fitted to the ^{4}He energy and size and to
nuclear matter. It was however fitted to what is now regarded as a
very large k_F = 1.45 fm^{-1} and has a small incompressibility
K = 190 MeV. It gives too little binding energy for the finite
nuclei ^{16}O and ^{40}Ca and is not in the same class as the Skyrme forces
so far as fitting the data is concerned.

    Gogny[11] defined an interaction, called D1, which has two
gaussians plus a δ- function density dependent term ρ^α and a δ-
function spin-orbit term. This interaction contains a large number
(thirteen) of parameters all of which are not well determined. He
chose $\alpha = 1/3$, ranges μ= 1.2 and 0.7fm and used ^{16}O and ^{90}Zr to fit
the remaining parameters. In nuclear matter D1 gives a large k_F= 1.36
and small K = 228 MeV. Both gaussians are attractive in even states
and there is no strong p-state repulsion.

    Campi and Sprung[17] use a sum of five gaussians plus a density
dependent δ- function to define an effective interaction called G0.
The reason for such complexity is that the interaction is first fitted
to the G-matrix calculated from the realistic Reid[18] potential but
since this does not reproduce the saturation properties of nuclear
matter, a number of parameters are substantially adjusted to fit the
data. This approach is similar to the earlier work of Negele[19].

    One general conclusion of these different HF calculations with
effective forces is that those with little density dependence give
a larger value of k_F from 1.32 to 1.36 fm^{-1} and a smaller K <300MeV
while those with strong density dependence give a small k_F <1.32 fm^{-1}

and large K>300MeV. There is thus an apparent inconsistency with the "experimental" values[6,11] of $k_F = 1.30$ fm^{-1} and K = 210$\pm$ 30 MeV. This difficulty has been discussed by Pearson[7] and the comments of Jennings and Jackson[20] do not appear to affect the conclusion.

For nuclear matter and for LS-coupled (i.e. containing both $j = \ell \pm \frac{1}{2}$ shells) N = Z closed shells only the spin-isospin average of the central force can contribute so that the spin-dependence of the effective interaction is irrelevant, in the independent particle approximation. When N $\neq$ Z and when the closed shell contains only the lowest $j = \ell + \frac{1}{2}$ member of a doublet both spin-orbit and tensor forces may contribute. The spin-orbit force is essential but although in his original paper[21] Skyrme included a tensor component this has been excluded from subsequent more detailed HF work without any ill effect.

EFFECTIVE INTERACTIONS IN SHELL-MODEL CALCULATIONS

In some ways the study of effective interactions for shell-model calculations is simpler than the corresponding problem for closed shell energies because one is concerned with few particles and the self-consistency requirement is absent. However there is more complexity because of the greater freedom in defining a model space and, away from closed shells, there is degeneracy for the many-particle wave function so that unless one is very restrictive in defining the model space, it is necessary to diagonalise large matrices. There will be an inevitable conflict between the desire to keep a small model space and the hope to find an effective interaction with a wide range of applicability.

Before discussing any detailed calculations I want to make a few general remarks. The real interaction contains a strong two-body spin-orbit force but its main role is in generating the familiar one-body spin-orbit splitting in the shell-model potential. Thus if we use empirical single-particle energies the major part of any two-body spin-orbit force will have been taken into account. In the same way it is known that the tensor force, which is undoubtedly present in the real interaction, has a large second-order contribution which is roughly equivalent to an increased central attraction in triplet-even states. It is therefore generally accepted that the effective interaction in shell-model calculations can reasonably be assumed to be central. It will be interesting to see if there is any real evidence that this assumption is inadequate.

The model space is usually defined by specifying the configuration j^k for the k valence nucleons or more generally by allowing the nucleons to be arranged in any configuration obtained from a finite set of close-lying levels j_1, j_2, j_3... One may perhaps impose a restriction that not more than so many nucleons may lie in orbits other than the lowest, because such configurations would be too high in energy to be appreciably mixed into the ground state. The effective interaction therefore enters as a finite set of two-body matrix elements $\langle jj'|V|j''\,j'''\rangle$. In many calculations these are simply treated as free parameters to be adjusted so that the low eigenvalues of the k-particle matrix in the chosen model-space agree with the observed energy levels for a range of k. Least squares fitting is usually employed, perhaps giving greater weight to the more reliable data. Having deduced a set of two-body matrix elements in this way what does this tell us about the effective interaction? One approach is to assume some form for the interaction containing a small number of parameters and then to fit them to the larger number of two-body matrix elements. Another approach, which is less used but which I shall discuss in some detail, is to reduce the two-body matrix elements to relative matrix elements using a Moshinsky transformation. This assumes that the wave functions have oscillator shape but since we are only concerned here with valence nucleons this is a good approximation at the optimum value of b. (The transformation is actually independent of b.) There is also the implicit assumption that the effective interaction is a relative one and although this is not necessarily true - for example we have seen density dependence in the earlier discussion, - it may well be an adequate approximation for the few interacting valence nucleons. One could imagine taking an average density appropriate for these nucleons. In making this transformation we are distinguishing between what might be called geometrical features, due to the trans-formation but independent of the interaction, (except for the two general assumptions mentioned above,) and physical features, which depend on the relative matrix elements of the interaction.

Small Model Space

For simplicity let us consider a single orbit j_1 so that there is a set of two-body matrix elements for even J with T = 1 and odd J with T = 0. The transformation involves first a 9j-symbol to go from jj to LS-coupling and then the Brody-Moshinsky transformation bracket.

$$|j_1^2 J\rangle = \sum_{LSn\ell N\pounds} \sqrt{(2L+1)(2S+1)}\,(2j_1+1) \begin{Bmatrix} \ell_1 & \ell_1 & L \\ \tfrac{1}{2} & \tfrac{1}{2} & S \\ j_1 & j_1 & J \end{Bmatrix} \times$$

$$\times \langle n_1\ell_1, n_1\ell_1; L \mid n\ell, n\pounds; L\rangle \mid n\ell, N\pounds; L, S; J\rangle$$

where $n_1\ell_1$ refer to the orbit j_1, $n\ell$ to the relative co-ordinate $\underline{r}_1 - \underline{r}_2$ and $N\pounds$ to the centre-of-mass co-ordinate $\tfrac{1}{2}(\underline{r}_1 + \underline{r}_2)$. A final transformation with a 6j-symbol will couple the relative angular momentum ℓ of the pair to their spin S giving a resultant g with the

same role as the j used in describing the nucleon-nucleon channels
in an earlier section.

$$|n\ell,N\pounds;L,S;J> = \sum_{g}(-1)^{\pounds+S+L+g}\sqrt{(2L+1)(2g+1)} \text{ x}$$

$$\text{x } \left\{\begin{matrix} \pounds & \ell & L \\ S & J & g \end{matrix}\right\} | n\ell,S;g,\pounds;J>$$

In this way the matrix element may be written

$$<j_1^{\,2}J\,|\,V\,|\,j_1^{\,2}J> = \sum_{nn'\ell\ell'SS'g} A(nn'\ell\ell'SS'g\,j,J\,) \text{ x}$$

$$\text{x } <n\ell Sg\,|\,V\,|\,n'\ell'S'g>$$

in terms of the relative matrix elements of V with the "geometrical"
coefficients A known from the previous two equations. For a central
force and for states with the same total oscillator energy there is
the further simplification that n = n', 1 = 1', S = S' and there is no
dependence on g. As a starting point for the discussion of this
expression let us take the particular case of the $f_{7/2}$ shell and assume
that the relative matrix elements are independent of n. The summation
then runs over the single index ℓ with S determined automatically from
antisymmetry. Table 3 shows the coefficients A for the different
J and T. For each J T the sum of coefficients is one so that they
represent the fraction of time spent by the two particles in that
particular relative ℓ state. Since both real and effective nuclear
interactions are dominated by their s-state components one can hope
to learn something already from the first column of table 3 which is
shown again in figure 1 by crosses with a smooth curve passed through
them. Also shown by circles in the figure are the experimental
points taken from the analysis of Schiffer and True[2] and the numbers

Table 3.   The fractional relative ℓ-state
composition of the states of $(f_{7/2})^2$.

	s	p	d	f	g	h	i	p'
T=1, J=0	.46	.41	.11	.02	–	–	–	.41
J=2	.11	.40	.40	.09	–	–	–	.39
J=4	.06	.26	.13	.37	.17	.01	–	.16
J=6	.04	.32	.03	.22	.03	.32	.04	–
T=0, J=1	.24	.53	.21	.02	–	–	–	.53
J=3	.12	.19	.40	.26	.03	–	–	.12
J=5	.13	.10	.26	.07	.34	.10	–	–
J=7	.31	–	.19	–	.19	–	.31	–

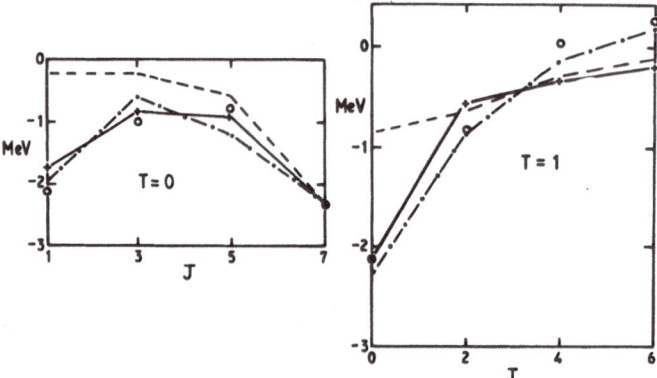

Fig. 1. Comparison of n-independent s-state interaction predictions
         (solid line) with experiment (circles) for $0f^2_{\frac{7}{2}}$

from table 3 have been multiplied by the values 7.35 MeV and 4.63
MeV for the effective ^{3}s and ^{1}s matrix elements chosen to fit the
T = 0, J = 7 and T = 1, J = 0 states respectively. One sees that the
trend of the experimental points is given very well both for T = 0
and for T = 1 by the elementary coefficients from table 3 which were
essentially geometrical. The curves - - - and -·--·- are obtained[32]
respectively with the bare and perturbed realistic matrix elements of
Kuo. In using table 3 we have assumed the effective interaction to
have matrix elements independent of n. How good is that assumption?
Figure 2 shows the ratio $\langle n|V|n\rangle/\langle o|V|o\rangle$ for a gaussian exp $(-r^2/a^2)$
in oscillator wave functions as a function of n for different choices
of the range-to-size ratio a/b. Curves are shown for both s and p
states. Especially for s-states the figure shows very little change
with n except at extremely short range. It is interesting to see how,
starting at infinite range, the curves first drop below the value 1.0
and then climb above. Hence at about a/b = $\frac{1}{2}$ the curve is again

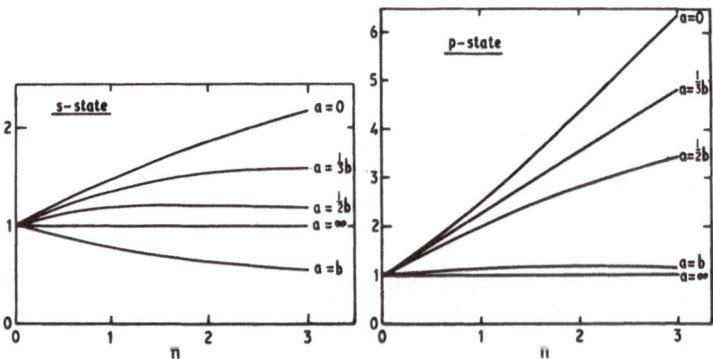

Fig. 2. The ratio $\langle n|V|n\rangle / \langle 0|V|0\rangle$ for a gaussion exp $(-r^2/a^2)$.

almost horizontal. In practice, light nuclei have b = 1.6fm and
heavy nuclei b = 2.2 fm so that a/b = $\frac{1}{2}$ represents a sensible range
of about 1 fm. The p-matrix elements follow the same trend but increase more sharply with n for short range.

In their review article[2] Schiffer and True analyse many simple
j^2 configurations and find the pattern in figure 1 to be typical of
them all. From their overall analysis they conclude that although
there is some evidence for non-central forces the significance is
low and a very acceptable fit may be obtained with central forces.
Indeed in the T = 0 channel a single even-state attractive potential
is sufficient with no odd-state force. For T=1 they require, in
addition to an even-state attraction, an odd-state force which is
attractive at short range and repulsive at long range. Indeed
these features are already apparent from figure 1 where the simple
s-state interaction fits well for T = 0 but is less good for T = 1.
The difference in the T = 1 energies cannot be corrected simply by
adding a long range p-component, the number in column 2 of table 3.
However, in the last column of that table, labelled p' we give the
expansion coefficients including a weighting factor n. As shown
in figure 2 this reproduces the linear increase of a short range
p-component.   One sees that a suitable combination of the three
columns s, p and p' can account for the T = 1 pattern. Very roughly
the combination of a long range s-attraction $< ns|V|ns> = - 3.8$ MeV
with a long range p-repulsion $< np|V|np > = 2.9$ and short range p-
attraction $< np|V|np > = - 1.6(1+1.5n)$ would fit the data.

As a further illustration of the geometrical effects, figure
3 shows the experimental points, marked by circles, for the $h_{9/2}\ i_{13/2}$
configuration[2]. The calculated points marked by crosses, refer to
the same assumption as in figure 1, s-state forces with no n-depen-
dence. The irregular pattern for these $j_1 \neq j_2$ spectra is hence seen
to be another consequence of the geometry rather than of any detailed
property of the interaction.

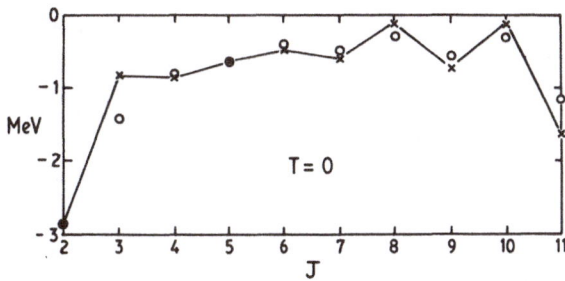

Fig. 3.   As for Fig.1 but for the T=0 states of $0h_{\frac{9}{2}}$ $0i_{\frac{13}{2}}$

Large Model Space

For the light nuclei $4 < A < 16$ there have been a number of successful shell-model calculations in which the nucleons are confined to the Op-shell, i.e. $p_{3/2}$ or $p_{1/2}$. Cohen and Kurath[22] have deduced the effective interaction matrix elements for a best fit to the nuclei $8 \leqslant A < 16$, arguing that it is difficult to include also the very light nuclei with $4 < A < 8$ with the same interaction. In table 4 we show their relative matrix elements, which they refer to as (8 - 16) POT. It is interesting that for s-states, only the sum of 0s and 1s matrix elements occurs and that the central component in d-states occurs only in combination with the 0s matrix elements. For triplet states, where non-central forces contribute, we give the central (C), vector (V) and tensor (T) component for the state of least J. This is sufficient, because the contributions for other J are then determined by a 6j-symbol. For p-states the ratios in $J = 0,1$ and 2 are 1:1:1 (Central), -2:-1:1 (Vector) and 10:-5:1(Tensor) while for d-states with $J = 1$, 2 and 3 the ratios are 1:1:1: (Central), 3:1:-2 (Vector) and 7:-7:2 (Tensor). For comparison we give the Sussex matrix elements[23] (SME) which were deduced from the nucleon-nucleon phase shifts and hence contain many of the features of the real interaction. For T=1 one concludes that (8-16) POT has (a) a shorter range in singlet-even states (b) a stronger triplet p-state repulsion (SME is weakly attractive) (c) a stronger vector force. In $T = 0$ states (8-16)POT is (a) slightly stronger in triplet-even states (b) has a much larger vector force and a d-state tensor force of opposite sign. Most of these differences are also apparent in a qualitative comparison between (8-16)POT and a realistic interaction like that of Hamada and Johnston[24]. Some of them, like the non-central d-state numbers, may not be well-determined in the p-shell fit but others, like the p-state repulsion, probably represent a genuine difference between the real and effective interactions.

Table 4.   Comparison of the relative matrix elements for
             the Cohen-Kurath(8-16)POT and Sussex
             potentials, the latter evaluated for b = 1.6fm.

T = 1	0s + 1s	0s + 0d	p_0(C)	p_0(V)	p_0(T)
(8-16)POT	-13.80	-4.74	1.40	1.07	-3.75
S M E	-11.99	-8.25	-0.25	0.57	-3.89

T = 0	0s + 1s	0s + 0d(C)	d_1(V)	d_1(T)	p	$\langle 1s|V|0d\rangle$
(8-16)POT	-16.77	-11.74	3.09	-2.59	-0.29	-1.47
S M E	-14.83	- 9.70	0.38	2.79	2.28	-1.28

It is encouraging to see agreement for the tensor force in the important $<1s|V|0d>$ coupling and in the triplet p-state. The slight weakness of the SME in s-states is a known deficiency which is at least partially removed when second order corrections are added[25] to the SME.

Corresponding large model space calculations have been performed by a number of authors for the sd-shell nuclei $16 < A < 40$ allowing all possible mixing of configurations in which the nucleons occupy the $0d_{5/2}$, $1s_{1/2}$ and $0d_{3/2}$ orbits. In the middle of the shell this is a very large model space, with matrices of dimension several thousand and is at the limit of present day computing capability. The most elaborate least-squares fitting of an effective interaction in the sd-shell seems to be that of Chung[26] and Wildenthal which is tabulated by Kelvin[27] et al and is more accurate than the earlier interaction of Preedom and Wildenthal[28]. Kelvin[27] proposes a refinement to include effective Coulomb force matrix elements but although this is a valid step it is a comparatively small correction and I shall not discuss it here.

An approximate analysis of the Chung-Wildenthal (CW) interaction, similar to that for (8-16)POT in table 4, is given in table 5. There are many features in common with the numbers in table 4.

The Sussex Potential Matrix Elements

As a final topic I shall describe work with the effective interaction which has been referred to as the Sussex potential, and include a report on some recent developments[29]. The interaction was deduced, in the form of a set of matrix elements (SME) in an oscillator basis for low n, from the nucleon-nucleon phase shifts by a DWBA approximation in each channel starting from a simple auxiliary potential. It was not intended as an effective interaction for immediate use in a small model space but was designed to avoid the difficulties of hard core potentials and to enable perturbation corrections to be calculated. The original[23] matrix elements nevertheless gave a reasonable first approximation in shell-model calculations and in particular gave a good account of single-particle spin-orbit splittings[30]. Second order perturbation corrections were later

Table 5.   An approximate analysis of the CW
interaction into relative matrix elements
and comparison with SME at b=1.8 fm.

T=1	0s+1s	0s+0d	0p (C)	$0p_0$(V)	$0p_0$(T)	$1p_0$(C)	$1p_0$(V)	$1p_0$(T)
CW	-12.61	-5.94	1.57	2.09	-2.31	-0.20	0.96	-1.88
SME	-10.35	-6.47	-0.19	0.98	-2.83	-0.12	2.15	-3.42

calculated[25] and found to improve agreement in spectroscopic calculations, giving results similar to those of Kuo and Brown[31] who constructed a g-matrix from the Hamada-Johnston potential. In a
calculation for the total binding energy of a nucleus the SME did not
have the correct saturation requirements and an attempt was made[25] to
overcome this deficiency by computing a small g-matrix correction.
In retrospect therefore, although the SME had many of the features
of the real interaction it was necessary to calculate corrections,
which were necessarily uncertain, of two kinds (a) from the short
range repulsion and (b) from perturbation corrections in the central
field. Realistically one cannot hope to calculate the latter much
beyond second order. Although the original philosophy of the SME was
that they should be free of adjustable parameters we have now asked
the question "What needs to be added to the original SME to ensure
correct saturation properties?" One could then introduce a few-
parameter additional interaction to replace (and to represent) the
two corrections referred to above which are uncertain and very
difficult to calculate.

We write the effective interaction as $V_{eff} = V_{SME} + W$ where
V_{SME} was tabulated[23] and the additional piece W has the form

$$W = \sum_{i<j} A \delta(\underline{r})\rho^{\alpha}(\underline{R}) + \tfrac{1}{2}\{B(1+P_{ij}) + C(1-P_{ij})\} e^{-r^2/a^2}$$

where $\underline{r} = \underline{r}_i - \underline{r}_j$, $\underline{R} = \tfrac{1}{2}(\underline{r}_i + \underline{r}_j)$, ρ is the density and P_{ij} the space
exchange operator. The additional interaction W is thus density
dependent with five parameters. In practice we find little sensitivity to the range a and hence fix a=1.0fm. We fit the remaining four
parameters α, A, B, and C to the energy and density of nuclear matter
and to the energy and size of ^{16}O in simple independent-particle
calculations, i.e. using the simplest model space. The calculations
are particularly sensitive to the two size parameters, which are the
least certain experimental numbers. Instead of fixing k_F and b we
simply confine them to the ranges $1.30 < k_F < 1.36$ fm^{-1} and $1.76 < b < 1.81$ fm.
We can use the two binding energies, for ^{16}O and nuclear matter, to
eliminate two of the parameters and the ranges for k_F and b then
define an acceptable region in the plane of the remaining two parameters α and C. This region is shown in figure 4 and it extends from
beyond $\alpha = 2.5$ back to $\alpha=0$, at which point there is no density
dependence. Five typical points, denoted by W1 to W5, are chosen and
the corresponding parameters are given in table 6. One finds that
the nuclear matter incompressibility increases with α from about
230 MeV at $\alpha = 0$ to 380 MeV at $\alpha = 2$. In figure 5 and table 7 we
compare our V_{eff} with other effective interactions but this is done
only for the spin-isospin averages since this is the only combination
which is involved in the N = Z closed LS-coupled shells considered
here. These averages are $\tfrac{1}{2}V(T=1, S=0) + \tfrac{1}{2} V(T=0, S=1)$ for even ℓ and
$\tfrac{9}{10}V(T=S=1) + \tfrac{1}{10} V(T=S=0)$ for odd ℓ. The figure 5 gives the matrix
elements in s-states as a function of density and it is interesting
to note that all lines intersect near $\rho = 0.8$ fm^{-3} which is the mean
density of ^{16}O. It is in fact the Os matrix element which dominates

Table 6.   Parameter sets for the interaction W.(B and C
            are in units of MeV and A in units of MeV fm$^{3+3\alpha}$.)

	α	C	B	A	A$(0.1)^{\alpha}$
W1	2	79.5	−50.7	3746	37
W2	1	68.4	−67.2	1189	119
W3	1	83.6	−60.5	827	83
W4	$\frac{1}{3}$	90.6	−101.7	568	264
W5	0	142.4	−227.5	782	782

Table 7.   Spin-isospin averaged relative matrix elements
            in an oscillator basis at b = 1.8fm.
   (The symbols +W1 are abreviations for V_{SME} + W1 etc.)

	V_{SME}	+W1	+W2	+W3	+W4	+W5	SIII	SV	D1	B1
0p	−0.03	0.49	0.42	0.52	0.56	0.90	−0.32	0.36	0.04	0.95
1p	0.17	1.15	1.02	1.20	1.29	1.93	−0.80	0.90	0.02	2.13
0d	−0.53	−0.57	−0.59	−0.58	−0.62	−0.73	−	−	−0.27	−0.81

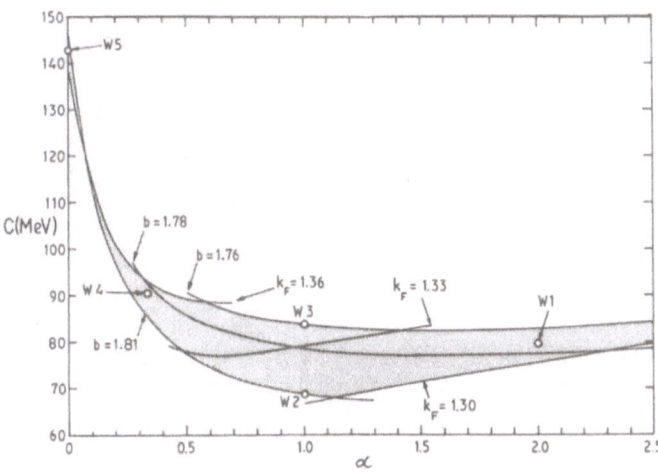

Fig. 4.   The acceptable region for the parameters α and C.

the binding energy of ^{16}O. For heavier nuclei, higher n are involved
and for these the intersections take place at somewhat higher
densities.

    One general feature of the additional interaction is the repul-
sive p-state which is a conclusion that has already been arrived at
from the work of Schiffer and True[2] and from the p-shell work of Cohen
and Kurath[22]. This p-state repulsion (measured by the magnitude of
C in figure 4) is seen to increase as α decreases. The extreme
solution is remarkably close to the B1 interaction of Brink and
Boeker[16] in both the p-state and s-state.

    Since the acceptable region extends over a wide range of α one
must look for   alternative criteria. As we discussed earlier the
incompressibility is such a quantity and the rather uncertain current
experimental value points to small α. It may also be significant
that first principles calculation of density dependence[19] also seem to
favour a small value of $\alpha = \frac{1}{3}$.

    Clearly more detailed work needs to be done but it does seem
that this comparatively small addition to the SME could produce an
effective interaction which has the detailed structure needed in a
spectroscopic calculation and also produces the correct saturation
properties.

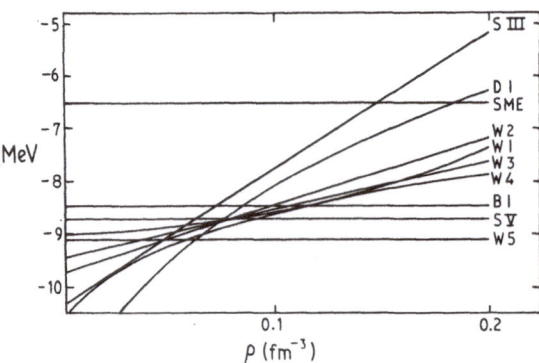

Fig. 5.   Spin-isospin averaged relative matrix element. <0s|V|0s> as
          a function of density for different interactions.

REFERENCES

1. P.J. Ellis and E. Osnes, Rev. Mod. Phys. 49:777 (1977)
2. J.P. Schiffer and W.W. True, Rev. Mod. Phys. 48:191 (1976)
3. T.T.S. Kuo, Ann. Rev. Nucl. Sci. 24:101 (1974)
4. J.P. Svenne, Adv. Nucl. Phys. 11:179 (1979)
5. H. Feldmeier, P. Manakos and T. Wolff, Z. Physik. 258:81 (1973)
6. W.D. Myers and W.J. Swiatecki, Ann. Phys. 55:395 (1969), 84:186 (1974)
7. J.M. Pearson, Phys. Lett. 91B:325(1980)
8. T.H.R. Skyrme, Nucl. Phys. 9:615(1959)
9. D. Vautherin and D.M. Brink, Phys. Rev. C5: 626 (1972)
10. M. Beiner, H. Flocard, Nguyen van Giai & P. Quentin, Nucl. Phys. A238:29(1975)
11. J.P. Blaizot, D. Gogny and B. Grammaticos, Nucl. Phys. A265:315 (1976)
12. H.S. Köhler, Nucl. Phys. A 258:301 (1976)
13. K.F. Liu and G.E. Brown, Nucl. Phys. A 265:385(1976)
14. S. Krewold, V. Klemt, J. Speth and A. Faessler, Nucl. Phys. A281: 166(1977)
15. M. Waroquier, J. Sau, K. Heyde, P.van Isacker and H. Vincx, Phys. Rev. C19: 1983(1979)
16. D.M. Brink and E. Boeker, Nucl. Phys. A 91:1(1967)
17. X. Campi and D.W. Sprung, Nucl. Phys. 194:401 (1972)
18. R.V. Reid, Ann. Phys. 50:411 (1968)
19. J.W. Negele, Phys. Rev. C1: 1260 (1970)
20. B.K. Jennings and A.D. Jackson, Nucl. Phys. A342:23(1980)
21. T.H.R. Skyrme, Nucl. Phys. 9:615(1959)
22. S. Cohen and D. Kurath, Nucl. Phys. 73:1 (1965)
23. J.P. Elliott, A.D. Jackson, H.A. Mavromatis, E.A. Sanderson and B. Singh, Nucl. Phys. A121:241 (1968)
24. T. Hamada and I.D. Johnston, Nucl. Phys. 34: 382 (1962)
25. E.A. Sanderson, J.P. Elliott, H.A. Mavromatis and B. Singh, Nucl. Phys. A 219:190 (1974)
26. W. Chung, Thesis, Michigan State University (1976)
27. D. Kelvin, A. Watt and R.R. Whitehead, J. Phys. G 3: 1539 (1977)
28 B.M. Preedom and B.H. Wildenthal, Phys. Rev. C 6 : 1633 (1972)
29. N. Kassis, J.P. Elliott and E.A. Sanderson, to be published (1980)
30. J. Dey, J.P. Elliott, A.D. Jackson, H.A. Mavromatis, E.A. Sanderson and B. Singh, Nucl. Phys. A 134: 385 (1969)
31. T.T.S. Kuo and G.E. Brown, Nucl. Phys. 85:40 (1966)
32. B.R. Barrett, Phys. Rev. C. 20:1926 (1979)

AN INTRODUCTION TO THE INTERACTING BOSON MODEL

F. Iachello

Kernfysisch Versneller Instituut, Groningen, The Netherlands
and
Physics Department, Yale University, New Haven, Ct 06520

1. INTRODUCTION

Properties of nuclei with several particles outside the closed
shells have been traditionally discussed in terms of the geometric
model of Bohr and Mottelson[1,2]. In the last few years, an alternative,
algebraic, description of these properties has been introduced[3,4],
known as interacting boson model. The purpose of my lectures is to
provide an introduction to the interacting boson model. No attempt
will be made to discuss the latest developments and refinements
of the model, in particular the explicit introduction of proton
and neutron degrees of freedom and the treatment of odd-A nuclei.
Rather, the lectures will be confined to a description of those
properties which require either no numerical calculation at all,
or very simple calculations. The two descriptions, geometric and
algebraic, are by no means contradictory. Much work has been done
recently on the relationship between the two models. This work
will not be discussed here. The study of the properties of the
interacting boson model makes use of group theory. Therefore, a
good fraction of these lectures will be concerned with the group
theory of the interacting boson model. I will try to keep the
lectures self-contained, introducing some basic concepts of group
theory, such as Casimir operators, etc., as they appear. For a
slightly more detailed account of the theory of Lie groups and of
their representations, the interested student may consult my
lecture notes from the Gull Lake Summer School[5] or any textbook
on group theory[6,7].

2. THE INTERACTING BOSON MODEL

In its simplest form, the interacting boson model assumes that the structure of low-lying levels of nuclei is dominated by excitations of the valence particles, i.e. the particles outside the major closed shells at 2, 8, 20, 28, 50, 82 and 126 (Fig. 1a). Furthermore, it assumes that the important particle configurations in the low-lying levels of even-even nuclei, are those in which identical particles are paired together in states with total angular momentum L = 0 and L = 2.

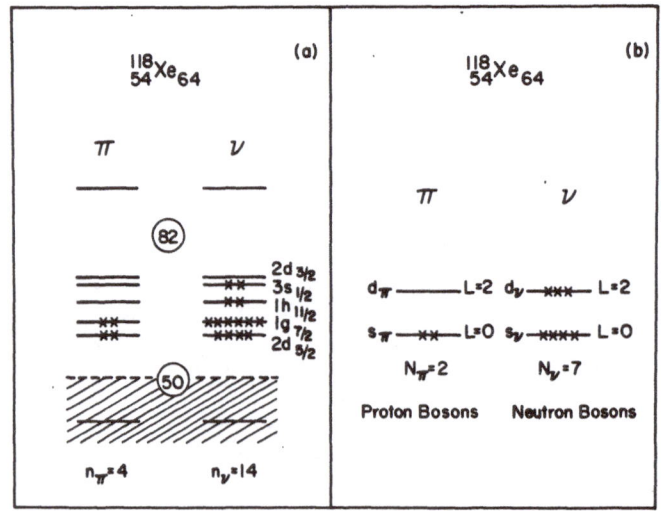

Fig. 1a. A schematic representation of the shell-model problem for $^{118}_{54}$Xe$_{64}$ (n_π and n_ν are the number of protons and neutrons outside the major closed shell at 50).
  b. The boson problem which replaces the shell-model problem for $^{118}_{54}$Xe$_{64}$.

Finally, the pairs are treated as bosons. Proton (neutron) bosons with angular momentum L = 0 are denoted by $s_\pi(s_\nu)$, while proton (neutron) boson with angular momentum L = 2 are denoted by $d_\pi(d_\nu)$, Fig. 1b. In order to take into account the particle-hole conjugation in the particle space, the number of proton, N_π, and neutron, N_ν, bosons is counted from the nearest closed shell, i.e. if more than half of the shell is full, $N_{\pi(\nu)}$, is taken as the number of hole pairs. Thus, for example, for $^{118}_{54}$Xe$_{64}$, Fig. 1, $N_\pi=(54-50)/2=2$, $N_\nu=(64-50)/2=7$, while for $^{128}_{54}$Xe$_{74}$, $N_\pi=(54-50)/2=2$, and $\bar{N}_\nu=(82-74)/2=4$. A bar is placed sometimes over the number $\bar{N}_{\pi(\nu)}$ in order to denote the fact that these are hole states. A detailed description of the properties of nuclei must treat separately proton and neutron pairs. This description, often referred as IBA-2, will not be discussed

here. We shall consider instead only the case in which no distinction
is made between proton and neutron bosons. In this approximation,
often referred as IBA-1, an even-even nucleus is treated as a system
of $N=N_\pi+N_\nu$ bosons.

In order to calculate properties of a given even-even nucleus,
one must first write down the relevant operators. For energies, one
needs the Hamiltonian, H. In the interacting boson model it is
assumed that H contains only one-body and two-body terms. Thus,

$$H = \sum_{i=1}^{N} \epsilon_i + \sum_{i<j}^{N} v_{ij} .$$

(2.1)

For calculations, it is convenient to make use of a second quantized
formalism, thus introducing creation ($s^\dagger, d_\mu^\dagger$) and annihilation (s, d_μ)
operators, where the index $\mu = 0, \pm 1, \pm 2$. These operators satisfy
Bose commutation relations

$$[s,s^\dagger] = 1 \quad , \quad [s,s] = 0 \quad , \quad [s^\dagger,s^\dagger] = 0 \quad ,$$

$$[d_\mu,d_{\mu'}^\dagger] = \delta_{\mu\mu'} \quad , \quad [d_\mu,d_{\mu'}] = 0 \quad , \quad [d_\mu^\dagger,d_{\mu'}^\dagger] = 0 \quad ,$$

$$[s,d_\mu^\dagger] = 0 \quad , \quad [s^\dagger,d_\mu^\dagger] = 0 \quad ,$$

$$[s,d_\mu] = 0 \quad , \quad [s^\dagger,d_\mu] = 0 \quad .$$

(2.2)

The six operators $s^\dagger, d_\mu^\dagger$ will be denoted altogether by $b_{\ell\mu}^\dagger$, $\ell = 0,2$.
It is well known that, while the creation operators $b_{\ell\mu}^\dagger$ transform
as spherical tensors under rotations, the annihilation operators
do not. However, one can easily construct spherical tensors by
introducing the operators $\tilde{b}_{\ell\mu} = (-)^{\ell+\mu}b_{\ell,-\mu}$. In particular, this
gives $\tilde{d}_\mu = (-)^\mu d_{-\mu}$ and $\tilde{s} = s$. Although there is no need to introduce $\tilde{s}$, since $\tilde{s} = s$, I will still do so in these lectures in order
to keep formulas symmetric.

With tensor operators one can form tensor products. The
tensor product of two operators $T_{\kappa_1}^{(k_1)}$, $T_{\kappa_2}^{(k_2)}$ is defined as[8]

$$T_{\kappa_3}^{(k_3)} = \sum_{\kappa_1,\kappa_2} < k_1\kappa_1 k_2\kappa_2 | k_3\kappa_3 > T_{\kappa_1}^{(k_1)} T_{\kappa_2}^{(k_2)}$$

(2.3)

and denoted by

$$T^{(k_3)} = [T^{(k_1)} \times T^{(k_2)}]^{(k_3)}. \tag{2.4}$$

The scalar product of two operators $T^{(k)}$ and $U^{(k)}$ is defined as

$$(T^{(k)} \cdot U^{(k)}) = (-)^k \sqrt{2k+1} \left[T^{(k)} \times U^{(k)} \right]^{(0)}_0 = \sum_\kappa (-)^\kappa T^{(k)}_\kappa U^{(k)}_{-\kappa}. \tag{2.5}$$

Thus, for example,

$$(d^\dagger \cdot \tilde{d}) = \sqrt{5} [d^\dagger \times \tilde{d}]^{(0)}_0 = \sqrt{5} \sum_{\mu_1, \mu_2} <2\mu_1 2\mu_2 | 00> d^\dagger_{\mu_1} \tilde{d}_{\mu_2} =$$

$$= \sum_{\mu_1} d^\dagger_{\mu_1} d_{\mu_1} = n_d, \tag{2.6}$$

the number operator for d-bosons.

We are now in a position to write down the second quantized form of the most general Hamiltonian which contains only one-body and two-body terms

$$H = \varepsilon_s (s^\dagger \cdot \tilde{s}) + \varepsilon_d (d^\dagger \cdot \tilde{d}) + \sum_{L=0,2,4} \frac{1}{2} (2L+1)^{1/2} c_L [[d^\dagger \times d^\dagger]^{(L)} \times [\tilde{d} \times \tilde{d}]^{(L)}]^{(0)}$$

$$+ \frac{1}{2^{1/2}} \tilde{v}_2 [[d^\dagger \times d^\dagger]^{(2)} \times [\tilde{d} \times \tilde{s}]^{(2)} + [d^\dagger \times s^\dagger]^{(2)} \times [\tilde{d} \times \tilde{d}]^{(2)}]^{(0)} \tag{2.7}$$

$$+ \frac{1}{2} \tilde{v}_0 [[d^\dagger \times d^\dagger]^{(0)} \times [\tilde{s} \times \tilde{s}]^{(0)} + [s^\dagger \times s^\dagger]^{(0)} \times [\tilde{d} \times \tilde{d}]^{(0)}]^{(0)}$$

$$+ u_2 [[d^\dagger \times s^\dagger]^{(2)} \times [\tilde{d} \times \tilde{s}]^{(2)}]^{(0)} + \frac{1}{2} u_0 [[s^\dagger \times s^\dagger]^{(0)} \times [\tilde{s} \times \tilde{s}]^{(0)}]^{(0)}.$$

This Hamiltonian is specified by 9 parameters, 2 appearing in the one-body terms, ε_s, ε_d, and 7 in the two body terms, $c_L(L=0,2,4)$, $\tilde{v}_L(L=0,2)$ and $u_L(L=0,2)$. However, since the total number of boson (pairs) is conserved, $N = n_s + n_d$, Eq. (2.7) can be rewritten as

$$H = \epsilon_s N + \frac{1}{2}u_o N(N-1)+\epsilon'(d^\dagger.\tilde{d})+ \sum_{L=0,2,4} \frac{1}{2}(2L+1)^{1/2}c_L'[[d^\dagger \times d^\dagger]^{(L)} \times [\tilde{d} \times \tilde{d}]^{(L)}]^{(0)}$$

$$+ \frac{1}{2^{1/2}} \tilde{v}_2[[d^\dagger \times d^\dagger]^{(2)} \times [\tilde{d} \times \tilde{s}]^{(2)} + [d^\dagger \times s^\dagger]^{(2)} \times [\tilde{d} \times \tilde{d}]^{(2)}]^{(0)} \qquad (2.8)$$

$$+ \frac{1}{2} \tilde{v}_0[[d^\dagger \times d^\dagger]^{(0)} \times [\tilde{s} \times \tilde{s}]^{(0)} + [s^\dagger \times s^\dagger]^{(0)} \times [\tilde{d} \times \tilde{d}]^{(0)}]^{(0)}$$

where

$$\epsilon' = (\epsilon_d - \epsilon_s) + \frac{1}{\sqrt{5}} u_2(N-1) - \frac{1}{2} u_0(2N-1),$$

$$c_L' = c_L + u_0 - 2u_2 . \qquad (2.9)$$

The first two terms in (2.8) contribute only to binding energies. They do not contribute to excitation energies. For a phenomenological analysis of excitation spectra within the framework of the interacting boson model one thus needs at most 6 parameters, ϵ', c_L' (L = 0,2,4) and $\tilde{v}_L$ (L = 0,2).

    There are several other equivalent ways of writing the Hamiltonian, H. Another form, often used in phenomenological analyses, is obtained by introducing the following operators

$$n_d = (d^\dagger.\tilde{d}) ,$$

$$P = \frac{1}{2}(\tilde{d}.\tilde{d}) - \frac{1}{2}(\tilde{s}.\tilde{s}) ,$$

$$L = \sqrt{10} [d^\dagger \times \tilde{d}]^{(1)} ,$$

$$Q = [d^\dagger \times \tilde{s} + s^\dagger \times \tilde{d}]^{(2)} - \frac{1}{2}\sqrt{7} [d^\dagger \times \tilde{d}]^{(2)} \qquad (2.10)$$

$$T_3 = [d^\dagger \times \tilde{d}]^{(3)} ,$$

$$T_4 = [d^\dagger \times \tilde{d}]^{(4)} .$$

In terms of these operators, the most general Hamiltonian, excluding terms which contribute only to binding energies, can be written as

$$H'=\epsilon'' n_d + a_0(P^\dagger.P) + a_1(L.L) + a_2(Q.Q) + a_3(T_3.T_3) + a_4(T_4.T_4) . \qquad (2.11)$$

The reason why it may be convenient to write H in this form is that
it has been found empirically that often only one or two terms in
(2.11) are sufficient to describe accurately the spectrum. Note
that Eq. (2.11) contains 6 independent parameters, ε'', a_i (i=0,1,...4)
as the corresponding part of Eq. (2.8), ε', c_L^J (L = 0,2,4) and
$\tilde{v}_L$ (L = 0,2).
Using some single algebra it is possible to convert (2.11) into
(2.7) and (2.8). The conversion coefficients are given in Table I.

Having written down the Hamiltonian, H, energy levels can now
be found by diagonalizing H in an appropriate basis. The construction of an appropriate basis for this problem is not at all trivial
and it is best done by making use of the powerful techniques of
group theory. We must therefore now turn to the study of the
group structure of the interacting boson model. In addition to
providing an orthonormal basis, this study will allow us to find
all those situations for which the eigenvalue problem for H can
be solved analytically. These situations, although rarely met in
actual nuclei, will shed considerable light into the structure of
low-lying states in nuclei.

Table I   Conversion coefficients from Eq. (2.11) to Eq. (2.7)

	$\varepsilon''n_d$	$a_0(\underset{\sim}{P}^\dagger\cdot\underset{\sim}{P})$	$a_1(\underset{\sim}{L}\cdot\underset{\sim}{L})$	$a_2(\underset{\sim}{Q}\cdot\underset{\sim}{Q})$	$a_3(\underset{\sim}{T}_3\cdot\underset{\sim}{T}_3)$	$a_4(\underset{\sim}{T}_4\cdot\underset{\sim}{T}_4)$
ε_s				$5\,a_2$		
ε_d	ε''		$6\,a_1$	$\frac{11}{4}\,a_2$	$\frac{7}{5}\,a_3$	$\frac{9}{5}\,a_4$
c_0		$\frac{5}{2}\,a_0$	$-12\,a_1$	$\frac{7}{2}\,a_2$	$-\frac{14}{5}\,a_3$	$\frac{18}{5}\,a_4$
c_2			$-\,6\,a_1$	$-\frac{3}{4}\,a_2$	$\frac{8}{5}\,a_3$	$\frac{36}{35}\,a_4$
c_4			$8\,a_1$	a_2	$\frac{1}{5}\,a_3$	$\frac{1}{35}\,a_4$
$\tilde{v}_2$				$-\sqrt{70}\,a_2$		
$\tilde{v}_0$		$-\frac{1}{2}\sqrt{5}\,a_0$		$2\sqrt{5}\,a_2$		
u_2				$2\sqrt{5}\,a_2$		
u_0		$\frac{1}{2}\,a_0$				

3. GROUP STRUCTURE OF THE BOSON HAMILTONIAN

I will keep the discussion here as short as possible, referring the reader to the more detailed description given in Ref. 5 and in the original articles. I begin by introducing the operators

$$G_\kappa^{(k)}(\ell\ell') = [b_\ell^\dagger \times \tilde{b}_{\ell'}]_\kappa^{(k)} \quad , \qquad \ell,\ell' = 0, 2 = s, d , \qquad (3.1)$$

with

$$[b_{\ell\alpha}, b_{\ell'\alpha'}^\dagger] = \delta_{\ell\ell'}\,\delta_{\alpha\alpha'} . \qquad (3.2)$$

The operators (3.1) satisfy the following commutation relations

$$\left[G_\kappa^{(k)}(\ell\ell'), G_{\kappa'}^{(k')}(\ell''\ell''')\right] = \sum_{k'',\kappa''} (2k+1)^{1/2}(2k'+1)^{1/2} \langle k\kappa k'\kappa'|k''\kappa''\rangle (-)^{k-k'} \times$$

$$\times \left[(-)^{k+k'+k''}\begin{Bmatrix} k & k' & k'' \\ \ell''' & \ell & \ell' \end{Bmatrix}\delta_{\ell'\ell''}G_{\kappa''}^{(k'')}(\ell\ell''') - \begin{Bmatrix} k & k' & k'' \\ \ell'' & \ell' & \ell \end{Bmatrix}\delta_{\ell\ell'''}G_{\kappa''}^{(k'')}(\ell''\ell')\right]. \qquad (3.3)$$

Operators, X, satisfying relations of the sort

$$[X_a, X_b] = \sum_c c_{ab}^c X_c , \qquad (3.4)$$

are said to form a Lie algebra with structure constants c_{ab}^c. One can verify that (3.3) defines the Lie algebra of the group U(6) of unitary transformations in six dimensions. Since the Hamiltonian H in (2.7) is built out of the operators $G_\kappa^{(k)}(\ell\ell')$ one says that it has the group structure of U(6). The operators $G_\kappa^{(k)}(\ell\ell')$ are called generators. There are a total of $36 = 6^2$, which written down explicitly read

$$G_0^{(0)}(ss) = [s^\dagger \times \tilde{s}]_0^{(0)} \qquad 1$$

$$G_0^{(0)}(dd) = [d^\dagger \times \tilde{d}]_0^{(0)} \qquad 1$$

$$G_\kappa^{(1)}(dd) = [d^\dagger \times \tilde{d}]_\kappa^{(1)} \qquad 3 \qquad\qquad (3.5)$$

$$G_\kappa^{(2)}(dd) = [d^\dagger \times \tilde{d}]_\kappa^{(2)} \qquad 5$$

$$G^{(3)}_{\kappa}(dd) \;=\; [\,d^{\dagger} \times \tilde{d}\,]^{(3)}_{\kappa} \qquad\qquad 7$$

$$G^{(4)}_{\kappa}(dd) \;=\; [\,d^{\dagger} \times \tilde{d}\,]^{(4)}_{\kappa} \qquad\qquad 9$$

$$G^{(2)}_{\kappa}(ds) \;=\; [\,d^{\dagger} \times \tilde{s}\,]^{(2)}_{\kappa} \qquad\qquad 5 \qquad\qquad (3.5)$$

$$G^{(2)}_{\kappa}(sd) \;=\; [\,s^{\dagger} \times \tilde{d}\,]^{(2)}_{\kappa} \qquad\qquad 5$$

$$\overline{}$$
$$36 \;=\; 6^2$$

Once the full algebraic structure of the problem has been identified, the next step is that of identifying all possible subalgebras of the full algebra. A subalgebra is a subset of generators which is closed under commutation. It turns out that in the present case there are three possible chains of subalgebras.

Subalgebras I

Delete from the 36 operators the 11 operators $G^{(0)}_{0}(ss)$, $G^{(2)}_{\kappa}(ds)$, $G^{(2)}_{\kappa}(sd)$. The remaining 25 operators close under the algebra $U(5)$, the group of unitary transformations in five dimensions.

A) U(5)

$$G^{(0)}_{0}(dd) \;=\; [\,d^{\dagger} \times \tilde{d}\,]^{(0)}_{0} \qquad\qquad 1$$

$$G^{(1)}_{\kappa}(dd) \;=\; [\,d^{\dagger} \times \tilde{d}\,]^{(1)}_{\kappa} \qquad\qquad 3$$

$$G^{(2)}_{\kappa}(dd) \;=\; [\,d^{\dagger} \times \tilde{d}\,]^{(2)}_{\kappa} \qquad\qquad 5 \qquad\qquad (3.6)$$

$$G^{(3)}_{\kappa}(dd) \;=\; [\,d^{\dagger} \times \tilde{d}\,]^{(3)}_{\kappa} \qquad\qquad 7$$

$$G^{(4)}_{\kappa}(dd) \;=\; [\,d^{\dagger} \times \tilde{d}\,]^{(4)}_{\kappa} \qquad\qquad 9$$

$$\overline{}$$
$$25 \;=\; 5^2$$

Delete from the 25 operators the 15 operators $G^{(0)}_{0}(dd)$, $G^{(2)}_{\kappa}(dd)$ and $G^{(4)}_{\kappa}(dd)$. The remaining 10 operators close under the algebra of $O(5)$, the orthogonal group in five dimensions.

B) O(5)

$$G^{(1)}_{\kappa}(dd) \;=\; [\,d^{\dagger} \times \tilde{d}\,]^{(1)}_{\kappa} \qquad\qquad 3$$

$$G^{(3)}_{\kappa}(dd) \;=\; [\,d^{\dagger} \times \tilde{d}\,]^{(3)}_{\kappa} \qquad\qquad 7 \qquad\qquad (3.7)$$

$$\overline{}$$
$$10 \;=\; 5 \times 4 / 2$$

Delete from the 10 operators the 7 operators $G_\kappa^{(3)}$ (dd). The remaining 3 operators close under the algebra of O(3), the ordinary rotation group.

C) O(3)

$$G_\kappa^{(1)}(dd) = [d^\dagger \times \tilde{d}]_\kappa^{(1)} \qquad\qquad 3 \qquad\qquad (3.8)$$

Finally, delete from the 3 operators the 2 operators $G_{+1}^{(1)}$ (dd) and $G_{-1}^{(1)}$ (dd). The remaining operator is the generator of O(2), the group of rotations around the z-axis.

D) O(2)

$$G_0^{(1)}(dd) = [d^\dagger \times \tilde{d}]_0^{(1)} \qquad\qquad 1 \qquad\qquad (3.9)$$

Thus, one possible chain of subalgebras is

$$U(6) \supset U(5) \supset O(5) \supset O(3) \supset O(2) \; . \qquad\qquad (I) \qquad (3.10)$$

Subalgebras II

A) U(3)

Consider the operators

$$G_0^{(0)}(ss) + \sqrt{5}\, G_0^{(0)}(dd) = [s^\dagger \times \tilde{s}]_0^{(0)} + \sqrt{5}[d^\dagger \times \tilde{d}]_0^{(0)} \qquad 1$$

$$G_\kappa^{(1)}(dd) = [d^\dagger \times \tilde{d}]_\kappa^{(1)} \qquad\qquad 3$$

$$G_\kappa^{(2)}(ds) + G_\kappa^{(2)}(sd) - \tfrac{1}{2}\sqrt{7} G_\kappa^{(2)}(dd) = [d^\dagger \times \tilde{s} + s^\dagger \times \tilde{d}]_\kappa^{(2)} - \tfrac{1}{2}\sqrt{7}[d^\dagger \times \tilde{d}]_\kappa^{(2)} \qquad 5$$

$$\overline{9=3^2}$$

$$(3.11)$$

These operators close under commutation and form the algebra of U(3). Obvious subalgebras are now

B) O(3)

$$G_\kappa^{(1)}(dd) = [d^\dagger \times \tilde{d}]_\kappa^{(1)} \qquad\qquad 3 \qquad\qquad (3.12)$$

and

C) O(2)

$$G_0^{(1)}(dd) = [d^\dagger \times \tilde{d}]_0^{(1)} \qquad\qquad 1 \qquad\qquad (3.13)$$

Thus, a second possible chain of subalgebras is

$$U(6) \supset U(3) \supset O(3) \supset O(2) \ . \qquad\qquad (II) \qquad (3.14)$$

Subalgebras III

A) O(6)

Consider the operators

$$G^{(1)}_\kappa (dd) = [\, d^\dagger \times \tilde{d}\,]^{(1)}_\kappa \qquad\qquad\qquad 3$$

$$G^{(3)}_\kappa (dd) = [\, d^\dagger \times \tilde{d}\,]^{(3)}_\kappa \qquad\qquad\qquad 7 \qquad\qquad (3.15)$$

$$G^{(2)}_\kappa (ds) + G^{(2)}_\kappa (sd) = [\, d^\dagger \times \tilde{s} + s^\dagger \times \tilde{d}\,]^{(2)}_\kappa \qquad 5$$

$$\overline{15} = 6 \times 5/2$$

These operators close under commutation, yielding the Lie algebra of O(6), the orthogonal group in six dimensions. Obvious subalgebras are now

B) O(5)

$$G^{(1)}_\kappa (dd) = [\, d^\dagger \times \tilde{d}\,]^{(1)}_\kappa \qquad\qquad\qquad 3$$

$$\qquad\qquad\qquad\qquad\qquad\qquad\qquad\qquad\qquad\qquad (3.16)$$

$$G^{(3)}_\kappa (dd) = [\, d^\dagger \times \tilde{d}\,]^{(3)}_\kappa \qquad\qquad\qquad 7$$

$$\overline{10}$$

C) O(3)

$$G^{(1)}_\kappa (dd) = [\, d^\dagger \times \tilde{d}\,]^{(1)}_\kappa \qquad\qquad\qquad 3 \qquad\qquad (3.17)$$

D) O(2)

$$G^{(1)}_0 (dd) = [\, d^\dagger \times \tilde{d}\,]^{(1)}_0 \qquad\qquad\qquad 1 \qquad\qquad (3.18)$$

Thus, a third possible chain is

$$U(6) \supset O(6) \supset O(5) \supset O(3) \supset O(2) \ . \qquad\qquad (III) \qquad (3.19)$$

It is possible to show that these are the only possible chains of

subgroups for this problem, if one insists that the angular
momentum L be a good quantum number (i.e. O(3) must be contained
in the chain). In fact, starting from U(6) we have considered U(5),
U(3) and O(6). But O(6) is locally isomorphic to SU(4). Thus, we
have considered all possible subgroups of U(6), namely U(5), U(4)
and U(3). In conclusion, there are three and only three possible
chains

$$U(6) \begin{cases} U(5) \supset O(5) \supset O(3) \supset O(2) & \text{I} \\ U(3) \supset O(3) \supset O(2) & \text{II} \\ O(6) \supset O(5) \supset O(3) \supset O(2) & \text{III} \end{cases} \quad (3.20)$$

4. CLASSIFICATION OF STATES

Once a group chain has been identified, its first, important
application is in constructing a basis in which the Hamiltonian H
can be diagonalized. In order to do this, we need to know the
labels which characterize the irreducible representations of the
various groups which appear in the chain. For example, it is well
known that the representations of the rotation group, O(3), are
characterized by the values of the angular momentum, L. For the
group U(6), U(5), etc. appearing in (3.20) the situation is more
complex. The general procedure by means of which one can find the
quantum numbers characterizing the irreducible representations of
the various groups is given in Refs. 5, 6 and 7. Here I give only
the results. Since we are dealing with a system of identical
bosons some simplifications occur as indicated below.

Group chain I

The labels needed to classify the states in this chain are

U(6)   [N]

U(5) (n_d)

O(5)   v

O(3)   L

O(2)   M

The representations of U(6), U(5) and O(5) are characterized
only by one number because they are totally symmetric. N is the
total number of bosons, n_d the number of d-bosons and v is called
seniority. One important problem is that of finding the represen-

tations of a subgroup G' contained in the larger group G. For example, for given n_d, what are the allowed values of v? This is one of the most difficult problems of group theory and it can be solved by means of a procedure known as building up process. For details the student is referred to Ref. 5. It should be noted incidentally that the same quantum numbers N and n_d label the representations of the special unitary groups SU(6) and SU(5). For this reason, this group chain is often referred as the SU(5) chain. The values of n_d contained in each [N] are

$$n_d = 0, 1, \ldots, N \quad . \tag{4.1}$$

The values of v contained in each (n_d) are

$$v = n_d, n_d-2, \ldots, 1 \text{ or } 0 \quad ; \quad n_d = \text{odd or even.} \tag{4.2}$$

When going from O(5) to O(3) one additional problem arises, namely that more than one state with a given value of L is contained in a given representation (v) of O(5). When this occurs one says that the step from O(5) to O(3) is not fully decomposable and an additional quantum number is needed to characterize the states. In this particular case the quantum number, denoted by n_Δ, is chosen as the number of boson triplets coupled to zero angular momentum. The values of L contained in each representation n_d of SU(5) are then given by the following algoritm. First, partition n_d as

$$n_d = 2n_\beta + 3n_\Delta + \lambda, \tag{4.3}$$

where

$$n_\beta = (n_d-v)/2 \quad ; \quad n_\beta = 0,1,\ldots, \frac{n_d}{2} \text{ or } \frac{n_d-1}{2} \tag{4.4}$$

Then,

$$L = \lambda, \lambda+1, \lambda+2, \ldots, 2\lambda-2, 2\lambda \tag{4.5}$$

[Note that $2\lambda-1$ is missing!]. This gives the following table.

Table II   Classification scheme for the group chain I

SU(5)	O(5)		O(3)
n_d	v	n_Δ	L
0	0	0	0
1	1	0	2

Table II cont'd

2	2	0	4,2
	0	0	0
3	3	0	6,4,3
		1	0
	1	0	2
4	4	0	8,6,5,4
		1	2
	2	0	4,2
	0	0	0

The complete classification scheme for chain I is thus[9]
$|[N](n_d) \, v \, n_\Delta \, L \, M >$

Group chain II

The labels needed to classify the states in this chain are

U(6)   [N]

SU(3) (λ,μ)

O(3)   L

O(2)   M

Two quantum numbers (λ,μ) are needed to characterize the representations of SU(3), here used instead of U(3). The values of (λ,μ) contained in each [N] are given by

$$[N] = (2N,0) \oplus (2N-4,2) \oplus (2N-8,4) \oplus \ldots \oplus \begin{Bmatrix} (0,N) \\ (2,N-1) \end{Bmatrix} \oplus \begin{Bmatrix} N=\text{even} \\ N=\text{odd} \end{Bmatrix}$$

$$\oplus (2N-6,0) \oplus (2N-10,2) \oplus \ldots \qquad \oplus \begin{Bmatrix} (0,N-3) \\ (2,N-4) \end{Bmatrix} \oplus \begin{Bmatrix} N-3=\text{even} \\ N-3=\text{odd} \end{Bmatrix}$$

$$\oplus (2N-12,0) \oplus (2N-16,2) \oplus \ldots \qquad \oplus \begin{Bmatrix} (0,N-6) \\ (2,N-7) \end{Bmatrix} \oplus \begin{Bmatrix} N-6=\text{even} \\ N-6=\text{odd} \end{Bmatrix}$$

$$\oplus \ldots \tag{4.6}$$

The step from SU(3) to O(3) is not fully decomposable. The simplest choice of the additional quantum number needed to classify the states is due to Elliott. The corresponding number is called K. The values of L contained in each (λ,μ) in Elliott basis are given by the following algoritm:

$$L = K, K+1, K+2,...., K+\max\{\lambda,\mu\}, \qquad (4.7)$$

where

$$K = \text{integer} = \min\{\lambda,\mu\}, \min\{\lambda,\mu\}-2,..., 1 \text{ or } 0 \qquad (4.8)$$

with the exception of K = 0 for which

$$L = \max\{\lambda,\mu\}, \max\{\lambda,\mu\}-2,..., 1 \text{ or } 0. \qquad (4.9)$$

Elliott basis has the drawback of not being orthogonal. For this reason, it is convenient to introduce another basis, called Vergados basis, which can be constructed from Elliott basis in the following way. Let $K_1, K_2,...K_n$ be the Elliott quantum numbers which occur in a given representation (λ,μ) with $K_1 < K_2 < ... < K_n$. The new basis is labelled by the quantum numbers $\chi_1, \chi_2,..., \chi_n$ with $\chi_1 < \chi_2 < .... < \chi_n$ and defined by

$$|(\lambda,\mu) \chi_1 L M> = |(\lambda,\mu) K_1 L M>_0 ,$$

$$|(\lambda,\mu) \chi_2 L M> = x_{21}|(\lambda,\mu) K_1 L M>_0 + x_{22}|(\lambda,\mu) K_2 L M>_0 ,$$

$$... \qquad (4.10)$$

$$|(\lambda,\mu) \chi_i L M> = \sum_{j=1}^{i} x_{ij}|(\lambda,\mu) K_j L M>_0 ,$$

where the states $|(\lambda,\mu) K L M>_0$ are related to Elliott states $|(\lambda,\mu) K L M>$ by the phase convention

$$|(\lambda,\mu) K L M>_0 = i^{\lambda+2\mu}|(\lambda,\mu) K L M> \qquad (4.11)$$

and the coefficients x_{ij} are obtained by the requirement

$$<(\lambda,\mu) \chi_i L M |(\lambda,\mu) \chi_j L M> = \delta_{ij} \qquad (4.12)$$

Thus, the sequence of quantum numbers $\chi_1, \chi_2,.., \chi_n$ is the same as $K_1, K_2,..., K_n$ but the values of L contained in each χ_i are different from those contained in K_i. In fact, from its definition,

it is clear that, if a given L occurs in a given representation only
once, it belongs to the lowest possible χ. If it occurs twice, it
belongs to the two lowest possible χ's, etc. The only exception is
when $\chi = 0$ for which the allowed L values are restricted to be even
or odd for λ even or odd, respectively. In the following, the
Vergados basis will be used. This gives the following table.

Table III Classification scheme for the group chain II

SU(6)	SU(3)		O(3)
N	(λ,μ)	χ	L
0	(0,0)	0	0
1	(2,0)	0	2,0
2	(4,0)	0	4,2,0
	(0,2)	0	2,0
3	(6,0)	0	6,4,2,0
	(2,2)	0	4,2,0
		2	3,2
	(0,0)	0	0

The complete classification for chain II is^{10}

$$|[N]\ (\lambda,\mu)\ \chi\ L\ M>.$$

Group chain III

The labels needed to classify the states in this chain are

U(6)   [N]

O(6) (σ)

O(5) τ

O(3)   L

O(2)   M

The values of σ contained in each [N] are given by

$$\sigma = N,\ N-2,\ ....,\ 0\ \text{or}\ 1\ ,\ \text{for}\ N = \text{even or odd}\ . \qquad (4.13)$$

The values of τ contained in each σ are given by

$$\tau = \sigma, \; \sigma-1, \ldots, 0 \; . \tag{4.14}$$

Once more, the step from O(5) to O(3) is not fully reducible. One needs a further quantum number, called ν_Δ. The values of L contained in each τ are found by partitioning τ as

$$\tau = 3\nu_\Delta + \lambda \quad , \qquad \nu_\Delta = 0, \; 1, \; \ldots \tag{4.15}$$

and taking

$$L = 2\lambda, \; 2\lambda-2, \; \ldots \; , \; \lambda+1, \; \lambda \tag{4.16}$$

[Note that $2\lambda-1$ is missing!]. This gives the following table.

Table IV   Classification scheme for the group chain III

SU(6)	O(6)	O(5)		O(3)
N	σ	τ	ν_Δ	L
0	0	0	0	0
1	1	1	0	2
		0	0	0
2	2	2	0	4,2
		1	0	2
		0	0	0
	0	0	0	0
3	3	3	0	6,4,3
			1	0
		2	0	4,2
		1	0	2
		0	0	0
	1	1	0	2
		0	0	0

The complete classification scheme for chain III is[11]

$$|[N] \; \sigma \; \tau \; \nu_\Delta \; L \; M >.$$

Having constructed a classification scheme of the states, we are now in a position to diagonalize the Hamiltonian H of Eq. (2.7).

Any of the three chains can be used in this diagonalization, since all three are complete. A computer program, called PHINT, has been written by Scholten for the diagonalization of H. This computer program makes use of the basis defined by chain I, and is available on request. However, before turning to the results of the numerical calculations, it is interesting to study those cases for which the eigenvalue problem can be solved analytically.

5. DYNAMICAL SYMMETRIES

The technique used to find analytic solutions of the eigenvalue problem of H is again based on group theory. One first introduces some operators C, called Casimir operators, with the property

$$[C, G^{(k)}_\kappa] = 0 \quad , \text{ any } k, \kappa \ , \tag{5.1}$$

i.e. the operators C commute with all operators of the algebra, $G^{(k)}_\kappa$. For example, the algebra of U(6) has a linear Casimir operator

$$C = G^{(0)}_0 (ss) + \sqrt{5}\, G^{(0)}_0 (dd) \ . \tag{5.2}$$

This operator commutes with all 36 operators, (3.5) and it is nothing but the total number of particles

$$C = [s^\dagger \times \tilde{s}]^{(0)} + \sqrt{5}\,[d^\dagger \times \tilde{d}]^{(0)} = n_s + n_d = N \tag{5.3}$$

The statement that C commutes with all $G^{(k)}_\kappa$ is the same as the statement that the total number of particles is conserved by the Hamiltonian, H. A linear Casimir operator of a group U(n) will be denoted by C_{1Un}. Note that only unitary groups have linear Casimir operators. Similarly, one has also quadratic Casimir operators. For example, the algebra of O(3), generated by the operators $G^{(1)}_\kappa(dd) = \tilde{G}^{(1)}$ of Eq. (3.8) has a quadratic Casimir operator

$$C = \tilde{G}^{(1)} \cdot \tilde{G}^{(1)} \tag{5.4}$$

Since $\tilde{G}^{(1)}$ is proportional to the angular momentum $\tilde{L}$, Eq. (2.10), the operator C is proportional to $\tilde{L} \cdot \tilde{L}$. This operator commutes with all components of $\tilde{L}$, as it is well known from simple angular momentum algebra. A quadratic Casimir operator of a group $O(n)$ will be denoted by C_{2On}.

It is now possible to show[12] that, apart from terms which contribute only to the binding energies, the most general Hamiltonian, H, can be written in terms of the Casimir operators of the groups $U(5)$, $O(5)$, $O(3)$, $SU(3)$ and $O(6)$. Since the Hamiltonian is at most two body, this expression will involve at most quadratic operators. [It should be noted that the group $O(2)$ does not play any role unless the nucleus is placed in an external magnetic field. Therefore it will be neglected henceforth]. Of the various groups mentioned above, only $U(5)$ has a linear Casimir operator. Thus, the most general Hamiltonian can be written as[5]

$$H' = \varepsilon''' \ C_{1U5} + \alpha' \ C_{2U5} + \beta' \ C_{2O5} + \gamma' \ C_{2O3} + \delta' \ C_{2SU3} + \eta' \ C_{2O6}.$$

$$(5.5)$$

As in (2.11) there are 6 independent parameters, ε''', α', β', γ', δ', η'. The Casimir operators appearing in (5.5) are clearly related to the operators of (2.11). Some care must be taken since Casimir operators are defined up to a constant factor and different authors use different values for these constant factors. If one uses the definitions of Ref. 5, the relations are as follows

$$C_{1U5} = n_d = (d^\dagger.\tilde{d}) \quad ,$$

$$C_{2U5} = n_d(n_d+4) = (d^\dagger.\tilde{d})(d^\dagger.\tilde{d}) + 4(d^\dagger.\tilde{d}) \quad ,$$

$$C_{2O5} = 4\left(\frac{1}{10}\ L.L+T_3.T_3\right)=4\left([d^\dagger\times\tilde{d}]^{(1)}.[d^\dagger\times\tilde{d}]^{(1)} +[d^\dagger\times\tilde{d}]^{(3)}.[d^\dagger\times\tilde{d}]^{(3)}\right),$$

$$C_{2O3} = 2(L.L) = 2\left(10[d^\dagger\times\tilde{d}]^{(1)}.[d^\dagger\times\tilde{d}]^{(1)}\right),$$

$$(5.6)$$

$$C_{2SU3} = \frac{2}{3}[2(Q.Q)+\frac{3}{4}(L.L)] = \frac{2}{3}\{2\left([d^\dagger\times\tilde{s}+s^\dagger\times\tilde{d}]^{(2)}-\frac{1}{2}\sqrt{7}[d^\dagger\times\tilde{d}]^{(2)}\right).$$

$$.\left([d^\dagger\times\tilde{s}+s^\dagger\times\tilde{d}]^{(2)} - \frac{1}{2}\sqrt{7}[d^\dagger\times\tilde{d}]^{(2)}\right) + \frac{15}{2}\left([d^\dagger\times\tilde{d}]^{(1)}.[d^\dagger\times\tilde{d}]^{(1)}\right),$$

$$C_{2O6} = 2N(N+4) - 8 \ (P^\dagger.P) = 2N(N+4)-2(d^\dagger.d^\dagger-s^\dagger.s^\dagger)(\tilde{d}.\tilde{d}-\tilde{s}.\tilde{s}).$$

Other operators used in Refs. 9, 10 and 11 are

$$C(\lambda,\mu) = \frac{3}{2} C_{2SU3} \quad ,$$

$$C_5 = \frac{1}{12} C_{2O5} \quad ,$$

$$C_3 = \frac{1}{2} C_{2O3} \quad ,$$

$$C_6 = C_{2O6} \quad . \tag{5.7}$$

Casimir operators have the important property of being diagonal in the representation provided by the corresponding group. This property gives the possibility to find all possible special cases for which the eigenvalue problem can be solved analytically. This will occur whenever the Hamiltonian can be written in terms only of Casimir operators of a complete chain of subgroups of U(6), for in that case H is diagonal. When this occurs, one says that the Hamiltonian H has a dynamical symmetry. Since there are three subgroup chains, there are three possible dynamical symmetries. As it is clear from Eq. (5.5), these symmetries correspond to the vanishing of some coefficients in (5.5).

Dynamical symmetry I

The group chain here is

$$U(6) \supset U(5) \supset O(5) \supset O(3) \supset O(2) \quad . \tag{5.8}$$

This symmetry corresponds to the vanishing of δ' and η' in (5.5). The corresponding Hamiltonian is

$$H^{(I)} = \varepsilon''' \, C_{1U5} + \alpha' \, C_{2U5} + \beta' \, C_{2O5} + \gamma' \, C_{2O3} \quad . \tag{5.9}$$

In order to find the expectation value of $H^{(I)}$ in the representation $|[N](n_d)v \, n_\Delta \, L \, M\rangle$ one needs to know the expectation values of the various Casimir operators appearing in (5.9). These are given by standard group theoretical techniques[5]. The resulting expression is

$$\langle H^{(I)}\rangle = \varepsilon''' \, n_d + \alpha' n_d(n_d+4) + 2\beta'v(v+3) + 2\gamma'L(L+1). \tag{5.10}$$

The structure of the spectrum for ε''', α', β', $\gamma' > 0$ is shown in

F. IACHELLO

Fig. 2. Several other combinations of the Casimir operators of the chain can be used to generate the solution. The combination used in Ref. 9 leads to the result

$$\langle H^{(I)} \rangle = \varepsilon n_d + \alpha \tfrac{1}{2} n_d (n_d - 1) + \beta [n_d (n_d + 3) - v(v+3)] + \gamma [L(L+1) - 6n_d] \,. \quad (5.11)$$

The parameters ε, α, β and γ in (5.11) are related in a simple way to the parameters ε''', α', β' and γ' in (5.10).

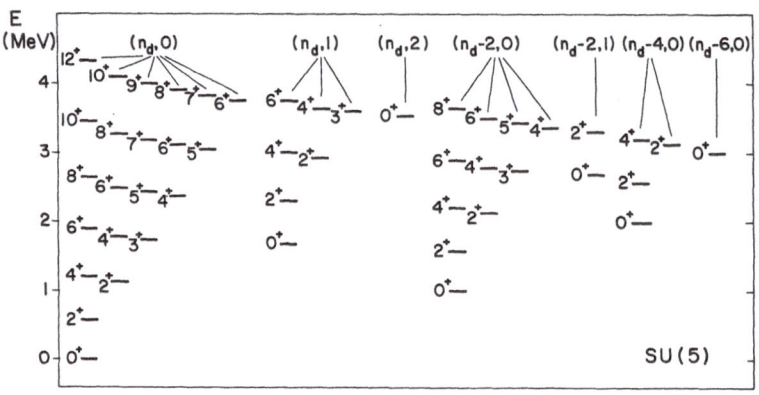

Fig. 2. A typical spectrum with U(5) symmetry and N=6. In parenthesis are the values of v and n_Δ.

Dynamical symmetry II

The group chain here is

$$U(6) \supset SU(3) \supset O(3) \supset O(2) \,. \qquad (5.12)$$

This dynamical symmetry corresponds to the vanishing of ε''', α', β' and γ' in (5.5). The corresponding Hamiltonian is

$$H^{(II)} = \delta' \, C_{2SU3} + \gamma' \, C_{2O3} \,. \qquad (5.13)$$

The expectation value of H in the representation $|[N](\lambda,\mu) \chi L M >$ is given by[5]

$$<H^{(II)}> = \delta' \frac{6}{9}[\lambda^2 + \mu^2 + \lambda\mu + 3(\lambda+\mu)] + \gamma' 2L(L+1). \qquad (5.14)$$

The structure of the spectrum when $\delta' < 0$, $\gamma' > 0$ is shown in Fig. 3. Again, several linear combinations of C_{2SU3} and C_{2O3} can be used to generate the spectrum. In Ref. 10 the operator

$$H^{(II)} = -\kappa \; 2 \; \underset{\sim}{Q} \cdot \underset{\sim}{Q} - \kappa' \; \underset{\sim}{L} \cdot \underset{\sim}{L} \qquad (5.15)$$

is used, with eigenvalues

$$<H^{(II)}> = (\frac{3}{4}\kappa - \kappa')L(L+1) - \kappa[\lambda^2 + \mu^2 + \lambda\mu + 3(\lambda+\mu)] \qquad . \qquad (5.16)$$

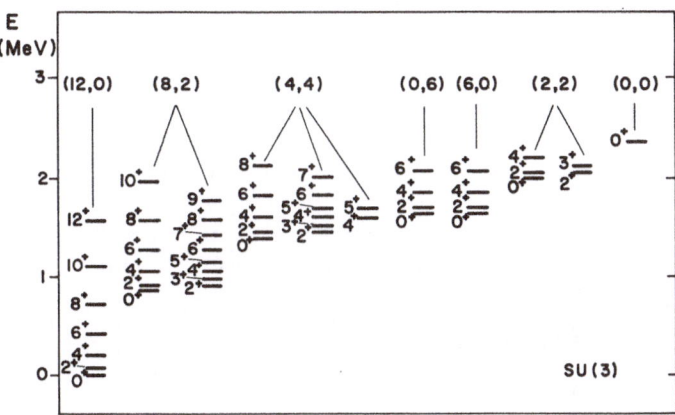

Fig. 3. A typical spectrum with SU(3) sym-
        metry and N=6. In parenthesis are
        the values of λ and μ.

Dynamical symmetry III

The group chain here is

$$U(6) \supset O(6) \supset O(5) \supset O(3) \supset O(2).$$ (5.17)

This symmetry corresponds to the vanishing of the coefficients ε''', α', δ' in (5.5). The corresponding Hamiltonian is

$$H^{(III)} = \beta' \, C_{205} + \gamma' \, C_{203} + \eta' \, C_{206} .$$ (5.18)

The expectation value of H in the representation $|[N](\sigma) \tau \, \nu_\Delta \, L \, M >$ is[5]

$$<H^{(III)}> = \beta' \, 2\tau(\tau+3) + \gamma' 2L(L+1) + \eta' 2\sigma(\sigma+4).$$ (5.19)

The corresponding structure of the spectrum for β', $\gamma' > 0$, $\eta' < 0$ is shown in Fig. 4
Another combination used in Ref. 11 to generate the spectrum is

$$H^{(III)} = A \, P_6 + B \, C_5 + C \, C_3 ,$$ (5.20)

where $P_6 = P^\dagger . P$ and C_5 and C_3 are given by (5.7). The eigenvalues of (5.20) are

$$<H^{(III)}> = A\frac{1}{4}(N-\sigma)(N+\sigma+4) + B\frac{1}{6}\tau(\tau+3) + C \, L(L+1).$$ (5.21)

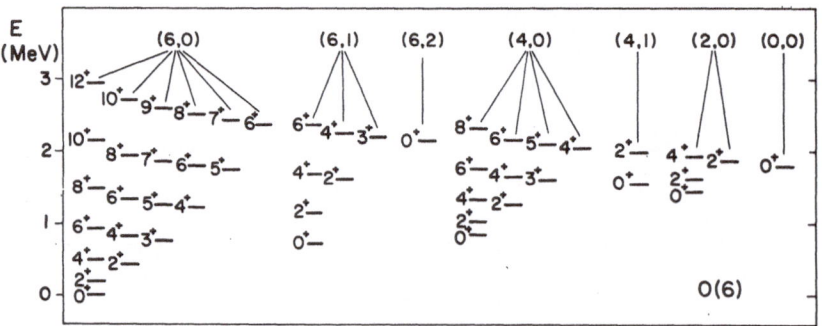

Fig. 4. A typical spectrum with O(6) symmetry
and N=6. In parenthesis are the values
of σ and ν_Δ.

6. EXAMPLES OF SPECTRA WITH DYNAMICAL SYMMETRY

    Having constructed the three analytic solutions, (5.11),
(5.16) and (5.21), it is interesting to see whether or not nuclear
spectra display examples of these three dynamical symmetries. It
turns out that among the known medium mass and heavy even-even
nuclei, there appear to be several nuclei whose spectrum is
relatively close to one of the three limiting cases, I), II) and
III). These nuclei thus provide evidence for the occurrence of
dynamical symmetries in nuclear physics. Three examples are
given in the following figures 5, 6 and 7.

<u>Dynamical symmetry I</u>

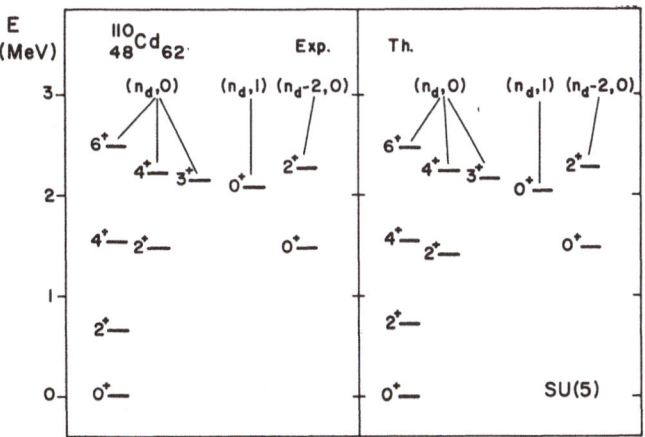

Fig. 5. An example of a spectrum with
        SU(5) symmetry: $^{110}_{48}Cd_{62}$, $N_\pi = 1$,
        $N_\nu = 6$, $N = 7$.

Dynamical symmetry II

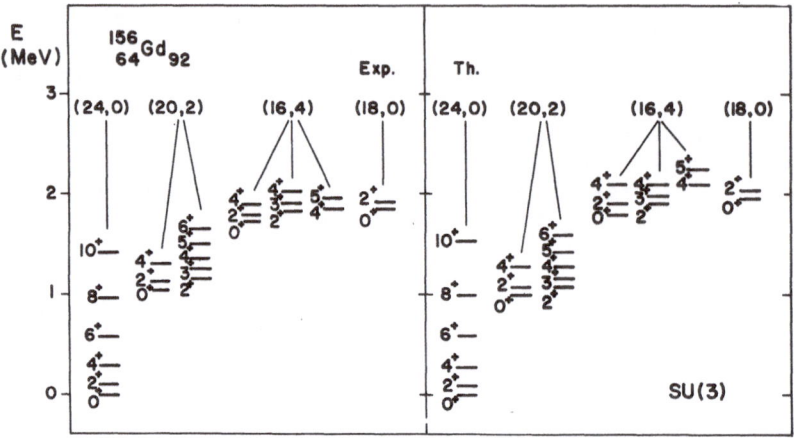

Fig. 6. An example of a spectrum with
SU(3) symmetry: $^{156}_{64}$Gd$_{92}$, N_π=7,
N_ν=5, N=12.

Dynamical symmetry III

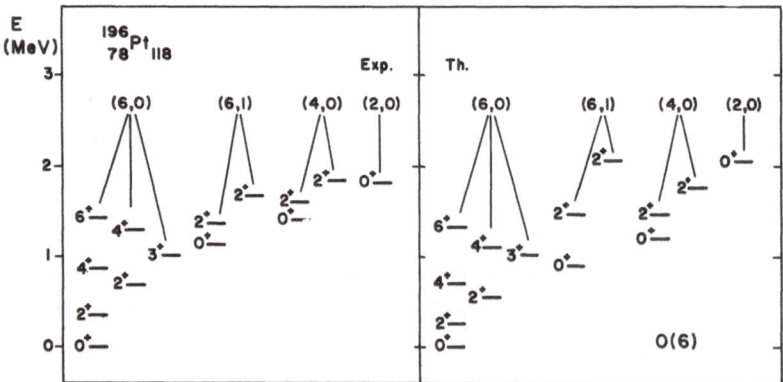

Fig. 7. An example of a spectrum with O(6)
symmetry: $^{196}_{78}$Pt$_{118}$, N_π=2, N_ν=4, N=6.

7. ELECTROMAGNETIC TRANSITION RATES

In order to calculate electromagnetic transition rates, one must specify the transition operators. In the simplest form of the interacting boson model, it is assumed that the transition operators are, at most, one-body operators

$$\underset{\sim}{T}^{(\ell)} = \sum_{i=1}^{N} \underset{\sim}{t}_i^{(\ell)} \ . \tag{7.1}$$

The second quantized form of (7.1) is then

$$\underset{\sim}{T}^{(\ell)}_m = \alpha_2 \delta_{\ell 2} [d^\dagger \times \tilde{s} + s^\dagger \times \tilde{d}]_m^{(2)} + \beta_\ell [d^\dagger \times \tilde{d}]_m^{(\ell)} + \gamma_0 \delta_{\ell 0} \delta_{m0} [s^\dagger \times \tilde{s}]_0^{(0)} . \tag{7.2}$$

Eq. (7.2) gives rise to the following operators

$$\underset{\sim}{T}^{(E0)}_0 = \beta_0 [d^\dagger \times \tilde{d}]_0^{(0)} + \gamma_0 [s^\dagger \times \tilde{s}]_0^{(0)} \ ,$$

$$\underset{\sim}{T}^{(M1)}_m = \beta_1 [d^\dagger \times \tilde{d}]_m^{(1)} \ ,$$

$$\underset{\sim}{T}^{(E2)}_m = \alpha_2 [d^\dagger \times \tilde{s} + s^\dagger \times \tilde{d}]_m^{(2)} + \beta_2 [d^\dagger \times \tilde{d}]_m^{(2)} \ , \tag{7.3}$$

$$\underset{\sim}{T}^{(M3)}_m = \beta_3 [d^\dagger \times \tilde{d}]_m^{(3)} \ ,$$

$$\underset{\sim}{T}^{(E4)}_m = \beta_4 [d^\dagger \times \tilde{d}]_m^{(4)} .$$

No multipole higher than four is possible if the operators $\underset{\sim}{T}^{(\ell)}$ are assumed to be at most one-body. Furthermore, the conservation of boson number $N = n_s + n_d$ allows one to rewrite $\underset{\sim}{T}^{(E0)}$ as

$$\underset{\sim}{T}^{(E0)} = \gamma_0 N + \left(\frac{\beta_0}{\sqrt{5}} - \gamma_0\right) n_d = \gamma_0 N + \tilde{\beta}_0 n_d . \tag{7.4}$$

The first term in (7.4) will not induce transitions, since it is diagonal in any representation. E0 transitions are thus only given in terms of the matrix elements of the operator n_d. Finally, it is easy to see from (2.10) that the M1 operator is proportional to the angular momentum operator

$$\underset{\sim}{T}^{(M1)} = \frac{\beta_1}{\sqrt{10}} \cdot \underset{\sim}{L}.$$ (7.5)

Since the angular momentum is a good quantum number in any coupling scheme, no M1 transition is allowed in this approximation. The M1 operator of (7.5) has only diagonal matrix elements, which give rise to magnetic moments.

Once the transition operators have been given, electromagnetic transition rates can be calculated in the usual way, by taking reduced matrix elements of $\underset{\sim}{T}^{(\ell)}$ between initial and final state $<J_f||\underset{\sim}{T}^{(\ell)}||J_i>$. The B(E$\ell$) and B(M$\ell$) values are then obtained as

$$B(\ell; J_i \rightarrow J_f) = \frac{1}{2J_i+1} |<J_f||\underset{\sim}{T}^{(\ell)}||J_i>|^2.$$ (7.6)

It is clear that in order to calculate electromagnetic transition rates in a phenomenological approach one needs, in addition to the wave functions of the initial and final states already obtained from the diagonalization of the Hamiltonian, one parameter for each of the various multipoles E0, M1, M3 and E4. For the E2 multipolarity, two parameters α_2 and β_2 are needed.

In general, the calculation of electromagnetic transition rates must be done numerically. The computer program PHINT mentioned above calculates, in addition to energies, also electromagnetic transition rates. However, as in the case of the energies, the calculations can be done analytically for the three limiting cases discussed in Sect. 5. Here I limit myself to quoting some results for E2 transitions.

Dynamical symmetry I

The operator

$$\underset{\sim}{T}^{(E2)}_m = \alpha_2 [d^\dagger \times \tilde{s} + s^\dagger \times \tilde{d}]^{(2)}_m + \beta_2 [d^\dagger \times \tilde{d}]^{(2)}_m ,$$ (7.7)

when taken between states $|[N](n_d) v n_\Delta L M>$ has selection rules

$$\Delta n_d = 0, \pm 1 .$$ (7.8)

Since the states of this group chain are characterized by a fixed number of d-bosons, n_d, the B(E2) values along the ground state band, defined by the quantum numbers n_d, $v=n_d$, $n_\Delta=0$ and $L=2n_d$ are given only by the first term in (7.7). Their explicit expression[9] is

$$B(E2; n_d+1,\ v=n_d+1,\ n_\Delta=0,\ L'=2n_d+2 \rightarrow n_d,\ v=n_d,\ n_\Delta=0,\ L=2n_d) =$$
$$= \alpha_2^2\ \frac{L+2}{2}\ \frac{2N-L}{2}\ . \tag{7.9}$$

Thus,

$$B(E2;\ 2_1^+ \rightarrow 0_1^+) = \alpha_2^2\ N\ , \qquad \text{in SU(5).} \tag{7.10}$$

The quadrupole moments, defined in the usual way

$$Q_L = \langle L,\ M=L | \sqrt{\frac{16\pi}{5}}\ \mathcal{T}_0^{(2)} | L,\ M=L \rangle \quad , \tag{7.11}$$

are instead given only by the second term in (7.7). For states belonging to the ground state band, one obtains

$$Q_L = \beta_2\ \sqrt{\frac{16\pi}{5}}\ \sqrt{\frac{1}{14}}\ L\ . \tag{7.12}$$

Thus,

$$Q_{2_1^+} = \beta_2\ \sqrt{\frac{16\pi}{5}} \cdot \sqrt{\frac{2}{7}}\ , \qquad \text{in SU(5).} \tag{7.13}$$

Dynamical symmetry II

For calculations in this limit, it is more convenient to rewrite the E2 operator as

$$\mathcal{T}_m^{(E2)} = \alpha_2\ \mathcal{Q}_m^{(2)} + \alpha_2'\ \mathcal{Q}_m'^{(2)}\ , \tag{7.14}$$

$$\alpha_2' = \beta_2 + \alpha_2 \cdot \frac{\sqrt{7}}{2}\ , \tag{7.15}$$

where $Q^{(2)}$ is the same operator as in (2.10) and

$$Q'^{(2)}_m = [d^\dagger \times \tilde{d}]^{(2)}_m .$$

                                                                (7.16)

It turns out that the first term in (7.14) is much larger than the second in regions where the symmetry II applies.
The selection rules of this term, when taken between states of the form $|[N]\ (\lambda,\mu)\ \chi\ L\ M >$ are

$$\Delta\lambda = 0, \qquad \Delta\mu = 0, \tag{7.17}$$

since $Q^{(2)}$ is a generator of SU(3) and thus cannot connect different SU(3) representations. The B(E2) values along the ground state band, defined by the quantum numbers $\lambda = 2N,\ \mu = 0,\ \chi = 0,\ L$ are given by[10]

$$B(E2;(\lambda=2N,\mu=0),\chi=0,L'=L+2 \to (\lambda=2N,\mu=0),\chi=0,L) =$$

                                                                     (7.18)

$$= \alpha_2^2\ \frac{3}{4}\frac{(L+2)(L+1)}{(2L+3)(2L+5)}\ (2N-L)(2N+L+3) .$$

Thus,

$$B(E2;\ 2_1^+ \to 0_1^+) = \alpha_2^2\ \frac{1}{5}\ N\ (2N+3), \quad \text{in SU(3).} \tag{7.19}$$

The quadrupole moments of the states in the ground state band are given by

$$Q_L = -\alpha_2\ \sqrt{\frac{16\pi}{40}}\ \frac{L}{2L+3}\ (4N+3) . \tag{7.20}$$

Thus,

$$Q_{2_1^+} = -\alpha_2\ \sqrt{\frac{16\pi}{40}}\ \frac{2}{7}\ (4N+3) , \qquad \text{in SU(3)} \tag{7.21}$$

Comparing (7.10) with (7.19) one sees a change from an N to an N^2 dependence when going from SU(5) to SU(3). The N^2 dependence in SU(3) is responsible for the large B(E2) values observed in the middle of the major shells where the symmetry II applies.

Dynamical symmetry III

It turns out that in regions where the symmetry III applies, the first term in the transition operator (7.7) is the dominant one. Thus, the appropriate E2 operator to discuss this limit is

$$T_m^{(E2)} = \alpha_2 [d^\dagger \times \tilde{s} + s^\dagger \times \tilde{d}]_m^{(2)}, \quad \beta_2 = 0 \tag{7.22}$$

This operator, when taken between states $|[N](\sigma)\tau \nu_\Delta L M >$, has selection rules

$$\Delta\sigma = 0, \quad \Delta\tau = \pm 1, \tag{7.23}$$

the first being a consequence of the fact that $T^{(E2)}$, Eq. (7.22), is a generator of $O(6)$, and thus cannot connect different $O(6)$ representations. The $B(E2)$ values along the ground state band, defined by the quantum numbers $\sigma = N$, τ, $\nu_\Delta = 0$, $L = 2\tau$ are given by[11]

$$B(E2;\sigma=N,\tau+1,\nu_\Delta=0,L'=2\tau+2 \rightarrow \sigma=N,\tau,\nu_\Delta=0,L=2\tau) =$$
$$\tag{7.24}$$
$$= \alpha_2^2 \frac{L+2}{2(L+5)} \frac{1}{4} (2N-L)(2N+L+8) \ .$$

Thus,

$$B(E2; 2_1^+ \rightarrow 0_1^+) = \alpha_2^2 \frac{1}{5} N(N+4) \ , \quad \text{in } O(6) \ . \tag{7.25}$$

Because of the second selection rule in (7.23), all quadrupole moments are zero in $O(6)$, if the E2 operator is strictly given by (7.22).

$$Q_L = 0 \tag{7.26}$$

In addition to $B(E2)$ values and quadrupole moments along the ground state band, it is possible to calculate analytically other $B(E2)$ values and quadrupole moments, as described in Refs. 9, 10 and 11. These analytic expressions can be tested against experiment. An example is shown in Fig. 8.

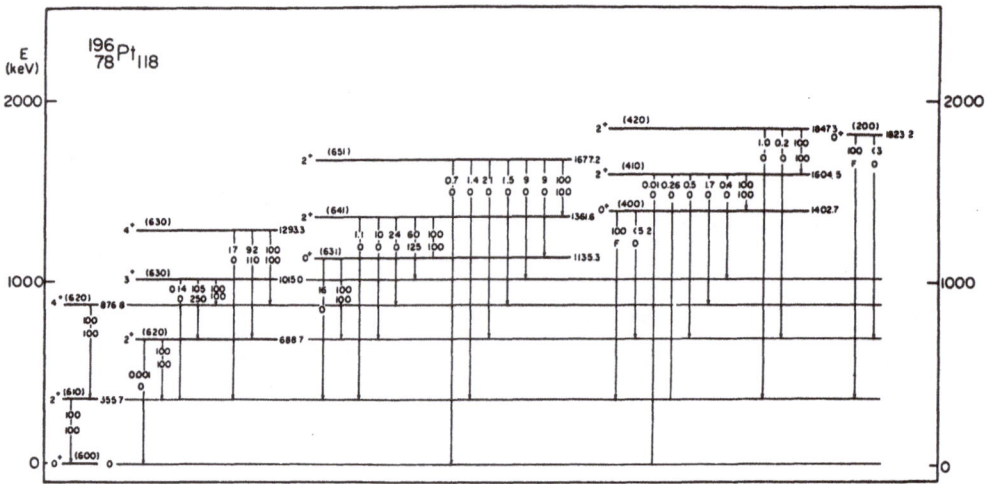

Fig. 8. Branching ratios for the decay of the
positive parity states in ^{196}Pt (Ref. 13).

8. TRANSITIONAL CLASSES

The three limiting cases discussed above are useful because
they provide a set of analytic relations which are easily tested
by experiment. However, only few nuclei can be describable by
the limiting cases. Most nuclei display spectra which are intermediate between them. In order to describe these, transitional,
nuclei, one must return to the full Hamiltonian H, Eq. (2.7), and
diagonalize it numerically. For the purpose of classification,
it is convenient to divide transitional nuclei into four classes:

A) nuclei with spectra intermediate between I) and II),
B) nuclei with spectra intermediate between II) and III),
C) nuclei with spectra intermediate between III) and I).
and
D) nuclei with spectra intermediate among all three limiting cases.

Nuclei in the transitional class D) are obviously the most
difficult to treat from a phenomenological point of view, because
they require the use of all the operators n_d, $P^+ \cdot P$, $L \cdot L$, $Q \cdot Q$,
$T_3 \cdot T_3$ and $T_4 \cdot T_4$ appearing in (2.11). Much simpler phenomenological
studies can be done for nuclei belonging to the transitional classes

A), B) and C). I will next discuss the major features of the transitional classes A) and B) which have been studied in detail both experimentally and theoretically.

Transitional class A.

This class is a mixture of the limits I) and II). Thus it can be studied by considering a mixture of the Casimir operators of both I) and II). Using the operators (5.6), the corresponding Hamiltonian can be written as

$$H^{(I)+(II)} = \varepsilon''' \, C_{1U5} + \gamma' \, C_{203} + \delta' \, C_{2SU3} \, .$$ (8.1)

Conversely, this Hamiltonian can also be written as[14]

$$H^{(I)+(II)} = \varepsilon \, n_d - \kappa \, 2Q \cdot Q - \kappa' L \cdot L \, .$$ (8.2)

It is clear from (8.2) that, when ε is large compared with κ and κ', the eigenfunctions of H will be those appropriate to symmetry I), while when ε is small, they will be those appropriate to symmetry II). For intermediate situations, they will be somewhat intermediate between the two limits. Suppose now, that one is studying a series of isotopes, variable N. It may happen that as N varies the parameters ε, κ, κ' change in such a way that one moves from a situation in which $\varepsilon > \kappa, \kappa'$ to a situation in which $\varepsilon < \kappa, \kappa'$, thus shifting from symmetry I) to II). This transitional class of nuclei is experimentally observed in many regions of the periodic table. A simple study of it can be done by letting ε, κ and κ' vary linearly with N

$$\varepsilon(N) = \varepsilon(N_0) + \frac{\partial \varepsilon}{\partial N} \Big|_{N=N_0} (N-N_0) + \dots \, ,$$

$$\kappa(N) = \kappa(N_0) + \frac{\partial \kappa}{\partial N} \Big|_{N=N_0} (N-N_0) + \dots \, ,$$ (8.3)

$$\kappa'(N) = \kappa'(N_0) + \frac{\partial \kappa'}{\partial N} \Big|_{N=N_0} (N-N_0) + \dots \, .$$

In particular, one could keep κ and κ' constant and let ε decrease with increasing N[14]

$$\varepsilon = \varepsilon_0 - \varepsilon_1 N.$$ (8.4)

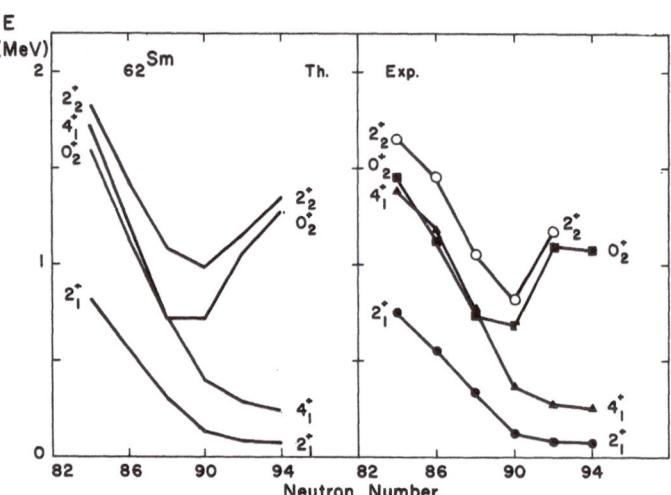

Fig. 9. Typical features of the transitional
class A. Energies.

The corresponding spectra have the properties shown in Fig. 9.
Similar changes occur in the electromagnetic transition rates.
Again, in order to study these changes in a simple way, it is
convenient to expand the coefficients appearing in the transition
operators as a function of N. For the E2 operator, this would
require

$$\alpha_2(N) = \alpha_2(N_0) + \frac{\partial \alpha_2}{\partial N}\bigg|_{N=N_0} (N-N_0) + \dots ,$$

$$\beta_2(N) = \beta_2(N_0) + \frac{\partial \beta_2}{\partial N}\bigg|_{N=N_0} (N-N_0) + \dots .$$

(8.5)

In particular, one could keep α_2 and $\beta_2 = -\frac{\sqrt{7}}{2}\alpha_2$ constant[14].
The corresponding transition rates are shown in Fig. 10.
In this figure, particularly important is the ratio

$$R = \frac{B(E2; \ 2_2^+ \rightarrow 0_1^+)}{B(E2; \ 2_2^+ \rightarrow 2_1^+)}$$

(8.6)

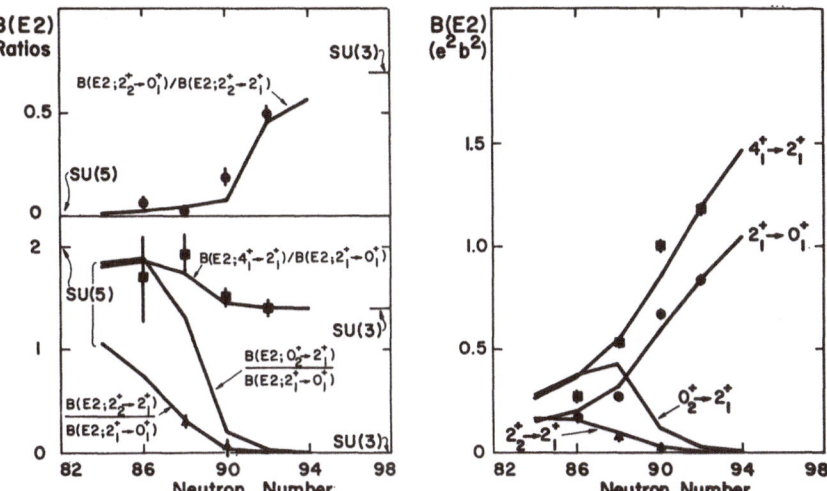

Fig. 10. Typical features of the transitional
class A. Electromagnetic transition
rates.

which changes from

$$R = 0, \text{ in I}), \qquad\qquad (8.7)$$

to

$$R = \frac{7}{10}, \text{ in II}). \qquad\qquad (8.8)$$

Transitional class B.

This transitional class is intermediate between II) and III)
and thus it can be studied by means of a mixture of Casimir
operators of both II) and III),

$$H^{(II)+(III)} = \beta' \, C_{205} + \gamma' \, C_{203} + \delta' \, C_{2SU3} + \eta' \, C_{206} \qquad (8.9)$$

Introducing the operators P_6, C_5, C_3 and $\underset{\sim}{Q}.\underset{\sim}{Q}$, this can also be
written as

$$H^{(II)+(III)} = AP_6 + BC_5 + CC_3 + \kappa 2\underset{\sim}{Q}.\underset{\sim}{Q} . \qquad (8.10)$$

Again, when κ is small the eigenfunctions of H are those appropriate to symmetry III), while when κ is large they are those appropriate to symmetry II). In studying a series of isotopes, one may expand A, B, C and κ as a function of N

$$A(N) = A(N_0) + \frac{\partial A}{\partial N}\Big|_{N=N_0} (N-N_0) + \dots ,$$

$$B(N) = B(N_0) + \frac{\partial B}{\partial N}\Big|_{N=N_0} (N-N_0) + \dots ,$$

$$C(N) = C(N_0) + \frac{\partial C}{\partial N}\Big|_{N=N_0} (N-N_0) + \dots , \qquad (8.11)$$

$$\kappa(N) = \kappa(N_0) + \frac{\partial \kappa}{\partial N}\Big|_{N=N_0} (N-N_0) + \dots .$$

In particular, one could keep A, B and C constant and let κ vary linearly with N^5

$$\kappa = \kappa_0 + \kappa_1 N . \qquad (8.12)$$

(This calculation is slightly different from that given in Ref. 15) The resulting spectra have the properties shown in Fig. 11.

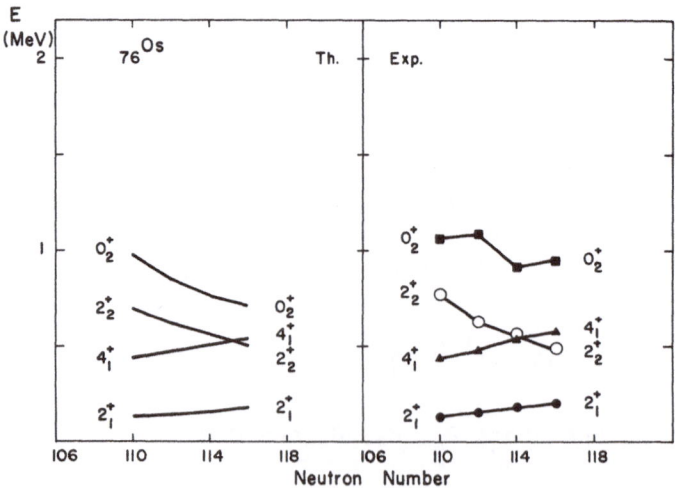

Fig. 11. Typical features of the transitional class B. Energies.

Similar changes occur in the electromagnetic transition rates.
If, for simplicity one keeps the coefficients α_2 and $\beta_2 = 0$
constant[5],[15] one obtains the results shown in Fig. 12. Also here
particularly important is the ratio R, defined in (8.6), which
changes from

$$R = 0, \quad \text{in III)} , \tag{8.13}$$

to

$$R = \frac{7}{10}, \quad \text{in II)} . \tag{8.14}$$

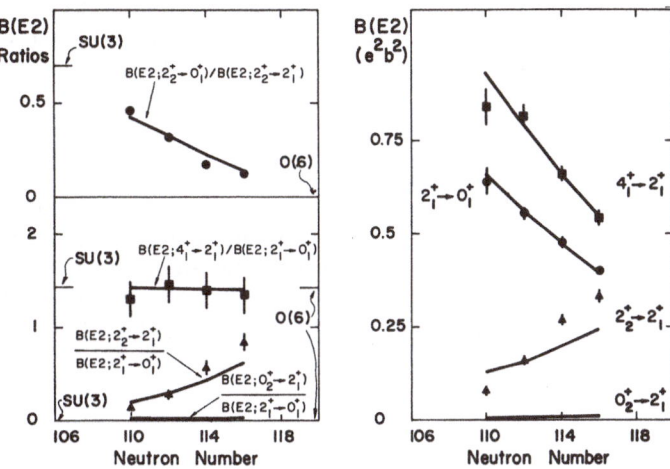

Fig. 12. Typical features of the transitional class B. Electromagnetic
transition rates.

9. GENERAL CASES

In some cases, the observed properties cannot be described
either by one of the limiting cases I), II) and III), or by one
of the transitional classes A), B) and C). For those cases, a
phenomenological description in terms of the interacting boson
model may proceed as follows:

(i)   start from the Hamiltonian, H', Eq. (2.11), and determine
        the parameters ε'', $a_i (i=0,...4)$ from a fit to some experi-
        mental levels. In doing this, it is convenient to enlarge
        the parameter space step by step. An appropriate order is
        ε'', a_2, a_0, a_1, a_3 and a_4. It has been found empirically

that a_3 and a_4 are rarely needed, while ε'', a_2 and a_0 play a
dominant role in almost all nuclei. In the fitting procedure
it is appropriate to include those states whose character is
experimentally well known. For example, the states 2_1^+, 4_1^+,
2_2^+ and 0_2^+.

(ii)   consider now the E2 transition operator, Eq. (7.7) and deter-
        mine the coefficients α_2 and β_2 by fitting two electromagnetic
        transition rates. In doing this, it is convenient to use,
        instead of α_2 and β_2, the parameters α_2 and $\chi_2 = \beta_2/\alpha_2$. Then,
        α_2 determines only the absolute magnitude of the transition
        rates, and it can be fixed by a fit to the B(E2; $2_1^+ \rightarrow 0_1^+$)
        value, which is quite often measured. The parameter χ_2 can be
        determined from branching ratios.

(iii)   if other properties, such as E0 and E4 transition rates are
        known, determine the coefficients $\tilde{\beta}_0$ in (7.4) and β_4 in (7.3)
        from a fit to one transition.

In doing these fits for a series of isotopes, it should be
remembered that it has been found empirically that the parameters
ε'', a_i (i=0,...,4); α_2, χ_2; $\tilde{\beta}_0$; β_4; etc. vary very smoothly with
mass number. Thus, if previous fits have been done in that mass
region, it is convenient to start with those values. It is suggested
to students to try this procedure by taking an arbitrary even-even
nucleus, with exception of magic or semimagic nuclei, and attempt
a description of its properties using the program PHINT mentioned
above.

10. CONCLUSIONS

I have discussed in these three lectures the major features
of the interacting boson model. My purpose has been that of providing an introduction to the model in its simplest form. For this
reason, I have omitted a discussion of the latest developments,
in particular the explicit introduction of proton and neutron
degrees of freedom, both at the phenomenological and at the fundamental level. However, for completeness, I include here a list
of references[16-20] to this work. I would suggest to the interested
student to consult these references. I have also omitted a discussion of the spectra of odd-A nuclei and of the related interacting
boson-fermion model. This model is still in the process of being
developed and I include here some preliminary references[21-24].
Finally, it is worthwhile mentioning once more the close connection
between the geometric and algebraic description. This connection
has been recently investigated in detail[25,26].

The major new feature of the interacting boson model is the
introduction and systematic exploitation of algebraic techniques,
which allows a simple and yet detailed description of many nuclear
properties. I hope that in the course of these three lectures I

have given you at least a flavor of these algebraic techniques and of their use in the solution of complex spectroscopic problems.

REFERENCES

1. A. Bohr, Mat. Fys. Medd. Dan. Vid. Selsk. 26, No. 14 (1952); A. Bohr and B.R. Mottelson, Mat. Fys. Medd. Dan. Vid. Selsk. 27, No. 16 (1953).
2. A. Bohr and B.R. Mottelson, "Nuclear Structure", Vol. II, W.A. Benjamin, Reading, Mass., 1975.
3. A. Arima and F. Iachello, Phys. Rev. Lett. 35 (1975), 1069.
4. F. Iachello ed., "Interacting Bosons in Nuclear Physics", Plenum Press, New York, 1979.
5. F. Iachello, Group Theory and Nuclear Spectroscopy, in: Lecture Notes in Physics, Vol. 119, "Nuclear Spectroscopy", Springer-Verlag, Berlin, 1980.
6. M. Hamermesh, "Group Theory", Addison-Wesley Publ. Co., Reading, Mass., 1962.
7. B.G. Wybourne, "Classical Groups for Physicists", J. Wiley and Sons, New York, 1974.
8. A. de-Shalit and I. Talmi, "Nuclear Shell Theory", Academic Press, New York, 1963.
9. A. Arima and F. Iachello, Ann. Phys. (N.Y.) 99 (1976), 253.
10. A. Arima and F. Iachello, Ann. Phys. (N.Y.) 111 (1978), 201.
11. A. Arima and F. Iachello, Ann. Phys. (N.Y.) 123 (1979), 468.
12. O. Castaños, E. Chacon, A. Frank and M. Moshinsky, J. Math. Phys. 20 (1979), 35.
13. J.A. Cizewski, R.F. Casten, G.J. Smith, M.L. Stelts, W.R. Kane, H.G. Börner and W.F. Davidson, Phys. Rev. Lett. 40 (1978), 167.
14. O. Scholten, F. Iachello and A. Arima, Ann. Phys. (N.Y.) 115 (1978), 325.
15. R.F. Casten and J.A. Cizewski, Nucl. Phys. A309 (1978), 477.
16. A. Arima, T. Otsuka, F. Iachello and I. Talmi, Phys. Lett. 66B (1977), 205; T. Otsuka, A. Arima, F. Iachello and I. Talmi, Phys. Lett. 76B (1978), 139.
17. T. Otsuka, A. Arima and F. Iachello, Nucl. Phys. A309 (1978), 1.
18. T. Otsuka, Ph.D. Thesis, University of Tokyo, Japan, 1979.
19. J. Ginocchio, Phys. Lett. 79B (1978) 173; Phys. Lett. 85B (1979), 9; Ann. Phys. (N.Y.) 126 (1980), 234.
20. F. Iachello, G. Puddu, O. Scholten, A. Arima and T. Otsuka, Phys. Lett. 89B (1979), 1.
21. F. Iachello and O. Scholten, Phys. Rev. Lett. 43 (1979), 679.
22. F. Iachello and O. Scholten, Phys. Lett. 91B (1980), 189.
23. F. Iachello, Phys. Rev. Lett. 44 (1980), 772.
24. O. Scholten, Ph.D. Thesis, University of Groningen, The Netherlands, 1980.
25. J. Ginocchio and M.W. Kirson, Phys. Rev. Lett. 44 (1980), 1744.
26. A.E.L. Dieperink, O. Scholten and F. Iachello, Phys.Rev.Lett. 44 (1980), 1747.

PION-NUCLEAR MANY BODY PROBLEMS

W. Weise

CERN, Geneva, and
Institute of Theoretical Physics
University of Regensburg, W. Germany

1. INTRODUCTION

Pionic modes of excitation in nuclear systems have received considerable experimental and theoretical interest in recent years. Vast amounts of pion-nucleus scattering data[1] have been produced at the meson factories, the analysis of which has provided information about the "optical" branches of the pion-nuclear excitation spectrum, where the propagation of $\Delta(1232)$ isobars and their interaction with the nuclear environment is one of the outstanding issues[2-6], and will be one of the subjects of these lectures.

The low frequency, or "acoustic" branches of the pion spectrum, those related to low-lying particle-hole excitations carrying the quantum numbers of a pion, have likewise been subject of various investigations. The interest there is motivated by the possibility of a phase transition of nuclear matter into a pion condensate at sufficiently high density[7-9]. Most of the conclusions or expectations derived from the possible existence of a pion condensate rest very strongly on our (limited) knowledge of spin-isospin dependent correlations operating in pionic modes. It is therefore an important task to learn as much as possible about this interaction from "standard" experiments, both at low and intermediate energies. Of particular interest are the high momentum transfer ($q \sim 2\text{-}3\ m_\pi$) properties of these correlations, because this is the domain where the tensor force from one-pion exchange, the driving interaction to form a pion condensate, is strongest.

Even if the threshold for pion condensation is far from being reached at normal nuclear densities, certain precritical effects might appear as indicators of the proximity of the pion condensation threshold. Ways to observe such pionic opalescence phenomena (see refs.[10,11]), or precritical enhancement effects[12,13], in actual experiments are presently subject of theoretical activities as well as experimental proposals. One chapter of these lectures will be devoted to this question.

Before facing the full complexity of the pion-nuclear many-body problem, however, it is necessary first to specify the elementary input: mesons, nucleons, $\Delta(1232)$ isobars, and their mutual inter-actions.

2. BASIC MESON-BARYON EFFECTIVE LAGRANGIANS AND ELEMENTARY PROCESSES

2.1 πN Scattering

In our kinematical region of interest (pion energies $\omega \lesssim 3\, m_\pi$, pion momenta $|\vec{k}| \lesssim 4m_\pi$, where $m_\pi = 139$ MeV is the pion mass), the pion-nucleon scattering amplitude, Fig. 1, is dominated by partial waves with $\ell = 0,1$. The p-wave interactions are of primary impor-tance, because of the presence of the $\Delta(1232)$ resonance in the spin-isospin-3/2 channel. The s-wave interactions play only a minor role in systems with equal numbers of protons and neutrons. The isovec-tor part of the s-wave πN interaction can be thought of as being generated by the exchange of a ρ meson, as shown in Fig. 2a, while the isoscalar s-wave amplitude is small due to cancellations between isoscalar (scalar) 2π exchange and short range repulsive contri-butions which, among other mechanisms, involve the virtual excitation of intermediate nucleon-antinucleon pairs (see Fig. 2b). We should point out, however, that this cancellation holds only for on-shell pions, and can be unbalanced if the pions move off shell[14]. We shall not specify here the details of the model illustrated in Fig. 2, but refer to it as a useful way to extrapolate off-shell, e.g. in the πd ↔ NN process, to be discussed later. We would like to mention some alternative models in the literature, among which in particular, is the one of ref.[15], which does not make explicit use of ρ exchange. Let us now concentrate on a somewhat more detailed description of the p-wave πN interaction. We wish to construct it from the known intermediate states (the nucleon and its excited states) in the πN s-channel (cf. Fig. 1). A useful way to proceed[16] is to develop a crossing symmetric K-matrix (reaction matrix), according to Fig. 3, in the form

$$\langle \pi(q')N | K(s) | \pi(q)N \rangle = \frac{1}{4\pi} \sum_X \frac{\langle q' | \delta H_{\pi Nx} | X \rangle \langle X | \delta H_{\pi Nx} | q \rangle}{M_X - \sqrt{s}} + \text{crossed terms}, \quad (1)$$

where the kinematic variables are explained in Fig. 1 ($\sqrt{s}$ is the πN c.m. energy), and the crossed terms are obtained essentially by the replacements $\sqrt{s} \to \sqrt{u}$, $q \leftrightarrow -q'$. Here X = N(939), $\Delta(1232)$, N*(1470), etc. refer to the nucleon and the important πN resonances, M_X are the corresponding physical masses. The $\delta H_{\pi NX}$ are the effective Hamiltonians which couple the πN system to a given intermediate state. Treating these non-relativistically, we have, for example

$$\delta H_{\pi NN}(q) = i \frac{f(q^2)}{m_\pi} \vec{\sigma} \cdot \vec{q} \ \tau_\lambda [\pi_\lambda(q) + \pi^+_{-\lambda}(-q)], \qquad (2a)$$

$$\delta H_{\pi N\Delta}(q) = i \frac{f^*(q^2)}{m_\pi} \vec{S}^+ \cdot \vec{q} \ T^+_\lambda [\pi_\lambda(q) + \pi^+_{-\lambda}(-q)] + \text{h.c.} \qquad (2b)$$

where $\pi_\lambda(q)$ annihilates a pion of charge λ and four-momentum $q=(\omega,\vec{q})$. In eq. (2b), $\vec{S}^+$ and $\vec{T}^+$ are transition spin and isospin operators, respectively, connecting spin-isospin 1/2- and 3/2-states[16]. The quantities $f(q^2)$ and $f^*(q^2)$ are πNN or $\pi N\Delta$ vertex form factors, to be specified later. Their values at $q^2=m_\pi^2$ define the relevant π-baryon coupling constants, which we denote by f and f^*, respectively.

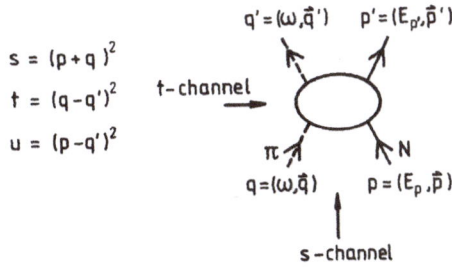

$s = (p+q)^2$

$t = (q-q')^2$

$u = (p-q')^2$

Fig. 1: Kinematic variables in the πN scattering amplitude

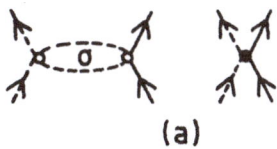
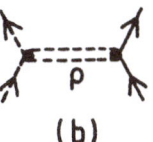

(a)   (b)

Fig. 2: Mechanisms for s-wave πN scattering: (a) isoscalar part through σ exchange and short-range repulsion; (b) isovector part through ρ exchange.

Fig. 3: Direct and crossed p-wave πN K-matrix. The intermediate states X refer to N(939), Δ(1232), N*(1470) etc.

The p-wave K matrix, eq. (1), has to be projected onto the different eigen channels $\alpha = (2S, 2I)$ of given spin S and isospin I(1/2 or 3/2 each). The corresponding channel K matrix, denoted by K_α, is related to the partial wave amplitude, expressed in terms of the phase shifts δ_α,

$$f_\alpha(\omega) = \frac{1}{|\vec{q}|} e^{i\delta_\alpha} \sin\delta_\alpha = \frac{1}{2i|\vec{q}|}(S_\alpha - 1), \tag{3a}$$

by the unitarity condition

$$S_\alpha = \frac{1 + i|\vec{q}|K_\alpha}{1 - i|\vec{q}|K_\alpha} \tag{3b}$$

or simply

$$K_\alpha = \frac{1}{|\vec{q}|} \tan\delta_\alpha. \tag{3c}$$

To illustrate how this scheme works, let us derive the spin-isospin 3/2-p-wave amplitude f_{33} under the simplifying assumption that only nucleon and $\Delta(1232)$-isobar intermediate states contribute, and that the baryons are treated in the static limit (i.e. $\sqrt{s} = M + \omega$). Except for isospin factors which we omit here for simplicity, the K-matrix of eq. (1) is then given by

$$\langle q'N|K(s)|qN\rangle = \frac{f^2}{4\pi m_\pi^2}\left[\frac{\vec{\sigma}\cdot\vec{q}'\vec{\sigma}\cdot\vec{q}}{-\omega} + \frac{\vec{\sigma}\cdot\vec{q}\vec{\sigma}\cdot\vec{q}'}{\omega}\right] + \frac{f^{*2}}{4\pi m_\pi^2}\left[\frac{\vec{S}\cdot\vec{q}'\vec{S}^+\cdot\vec{q}}{\omega_\Delta - \omega} + \frac{\vec{S}\cdot\vec{q}\vec{S}^+\cdot\vec{q}'}{\omega_\Delta + \omega}\right], \tag{4}$$

where $\omega_\Delta = M_\Delta - M_N = 2.1\, m_\pi$. If projected into the spin-isospin-3/2 channel, and neglecting the small crossed Δ-isobar piece we obtain

$$K_{33}(\omega) = \frac{1}{|\vec{q}|}\tan\delta_{33} = \frac{1}{3}\frac{|\vec{q}|^2}{4\pi}\left[\frac{4f^2}{\omega} + \frac{f^{*2}}{\omega_\Delta - \omega}\right]. \tag{5}$$

Following eqs. (3) and employing the Chew-Low value $f^* = 2f$ for the $\pi N\Delta$ coupling constant, we find, with $\lambda = (4/3)\, f^2/4\pi m_\pi^2$:

$$f_{33}(\omega) = \frac{\lambda\,\vec{q}^2(\omega_\Delta/\omega)}{\omega_\Delta - \omega - (i/2)\Gamma_\Delta(\omega)}, \qquad \Gamma_\Delta = \frac{2}{3}\frac{f^{*2}}{4\pi}\frac{|\vec{q}|^3}{m_\pi^2}\frac{\omega_\Delta}{\omega}, \tag{6}$$

which is identical with the result from Chew-Low theory. Note that the decay width Γ_Δ of the $\Delta(1232)$ has a factor ω_Δ/ω which comes from the crossed nucleon pole term. This factor, which is proportional to $(\vec{q}^2 + m_\pi^2)^{-1/2}$, is sometimes misinterpreted as a "long range $\pi N\Delta$ vertex factor", but it has obviously nothing to do with the $\pi N\Delta$ form factor $f^*(q^2)$.

For a quantitative description of the p-wave πN phase shifts, the full relativistic kinematics has to be treated appropriately, and inclusion of the $N^*(1470)$ is required to reproduce the correct energy dependence of δ_{11}. The phase shifts calculated with the model are then shown in Figs. 4, while the explicit form of the partial wave K-matrices, including all factors, is given in table 1. This model, based entirely on π-baryon effective Lagrangians and the physical masses of nucleon, Δ and N^*, is obviously a good starting point for subsequent discussions of pion-nucleus interactions. We emphasize that the separation of the p-wave π-nucleon amplitude into a propagator part and a π-baryon vertex piece will turn out to be useful in the many-body problem, since both ingredients will be renormalized in a nuclear medium.

Table 1.   Explicit forms of the isobar-model K matrices $K_\alpha = (1/q)\tan\delta_\alpha$ in the p-wave eigenchannels, employing $N(939)$, $\Delta(1232)$ and $N^*(1470)$ intermediate states. Relativistic kinematic factors are included here. Note that $\sqrt{s} = E + \omega$, $E = \sqrt{M^2 + \vec{q}^2}$, in terms of c.m. variables, and $\bar{u} = (E - \omega)^2 - 2q^2$. The masses and coupling constants are $M = 939$ MeV, $M_\Delta = 1232$ MeV, $M_{N^*} = 1450$ MeV; $f^2/4\pi = 0.08$, $f_\Delta^2/4\pi \equiv f^{*2}/4\pi = 0.37$, $f_{N^*}^2/4\pi = 0.02$.

$$K_{11} = \frac{1}{3} \; \frac{q^2}{m_\pi^2} \; \frac{M}{\sqrt{s}} \left[\frac{2Mf^2}{4\pi} \left(\frac{9}{M^2-s} + \frac{1}{M^2-\bar{u}} \right) + \right.$$

$$\left. + \frac{2M_{N^*}f_{N^*}^2}{4\pi} \left(\frac{9}{M_{N^*}^2-s} + \frac{1}{M_{N^*}^2-\bar{u}} \right) + \frac{2M_\Delta f_\Delta^2}{4\pi} \; \frac{16}{9} \; \frac{1}{M_\Delta^2-\bar{u}} \right]$$

$$K_{13} = K_{31} = \frac{1}{3} \; \frac{q^2}{m_\pi^2} \; \frac{M}{\sqrt{s}} \left[- \frac{2Mf^2}{4\pi} \; \frac{2}{M^2-\bar{u}} - \right.$$

$$\left. - \frac{2M_{N^*}f_{N^*}^2}{4\pi} \; \frac{2}{M_{N^*}^2-\bar{u}} + \frac{2M_\Delta f_\Delta^2}{4\pi} \; \frac{4}{9} \; \frac{1}{M_\Delta^2-\bar{u}} \right]$$

$$K_{33} = \frac{1}{3} \; \frac{q^2}{m_\pi^2} \; \frac{M}{\sqrt{s}} \left[\frac{2Mf^2}{4\pi} \; \frac{4}{M^2-\bar{u}} + \frac{2M_{N^*}^2 f_{N^*}^2}{4\pi} \; \frac{4}{M_{N^*}^2-\bar{u}} \right.$$

$$\left. + \frac{2M_\Delta f_\Delta^2}{4\pi} \left(\frac{1}{M_\Delta^2-s} + \frac{1}{9} \; \frac{1}{M_\Delta^2-\bar{u}} \right) \right]$$

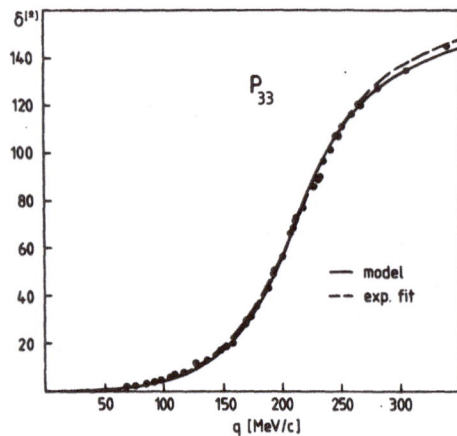

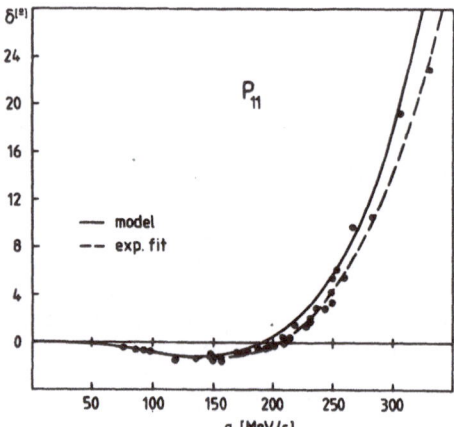

Fig. 4: p-wave πN phase shifts
calculated within the
model presented in
table 1. The data are
taken from ref. [17].
(The autor is grateful
to G. Bienek for pre-
paration of these
figures.)

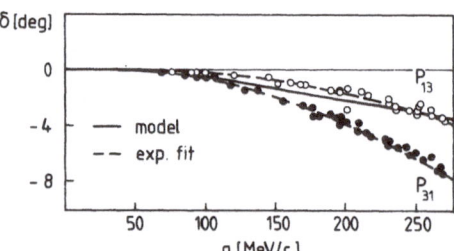

2.2 Pion-Baryon Vertex Formfactors

The K-matrix scheme developed in the previous chapter deter-
mines only the on-shell properties of the pion-baryon vertex functions. Their off-shell extrapolation, which is required in the pion-
nuclear many-body problem, has to be obtained from other sources. We
would like to parametrize the π-baryon vertex factors, appearing in
eqs. (2), in the form, e.g.,

$$f(q^2,s) = f \left(\frac{2M}{M + \sqrt{s}} \right)^{1/2} \left(\frac{\Lambda^2 - m_\pi^2}{\Lambda^2 - q^2} \right)^n , \qquad (7a)$$

$$f^*(q^2,s) = f^* \left(\frac{2M_\Delta}{M_\Delta + \sqrt{s}} \right)^{1/2} \left(\frac{\Lambda^{*2} - m_\pi^2}{\Lambda^{*2} - q^2} \right)^n , \qquad (7b)$$

where $q^2 = q_0^2 - q^2$. The additional kinematic correction factor depending on $\sqrt{s}$ becomes obvious from table 1. The coupling constants f, f*
are determined at $q^2 = m_\pi^2$ and $s = M^2$ or $s = M_\Delta^2$, respectively

($f^2/4\pi = 0.08$, $f^{*2}/4\pi = 0.37$). We shall usually employ monopole formfactors with n = 1. The cutoff mass Λ, by its inverse, is related to
the range of the vertex and has to be determined phenomenologically.

It is clear that the parametrization of eq. (7) hides just an-
other many-body problem, namely the quark dynamics of the $\pi N \leftrightarrow N$
or $\pi N \leftrightarrow \Delta$ transitions. The conceptual hypothesis we have tacitly
introduced here is that the pion-nuclear many-body problem, in the
limited range of energies and momenta considered, can be described
entirely in terms of nucleons, Δ-isobars and mesons, disregarding
their underlying quark structure. This is by no means a trivial
assumption, and its range of validity is clearly connected with
questions about the size of the quark bag. The present status of the
problem[18,19] must be regarded as controversal. However, we would
like to emphasize that the picture developed here cannot be easily
accomodated with typical MIT-bag radii larger than 1 fm.

2.3 The spin-isospin dependent baryon-baryon interaction

Given the effective Lagrangians of eqs. (2) together with eqs. (7),
the one-pion exchange (OPE) NN interaction becomes (Fig. 5(a)):

$$W_\pi(\omega, \vec{q}) = \frac{f^2(q^2)}{m_\pi^2} \, V_\pi(\omega, q) \, \vec{\sigma}_1 \cdot \hat{q} \, \vec{\sigma}_2 \cdot \hat{q} \, \vec{\tau}_1 \cdot \vec{\tau}_2, \tag{8}$$

where

$$V_\pi(\omega, q) = \frac{\vec{q}^2}{\omega^2 - \vec{q}^2 - m_\pi^2 + i\delta} \tag{8a}$$

while the E $N\Delta \to \Delta N$ interaction can be written similarly as
(Fig. 5(b))

$$W_\pi^{N\Delta \to \Delta N}(\omega, \vec{q}) = \frac{f^{*2}(q)^2}{m_\pi^2} \, V_\pi(\omega, q) \, \vec{S}_1^+ \cdot \hat{q} \, \vec{S}_2 \cdot \hat{q} \, \vec{T}_1^+ \cdot \vec{T}_2. \tag{9}$$

These will obviously be the driving forces in excitation modes
carrying pion quantum numbers. At shorter range, the exchange of two
interaction pions in an isovector state has to be considered as well
(Fig. 5(c)). In the NN sector, this interaction can be represented as

$$W_{2\pi}(\omega, \vec{q}) = \int_{4m_\pi^2}^{\infty} dt \, \frac{\rho_{2\pi}(t)}{\omega^2 - \vec{q}^2 - t} \, (\vec{\sigma}_1 \times \vec{q}) \cdot (\vec{\sigma}_2 \times \vec{q}) \, \vec{\tau}_1 \cdot \vec{\tau}_2. \tag{10}$$

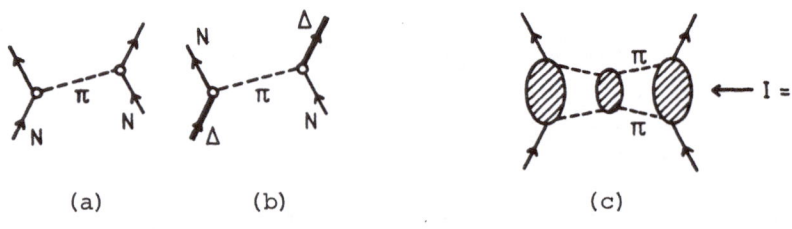

Fig. 5.   (a) and (b): one-pion exchange NN and ΔN $\rightarrow$ NΔ interaction
            (c) Two-pion exchange interaction in I = 1 t-channel.

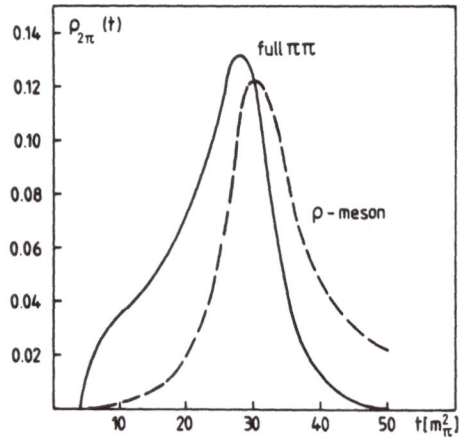

Fig. 6.   Spectral function (mass
          distribution) of isovec-
          tor two-pion exchange.
          Dashed curve: resonant
          part corresponding to the
          ρ meson.
          Solid curve: full spectral
          distribution, including
          non-resonant $\pi\pi$ interme-
          diate states.

Here $\rho_{2\pi}(t)$ is related to the square of the $\pi\pi \rightarrow \bar{N}N$ helicity amplitude f_1^-, as it appears for example in the Paris NN interaction[20]. Its
structure is shown in Fig. 6. It exhibits a resonant part related to
the ρ meson, but also a non-resonant background which contains,
among other pieces, the contribution of iterated OPE to the isovector
two-pion exchange. For low momentum transfers, it is useful to approximate the resonant part of $W_{2\pi}$ with zero-width ρ exchange,

$$W_\rho = \frac{f_{\rho NN}^2}{m_\rho^2} \frac{(\vec{\sigma}_1 \times \vec{q}) \cdot (\vec{\sigma}_2 \times \vec{q})}{\omega^2 - q^2 - m_\rho^2} \vec{\tau}_1 \cdot \vec{\tau}_2 \tag{11}$$

which defines the ρNN coupling constant. Typical values are $f_{\rho NN}^2/4\pi =$
= 4.5 - 5.0. An important property of ρ exchange is that its tensor
force component has opposite sign as compared to that from OPE, and

therefore provides a natural regularization of the pathologically large OPE tensor interaction at small distance. This is seen by remembering that $\vec{\sigma}_1 \cdot \hat{q} \, \vec{\sigma}_2 \cdot \hat{q} = \frac{1}{3} \vec{\sigma}_1 \cdot \vec{\sigma}_2 + \frac{1}{3} S_{12}(\hat{q})$, with $S_{12}(\hat{q}) = 3\vec{\sigma}_1 \cdot \hat{q} \, \vec{\sigma}_2 \cdot \hat{q} - \vec{\sigma}_1 \cdot \vec{\sigma}_2$, whereas $(\vec{\sigma}_1 \times \hat{q}) \cdot (\vec{\sigma}_2 \times \hat{q}) = \frac{2}{3} \vec{\sigma}_1 \cdot \vec{\sigma}_2 - \frac{1}{3} S_{12}(\hat{q})$.

If one of the nucleons in $W_{2\pi}$ is replaced by a Δ, the usual insertions $\vec{\sigma} \rightarrow \vec{S}$ and $\vec{\tau} \rightarrow \vec{T}$ have to be made. Not very much is known about the two-pion exchange NΔ -interaction. We assume a simple scaling:

$$V_{2\pi}(N\Delta \rightarrow \Delta N) = \alpha_{2\pi} \, V_{2\pi}(N\Delta \rightarrow NN) = \alpha_{2\pi}^2 \, V_{2\pi}(NN \rightarrow NN),$$

where $\alpha_{2\pi}$ measures the ratio of $\rho N\Delta$ and ρNN coupling strength. One of the best testing grounds for determining α_ρ, together with the cutoff mass Λ associated with the π-baryon formfactors, is the $\pi d \leftrightarrow pp$ reaction. This is so because the kinematics of this process requires the transfer of large momenta, but little energy, between the two nucleons. In the 3.3 resonance region, the excitation of a $\Delta(1232)$ with subsequent $\Delta N \rightarrow NN$ transition dominates, and the large momentum transfer in the latter process makes it sensitively dependent on $\alpha_{2\pi}$ and Λ. From a detailed analysis of $\pi d \leftrightarrow pp$ total and differential cross sections as well as polarization parameters[21], one finds that $\alpha_{2\pi} = 1.7 - 2.0$ and $\Lambda = 1.2 - 1.4$ GeV, where the same Λ has been used for both πNN and $\pi N\Delta$ form factors. The lower value of $\alpha_{2\pi}$ would also follow for the quark model ratio of $\rho N\Delta$ and ρNN coupling constants, and is consistent with the $\gamma N\Delta$ transition form factor within the ρ meson dominance model.

3. PIONS IN NUCLEAR MATTER

3.1 The spectrum of pion-like excitations

Consider now first the response function $R(\omega,q)$ of an infinite nuclear medium with respect to a pion field of frequency ω and momentum q. Following the observations of the previous chapter, the pion field will polarize the medium primarily by exciting nucleon-hole and $\Delta(1232)$-hole states. A first order picture of the response is then given by the pion self energy $\Pi^{(o)}$, as illustrated in Fig. 7. We write it - somewhat schematically - in the form

$$\Pi^{(o)}(\omega,q) = - \sum_{ph} \frac{|<\pi(q)|\delta H|ph>|^2}{E_p - E_h - \omega} + \text{ crossed terms}, \qquad (12)$$

where $|ph>$ denotes a nucleon- or Δ-hole state carrying pion quantum numbers and δH is the πNN or $\pi N\Delta$ vertex operator given by eqs. (2) (see table 1 for the relevant input parameters). We rewrite eq. (12) as

$$\Pi^{(o)}(\omega,q) = q^2 \chi(\omega,q), \tag{13}$$

which defines the elementary (lowest order) pionic susceptibility χ
of the medium. Because of the spin dependence of the p-wave πN interaction, it is very useful to point out [22-24] the analogy of the
pionic response problem with that encountered in magnetic materials.
In that sense, χ has a "diamagnetic" part χ_Δ which involves high-lying
Δ-hole excitations, and "paramagnetic" component χ_N, related to low
lying nucleon-hole excitations:

$$\chi = \chi_N + \chi_\Delta. \tag{14}$$

For symmetric nuclear matter at density ρ (Fermi momentum k_F), we find
at large frequencies ($\omega \gg \frac{q^2}{2M}$):

$$\chi_N(\omega,q) \overset{\sim}{=} - \frac{f^2(q^2)}{m_\pi^2} \left[\frac{\rho}{\frac{q^2}{2M^*} - \omega} + \frac{\rho}{\frac{q^2}{2M^*} + \omega} \right] \tag{15a}$$

and

$$\chi_\Delta(\omega,q) \simeq - \frac{f^{*2}(q^2)}{m_\pi^2} \frac{4}{9} \left[\frac{\rho}{\omega_\Delta - \omega - (i/2)\Gamma_\Delta(\omega)} + \frac{\text{crossed}}{\text{term}} \right], \tag{15b}$$

where M^* is the nucleon effective mass, and $\omega_\Delta = 2.1\, m_\pi$. Note that
$\chi_N(\omega,q) \to 0$ as $M^* \to \infty$, so that in this limit the response (for symmetric nuclear matter) is completely determined by $\Delta(1232)$ excitations.
At low frequency ω and $q < k_F$,

$$\chi_N(\omega=o,q) = - \frac{f^2(q^2)}{m_\pi^2} \frac{2M^* k_F}{\pi^2} \left[1 - \frac{1}{12} \frac{q^2}{k_F^2} + \dots \right] \tag{15c}$$

and

$$\chi_\Delta(\omega=o,q) = - \frac{8}{9} \frac{f^{*2}(q^2)}{m_\pi^2} \frac{\rho}{\omega_\Delta}. \tag{15d}$$

Here χ_N dominates, but the contribution from χ_Δ is still sizeable.

For $\omega \gtrsim m_\pi$, we have the "pion optics" domain, which is explored
by pion elastic scattering. It is convenient there to work with an
optical potential. For the lowest order p-wave optical potential, the

relation is simply

$$U_{opt}^{(o)}(\omega,q) = \frac{1}{2\omega} \Pi^{(o)}(\omega,q). \tag{16}$$

The response function in this simplest possible model is obtained (see Fig. 8) by the RPA iteration

$$R(\omega,q) = \Pi^{(o)}(\omega,q) + \Pi^{(o)}(\omega,q) \frac{1}{\omega^2-q^2-m_\pi^2} R(\omega,q). \tag{17}$$

It is convenient to define a "dimesic function"[13,24] $\varepsilon(\omega,q)$ by

$$R(\omega,q) = \frac{\Pi^{(o)}(\omega,q)}{\varepsilon(\omega,q)}. \tag{18}$$

From eqs. (17,13),

$$\varepsilon(\omega,q) = 1 - \frac{q^2}{\omega^2-q^2-m_\pi^2} \chi(\omega,q). \tag{19}$$

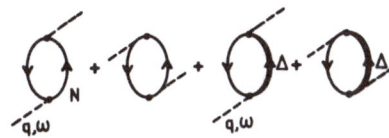

Fig. 7. Lowest order p-wave pion self-energy.

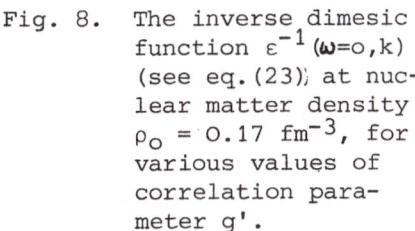

Fig. 8. The inverse dimesic function $\varepsilon^{-1}(\omega=o,k)$ (see eq.(23)) at nuclear matter density $\rho_o = 0.17$ fm^{-3}, for various values of correlation parameter g'.

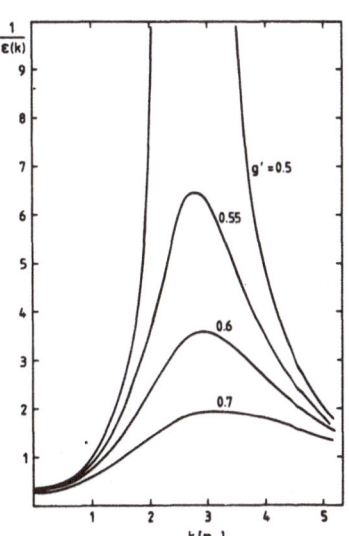

It is clear that the zeros of $\varepsilon(\omega,q)$ determine the spectrum $\omega(q)$ of pion-like excitations of the nuclear medium. Note also that, in the optical part of the spectrum, an index of refraction n for pions can be defined as

$$n^2 = \frac{q^2}{\omega^2 - m_\pi^2} = \left[1 + \chi(\omega,q) \right]^{-1}. \tag{20}$$

The T-matrix for pion scattering is simply given by $T=R/2\omega$.

Of particular interest is the question whether the nuclear medium can act as a strong amplifier for the pion field at no cost of energy, because of the strong attraction provided by the p-wave πN interaction in the spin-isospin-3/2-channel. To find this out, let us discuss the dimesic function at zero frequency. Clearly, $\varepsilon(\omega=0,q)$ can become small compared to unity at sufficiently large q and at sufficiently high density. In fact, as $\varepsilon \to 0$ for $\omega = 0$, a pionic soft mode instability occurs which indicates the onset of a pion condensate.

For the simple model described above with only OPE as particle-hole interaction, the OPE tensor force is sufficiently attractive at high momentum transfers ($q_c \sim 2\text{-}3\ m_\pi$) so that, in the absence of repulsive correlations, the condition $\varepsilon \to 0$ is actually met at critical densities ρ_c which are only a fraction of nuclear matter density $\rho_0 = 0.17\ \text{fm}^{-3}$. Clearly, the critical density ρ_c will depend sensitively on correlations accompanying OPE.

3.2 Nuclear spin-isospin correlations

Guided by the developments of chapter 2.3, we construct the spin-isospin dependent particle-hole interaction as follows:

$$W_{ph}(\omega,q) = \frac{f^2(q^2)}{m_\pi^2} \left[V_\pi(\omega,q) \vec{\sigma}_1 \cdot \hat{q}\, \vec{\sigma}_2 \cdot \hat{q} + V_{2\pi}(\omega,q)\, (\vec{\sigma}_1 \times \hat{q}) \cdot (\vec{\sigma}_2 \times \hat{q}) \right.$$

$$\left. + g'(\omega,q)\, \vec{\sigma}_1 \cdot \vec{\sigma}_2 + h'(\omega,q) S_{12}(\hat{q}) \right] \vec{\tau}_1 \cdot \vec{\tau}_2, \tag{21}$$

where $S_{12} = 3\, \vec{\sigma}_1 \cdot \hat{q}\ \vec{\sigma}_2 \cdot \hat{q} - \vec{\sigma}_1 \cdot \vec{\sigma}_2$. Here V_π is again the OPE interaction, eq. (8a), and

$$V_{2\pi}(\omega,q) = \int_{4m_\pi^2}^\infty dt\, \frac{q^2 \rho_{2\pi}(t)}{\omega^2 - q^2 - t} \left[\frac{f^2(q^2)}{m_\pi^2} \right]^{-1}, \tag{22}$$

is the isovector two-pion exchange, following eq. (10), including

ρ exchange. All other spin-spin and tensor correlations are incorporated by g' and h', respectively. The h' is quite small in our region of interest, since the leading tensor terms are already contained in the uncorrelated OPE and isovector two-pion exchange (TPE) interactions. On the other hand, $g'(\omega,q)$ represents important repulsive correlations. Note that $W_{ph}(\omega=o,\ q\to o)= \dfrac{f^2(o)}{m_\pi^2}\ g'(\omega=q=o)\vec{\sigma}_1\cdot\vec{\sigma}_2\ \vec{\tau}_1\cdot\vec{\tau}_2$,

so that $g'(0,0)$ is related, up to a factor, to the Landau parameter G_o' in Migdal's theory of finite Fermi systems. Values $g'(\omega=o,q=o)\gtrsim 0.7$ are deduced from recent phenomenological Migdal theory calculations[25], but this is not necessarily inconsistent with smaller values of $g'(\omega=o,q)$ at large momentum transfers q, which we are primarily interested in.

For $\Delta(1232)$ isobars, the replacements $f \to f^*$, $\vec{\sigma} \to \vec{S}$ and $\vec{\tau} \to \vec{T}$ are again to be made, together with the scaling factor $\alpha_{2\pi}$ or $\alpha_{2\pi}^2$ in $V_{2\pi}$. The repulsive correlations for Δ isobars are assumed to be the same as for nucleons. Such a universal g' is favoured by quark model considerations and receives some support by the role which these correlations play in the quenching of Gamow-Teller transitions in various nuclei[26]. We also note that $g'=1/3$ corresponds to the original value of the Lorentz-Lorenz correction[27].

The TPE interaction component $V_{2\pi}$ in W_{ph} of eq. (21) does actually not participate in the pionic response in infinite nuclear matter, because of its transverse nature of the coupling between spin and momentum, as compared to the longitudinal coupling imposed by the pion source operator. We include it here for completeness, both because it will play a role in finite nuclei due to surface effects, and because it becomes important in photon-induced excitations of pion-like states.

With inclusion of correlations, the dimesic function of eq. (19) turns into

$$\varepsilon(\omega,q) = 1 - \left[\frac{q^2}{\omega^2-q^2-m_\pi^2} + g'\right]\chi(\omega,q). \qquad (23)$$

The g' now plays a crucial role in the determination of the threshold for pion condensation. For reference [8], we mention that g' = 0.4 (in pion mass units) would place the critical density ρ_c at ρ_o, while g'=0.5 moves ρ_c to about $2\rho_o$.

It is instructive to discuss $\varepsilon(q,\omega=o)$ of eq. (23) at normal nuclear matter density as a function of the correlation parameter g'. From Fig. 8 it becomes obvious that the dimesic function at zero frequency carries signatures of the phase transition long before the critical density is actually reached. Note the strong precritical enhancement of $1/\varepsilon$ at momenta close to the critical momentum q_c even

for values of g' which would place ρ_c well above ρ_o. The question is
then whether such enhancement effects can actually be observed in
appropriate experiments. Their presence (or absence) would put constraints on the strength of spin-isospin correlations at high momentum transfer. Note also that the behaviour of $\varepsilon(q, \omega=o)$ in the long-
wavelength limit ($q \to o$) indicates a characteristic quenching (rather
than amplification), related to the repulsive correlations governed
by g'. This quenching is a well known phenomenon, e.g. in Gamow-Teller
and magnetic multipole transitions[28].

To summarize this section, we would like to present, in Fig. 9,
a schematic picture of the spectrum of pionic excitations in nuclear
matter, determined by $\varepsilon(\omega, q) = o$, i.e. by the singularities of the pion-
nuclear response function $R(\omega, q)$. Of particular interest in these lectures is the region of $\Delta(1232)$ propagation (I) and the domain of
possible pionic soft-mode behaviour (III). The area of low-energy
pion-nucleus scattering and pionic atoms (II) is an important topic
in itself, but will be touched here only occasionally.

The main objective is now to transcribe the conceptual nuclear
matter framework, developed so far, to finite nuclei.

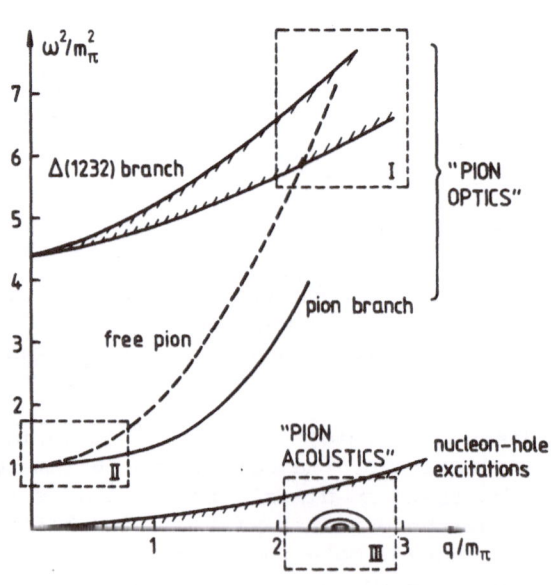

Fig. 9.   Schematic picture
of the various
branches of pionic
modes of excita-
tion in nuclear
matter.
Region I: domain
of $\Delta(1232)$ propa-
gation;
region II: domain
of low energy
pion-nucleus scat-
tering and pionic
atoms;
region III: high
momentum part of
low-frequency
pionic particle-
hole spectrum,
where pionic soft-
modes may develop
at sufficiently
high density.

4. PION-NUCLEUS SCATTERING AND PROPAGATION OF THE Δ(1232) IN NUCLEI

4.1 The Δ-hole model

We would like to discuss the optical potential, related to the pion self-energy Π by (cf. eq. (16)):

$$\langle K' | U(\omega) | \vec{k} \rangle \; = \; \frac{1}{2\omega} \; \langle \vec{k}' | \Pi(\omega) | \vec{k} \rangle \tag{24}$$

for the scattering of pion of frequency ω, with in- and outgoing momenta $\vec{k}$ and $\vec{k}'$, from a given finite nucleus. Instead of the simple geometric series, eq. (17), the pion-nuclear response function, or T matrix, is now determined by the integral equation

$$- \frac{2\pi}{\omega} \; F_{\pi A}(\Theta, \omega) \; = \tag{25}$$

$$\langle k' | T(\omega) | k \rangle \; = \; \langle \vec{k}' | U(\omega) | \vec{k} \rangle \; + \; 2\omega \int \frac{d^3 q}{(2\pi)^3} \; \frac{\langle \vec{k}' | U(\omega) | \vec{q} \rangle \langle \vec{q} | T(\omega) | \vec{k} \rangle}{\omega^2 - \vec{q}^2 - m_\pi^2 + i\delta} \; ,$$

where $d\sigma/d\Omega = |F_{\pi A}|^2$. It is convenient to split U into a Δ-resonant and a background part:

$$U \; = \; U_\Delta \; + \; U_{background}. \tag{26}$$

Let us neglect the background contribution for the moment and concentrate on the 3.3 resonance region, where the pionic response is dominated by the excitation of Δ-hole states,

$$|\alpha\rangle \; = \; |(\Delta h)J^\pi, \; T = 1\rangle$$

coupled to angular momentum $J^\pi = 0^-, \; 1^+, \; 2^-, \; \ldots$ etc.; we assume that a complete set of such states is chosen, given for example in a harmonic oscillator basis. Following eq. (12), the lowest order resonant optical potential $U^{(0)}$, illustrated in Fig. 10, can be written as

$$2\omega \langle \vec{k}' | U_\Delta^{(0)}(\omega) | \vec{k} \rangle \; = \; \sum_\alpha \langle \vec{k}' | \delta H_{\pi N \Delta} | \alpha \rangle \; \frac{1}{\omega - E_\alpha} \; \langle \alpha | \delta H_{\pi N \Delta} | \vec{k} \rangle \tag{27}$$

where

$$E_\alpha(\omega) \; = \; M_\Delta \; + \; \varepsilon_\Delta \; - \; \frac{i}{2} \Gamma_\Delta(\omega) \; - \; M_N \; - \; \varepsilon_h \tag{27a}$$

are the unperturbed Δ-hole energies: ε_Δ is the kinetic energy of the $\Delta(1232)$, and $\varepsilon_h(< 0)$ is the single particle energy of the bound nucleon (note that nucleon binding effects shift the Δ resonance slightly upward). Insertion of $U_\Delta^{(0)}$ into the integral equation, eq. (25), generates multiple scattering (Fig. 11). In the isobar-hole scheme, this is equivalent to the Tamm-Dancoff approximation. In fact the T matrix can be rewritten as

$$2\omega\langle\vec{k}'|T(\omega)|\vec{k}\rangle = \sum_{\alpha\beta}\langle\vec{k}'|\delta H_{\pi N\Delta}|\alpha\rangle\ G_{\alpha\beta}^{(0)}(\omega)\langle\beta|\delta H_{\pi N\Delta}|\vec{k}\rangle \tag{28}$$

with the Green's function

$$G^{(0)}(\omega) = \frac{1}{\omega - \mathcal{H}^{(0)}} \tag{28a}$$

where

$$\mathcal{H}_{\alpha\beta}^{(0)} = E_\alpha(\omega)\delta_{\alpha\beta} + \langle\alpha|W_\pi(\omega)|\beta\rangle\ . \tag{28b}$$

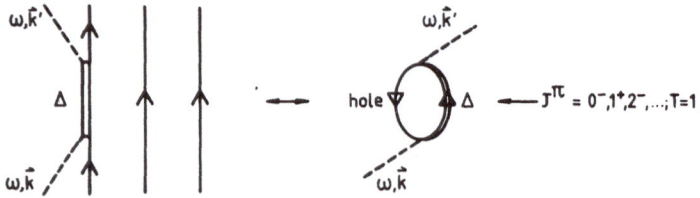

Fig. 10.   Resonant first order p-wave π-nucleus scattering and its interpretation in the Δ-hole model (right figure)

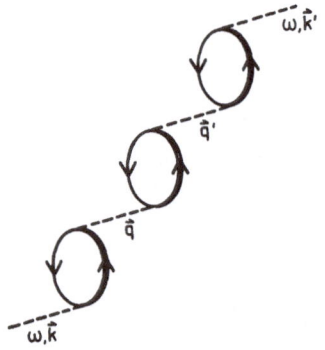

Fig. 11.   Multiple scattering in the isobar-hole model with one-pion exchange only

Here W_π is the OPE component of the Δ-hole interaction, described in section 3.2. Note that W_π is complex, because the pion can propagate on-shell. The imaginary part produces a sizeable increase of the Δ width (the well-known elastic broadening). The real part of OPE is attractive and generates a downward shift of the peak in all important partial waves. If small background pieces are added, then the procedure described here is equivalent (except for nucleon binding effects) to a standard first order optical potential calculation. The failure of such a calculation to reproduce the total cross section, and consequently the requirement to go beyond this first order treatment, becomes obvious from Fig. 14.

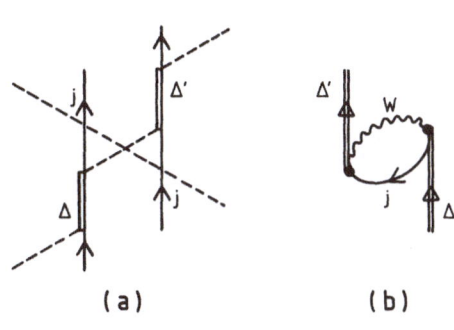

(a)   (b)

Fig. 12.   Effect of the Pauli principle on intermediate rescattering between two nucleons: the diagram (a) is Pauli forbidden, if j is an occupied state. (b): representation of the same process as a Fock term in the Δ self-energy.

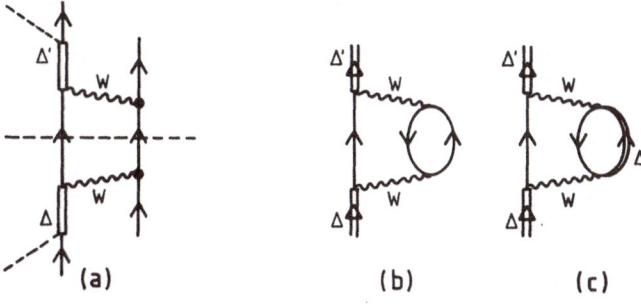

(a)   (b)   (c)

Fig. 13.

(a) coupling of the π-nucleus elastic channel to two-nucleon absorption channel. (b): representation of (a) as a Δ-isobar self-energy correction which couples the Δ to 2-nucleon-1-hole continuum configurations.

(c) additional contribution from resonant rescattering (reflection term) also incorporated in the Δ-hole model calculation.

In the isobar-hole model, higher order many-body corrections enter at two places:

(a) The non-static OPE Δ-hole interaction in the elastic channel must be replaced by the full particle-hole interaction described in section 3.2.

(b) The Δ(1232), once excited, experiences interactions which couple it to reaction channels (e.g. the quasifree $(\pi,\pi'N)$ channel, the two-body absorption $(\pi,2N)$ channel etc.) Such couplings modify the mass and width of the Δ-isobar in each individual wave ("Δ self-energy corrections").

The authors of refs.[2,6] have simulated the full diversity of these many-body corrections phenomenologically by a local complex isobar spreading potential. These results have provided clear evidence for the importance of true absorption channels. The position taken in refs.[3,29] was instead to calculate these corrections microscopically in order to explore their sensitivity with respect to the basic meson-baryon interactions. The discussion in ref.[3] is based on the full Δ-hole interaction described in section 3.2. The model used to calculate $g'(\omega,\vec{q})$ consists of π and 2π exchange, accompanied by a baryon-baryon correlation function. This gives vlues $g' = 0.5 - 0.6$ at $q = \omega = o$, depending on the precise value of the ρNΔ coupling strength.

We proceed now to discuss those parts of the Δ(1232) self energy in nuclei which have turned out to be the most important ones:

(i)   Pauli principle corrections
        As the pion propagates between pairs of nucleons, processes such as shown in Fig. 12a are obviously forbidden by the Pauli principle. In many-body language, such processes appear as isobar-Fock-terms, as shown in Fig. 12b. The effect of the Pauli blocking is to reduce the phase space for the decay $\Delta \rightarrow \pi N$, since parts of the nucleon states are occupied. This leads to a sizeable reduction of the Δ width[2,3], accompanied by a real part which depends on the details of the $\Delta N \rightarrow N\Delta$ interaction. For the model interaction W, this shift is attractive and strongly energy dependent[3].

(ii)   True absorption corrections
        The essential damping mechanism in the propagation of Δ-hole excitations is introduced by their strong coupling to two-nucleon-two-hole continuum configurations (Fig. 13). Let us illustrate briefly how such corrections increase the Δ width and shift its mass, by investigating, in a much simplified picture, the diagram Fig. 13b. Let $|\lambda\rangle = |2N1h\rangle$ be a two-nucleon-one-hole continuum state. Then the coupling of the Δ to this state leads to a complex shift of the Δ mass,

$$\delta E_\Delta = \sum_\lambda \frac{|<\Delta|W|\lambda>|^2}{E_O - E_\lambda - i\delta} \tag{29}$$

Here the sum over λ is understood to be also an integration over E_λ, and E_O is the starting energy, given essentially by the pion energy (except for binding corrections). Now, E_O is much larger than the threshold for two-nucleon emission. Using $[E_O - E_\lambda - i\delta]^{-1} = P/(E_O - E_\lambda) - i\pi \delta(E_O - E_\lambda)$, we see how the imaginary part gives an energy dependent absorptive width, $\Gamma_{abs} = -2Im\delta E_\Delta$, which adds to the Δ decay width. This absorptive damping is a substantial effect and overcompensates the reduction of the width by Pauli blocking effects.

Microscopic calculations of all the processes shown in Figs. 12-13 have been performed in ref.[3] for ^{4}He and ^{16}O, where the parameters of the interaction W have been chosen in accordance with a calculation of πd absorption which is a very important consistency requirement. We mention that Fig. 13c becomes important in the 3.3 resonance region (sometimes called "reflection term" because it incorporates the resonant rescattering of a pion in the quasi-elastic channel).

The pion-nuclear T-matrix with inclusion of these many-body corrections is now given by

$$2\omega<\vec{k}'|T(\omega)|\vec{k}> = \sum_{\alpha\beta} <\vec{k}'|\delta H_{\pi N\Delta}|\alpha>G_{\alpha\beta}(\omega)<\beta|\delta H_{\pi N\Delta}|\vec{k}>, \tag{30}$$

where

$$G(\omega) = (\omega - \mathcal{H}(\omega))^{-1}$$

is the full isobar-hole Green's function. For simplicity, we write only the direct part of it which defines the Tamm-Dancoff approximation (TDA), while inclusion of crossed terms (the step from TDA to RPA) is shown in detail in ref.[3]. The G is obtained by inversion of the complex matrix

$$\mathcal{H}_{\alpha\beta}(\omega) = \bar{E}_\alpha(\omega)\delta_{\alpha\beta} + \Sigma_{\alpha\beta}(\omega) + W_{\alpha\beta}(\omega), \tag{31}$$

where α and β refer to the Δ-hole basis states. Here $\bar{E}_\alpha$ is given by E_α of eq. (27a), except that the free Δ isobar kinetic energy can now be replaced by the single particle energies in an isobar Hartree potential, which describes additional binding effects. This Hartree potential is estimated to be roughly of the same (attractive) magnitude as the nucleon single particle potential[3]. The $W_{\alpha\beta}$ are matrix elements of the full isobar-hole interaction, as before, while

$\Sigma = \Sigma$ (Pauli) + Σ(2p2h) summarizes the self-energy corrections to the masses and widths of isobar-hole states in each partial wave J generated by the Fock terms (Σ(Pauli), Fig. 12) and the absorption and reflection contributions (Σ(2p2h), Fig. 13). We show in Fig. 14 the influence of higher order corrections on the total $\pi^{16}O$ cross secsion[3]. While Σ(Pauli) (through its positive imaginary part) leads to an appreciable narrowing of the isobar width in all partial waves, additional inclusion of Σ(2p2h) provides the necessary damping. At low energy, the absorptive width comes exclusively from two-nucleon-absorption, Fig. 13(b). The decomposition of σ_{tot} into partial waves is shown in Fig. 15. In the isobar-doorway picture [30,2], each one of these partial waves is interpreted as an isobar-doorway state of given J^{π}. Also shown in Fig. 16 are selected examples of differential cross sections.

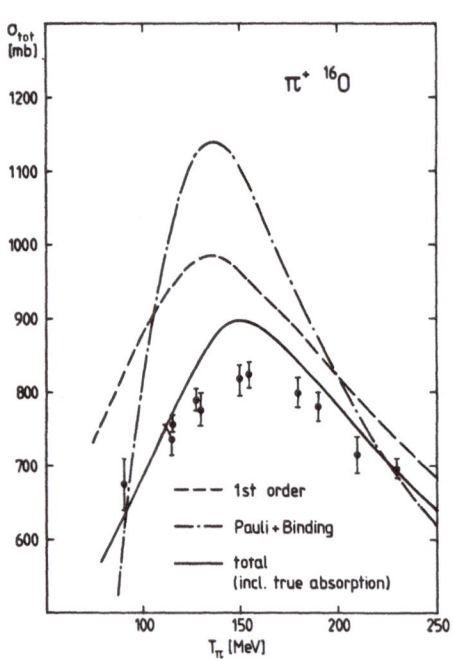

Fig. 14. Total $\omega^{16}O$ cross sections calculated at different levels of the π-nuclear many-body problem: shown is a standard first order momentum space calculation for reference. The other two curves are calculated within the RPA scheme of ref.[3].

Pauli + Binding:
incorporates isobar Hartree potential, Fock term Σ(Pauli) and full interaction W of eq. 21.

Total:
includes in addition true absorption and reflection terms Σ(2p2h)

Input:
π and ρ exchange plus baryon-baryon correlation function. (Cutoff in $\pi N\Delta$ vertex: Λ = 1.2 GeV; $\rho N\Delta$ coupling: $f_{\rho}^{*}/4\pi$ = 4.1 ("strong ρ").) Calculation includes all partial waves $J^{\pi} = 0^{-}$, 1^{+}, ... up to 8^{-}. For further details see ref.[3].

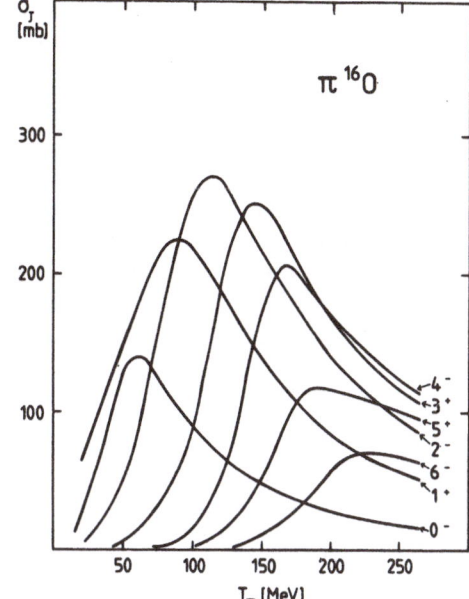

Fig. 15.   Decomposition of the
           $\pi^{16}O$ cross section
           into partial wave cross
           sections, once all
           higher order correc-
           tions are included
           (taken from [3]).

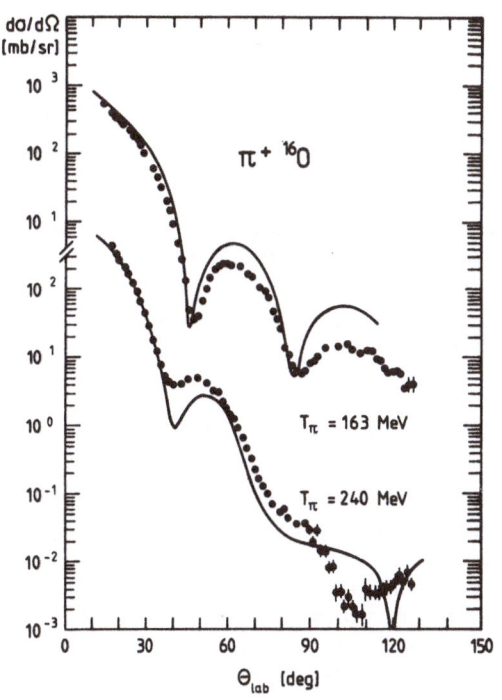

Fig. 16.   Examples of differen-
           tial cross sections
           for $\pi^{16}O$ including all
           medium corrections.
           Taken from ref.[3].

The apparent success of these and other isobar-hole model calculation, especially in reproducing the important damping mechanisms
which the Δ(1232) encounters in the nucleus, should not obscure some
of the basic uncertainties. The shifts and widths of Δ-hole states
depend sensitively on details of the NΔ -interaction which are not
very well under control (cutoff factors, ρNΔ coupling constant, short
range NΔ correlations etc.) It seems however that the quasifree and
absorptive widths are reasonably well reproduced once the parameters
of the basic baryon-baryon interactions are fixed in accordance
with the πd $\leftrightarrow$ pp reaction. On the other hand, the shifts in the positions of isobar-hole states are certainly not better determined than
within $\sim$ 30 MeV.

To summarize the picture obtained so far, we conclude that many
of the observed features of π-nucleus scattering in the 3.3 resonance
region can be understood in terms of the excitation of Δ(1232) isobars, followed by their interactions with surrounding nucleons. These
interactions can be figured by a modification of the Δ mass and width
in each partial wave. We have shown here to what extent microscopic
calculations can account for these modifications. On the phenomenological level, it is very useful to parametrize these features in
terms of a complex optical potential for Δ-isobars. A refined analysis[6] reveals that, if this potential is assumed to consist of central
and spin-orbit pieces, then the required real and imaginary parts of
the central potential are only smoothly energy dependent, the real part
being somewhat weaker than the nucleon single particle potential. The
strong dependence on the angular momentum of the Δ seems to be well
accounted for by the Δ spin-orbit potential.

The π-nucleus optical potential U is generally a highly non-local
quantity. However, it turns out that at low energy ($T_\pi \lesssim 50$ MeV), it
can be converted to a good degree of approximation into an equivalent
r-space potential of the Kisslinger type, $\vec{\nabla} F(r)\vec{\nabla}$. This means that
connections can be made between the many-body theoretical picture
and standard parametrizations of the p-wave optical potential,

$$2\omega U(\vec{r}) = - 4\pi \vec{\nabla} L(r) n(r)\vec{\nabla},$$

(32)

$$n(r) = c_o\rho(r) + C_o\rho^2(r), \quad L(r) = [1+4\pi g'n(r)]^{-1},$$

which have been used with considerable success to describe low energy
π-nucleus scattering data[31,32]. The isobar-hole model (for ^{16}O)
yields[33] Im C_O = 0.08 m_π^{-6}. The discussion of Re C_o includes a
variety of many-body effects like binding and Pauli corrections as
well as dispersive real parts from two-nucleon absorption. The isobar-hole model gives a positive (i.e. attractive) Re C_O of about the
same magnitude as Im C_O, the precise value depending on the treat-

ment of binding effects (but Re C_O remains positive for a wide variety of model assumptions). The value of the Ericson-Ericson-Lorentz-
Lorenz (EELL) correction is determined by the correlation part g' of
the full Δ-hole interaction W. A strong ρ exchange accompanied by
additional short-range correlations favours a large EELL correction,
g' $\gtrsim$ 0.5, as compared to the "classical" value g' = 1/3.

Most recent fits to shifts and widths of pionic atom levels[34] are
in fact consistent with a large EELL correction (g' $\simeq$ 0.6) in conjunction with large positive values of Re C_O (around 0.1 m_π^{-6}), where
at the same time Im C_O is required to be about twice as large as
the "canonical" value Im C_O = 0.07 m_π^{-6}, which has been used for many
years. The Δ-hole model, on the basis of the diagram Fig. 13(b),
gives only about one half of this absorption strength at threshold;
This is not surprising in the light of refs. 45,46, where it is
pointed out that a large fraction of Im C_O comes from processes
where the absorption takes place on correlated nucleon pairs, without primarily involving intermediate Δ(1232) states.

4.2 Photon-induced excitation of Δ-hole states

There is general agreement about the relevance of true absorption and other higher order processes. In pion-nucleus scattering,
however, much of the shift and broadening of Δ-hole states in the
dominant partial waves occurs simply because of coherent pion multiple scattering, which then combines with all other damping and
dispersive effects. It is therefore useful to investigate processes
complementary to pion induced ones, in particular those where the
elastic broadening is possibly reduced, in order to obtain independent information about the modification of positions and widths of
Δ-hole states. We claim[35] that total photonuclear cross sections in
the 3.3 resonance region[47,48] are a source of such information. In
addition, the coherent photoproduction of π^o on nuclei[48] has been
pointed out[36] as a useful process, because it is completely dominated
by Δ(1232) excitation.

The $\gamma N\Delta$ transition operator follows from the $\pi N\Delta$ vertex Hamiltonian simply by the replacements

$$ f \rightarrow f_{\gamma N\Delta}, \quad \vec{S}^+ \cdot \vec{q} \rightarrow \vec{S}^+ \cdot (\vec{k} \times \hat{\epsilon}), \quad T_\lambda^+ \pi_\lambda \rightarrow T_3, \qquad (33) $$

where $\vec{k}$ is the photon momentum, $\hat{\epsilon}$ the photon polarization vector, and
$f_{\gamma N\Delta}$ = 0.116. Note that the ratio $f_{\gamma N\Delta}/f^*$ for $f^{*2}/4\pi$ = 0.37 is close
to the canonical value $(e/g)\mu_v$, where $g^2/4\pi$ = 14, the πNN coupling
constant, and $\mu_v = (\mu_p - \mu_n)/2$ = 3.7 is the isovector magnetic moment.
The T-matrix for the coherent (γ, π^o) process can now be obtained at
once in the Δ-hole model, in analogy with eq. (30) (see Fig. 17(a)):

$$2\omega < \pi^{0}(\vec{q}) \,|\, T(\omega) \,|\, \gamma(\vec{k}) > \; = \; \sum_{\alpha\beta} <\vec{q}\,|\, \delta H_{\pi N\Delta} \,|\, \alpha> \; G_{\alpha\beta}(\omega) \; <\beta\,|\, \delta H_{\gamma N\Delta} \,|\,\vec{k}> \;\; , \qquad (34)$$

where G is again the full Δ-hole Green's function discussed in section 4.1.

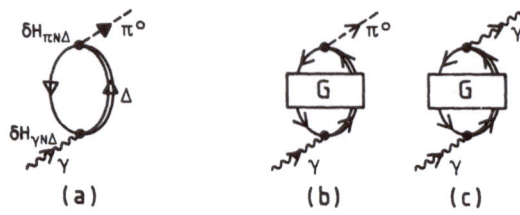

Fig. 17.   (a): lowest order coherent π^{0} photoproduction on a nucleus in the Δ-hole model. The $A(\gamma,\pi^{0})$ A_{gs} (b) and resonant photon scattering (c) processes in the Δ-hole model, employing the full Δ-hole Green's function.

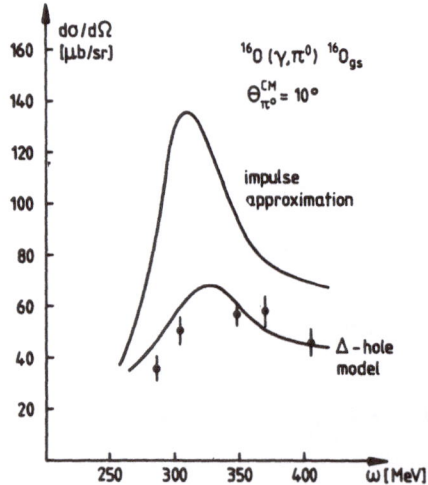

Fig. 18.   Differential cross section for coherent π^{0} photoproduction on ^{16}O at $\Theta = 10^{0}$ as function of photon energy. A comparison is shown between an impulse approximation calculation (Fig. 17(a)) and a full Δ-hole model calculation Fig. 17(b)). The experimental data are from ref.[48].

An immediate consequence of the tranverse nature of the $\gamma N\Delta$ coupling, as compared to the longitudinal $\pi N\Delta$ coupling, is the suppression of π^{O} propagation under extreme forward angles ($\Theta = O$). The $^{16}O(\gamma,\pi^{O})^{16}O_{gs}$ angular distribution peaks around $\Theta = 20-30$ degrees, depending on the photon energy. The damping of Δ-hole modes discussed in section 4.1 should be visible also in the (γ,π^{O}) reaction. This is evident from Fig. 18, where the impulse approximation result for $^{16}O(\gamma,\pi^{O})^{16}O_{gs}$ at $\Theta_{CM} = 10^{O}$ is compared with the full Δ-hole model calculation[37]. It appears that the experimental data, obtained at the Bonn synchroton [48], clearly require such a strong damping.

Let us now turn to photon scattering and total photon-nucleus cross section $\sigma_{\gamma A}(\omega)$ in the 3.3 resonance region. The amount of experimental data in this area is still quite limited. However, the measurements of the Mainz group[47] as well as the Bonn data on total cross sections for charged pion photoproduction[48], show again a considerable damping of the Δ, but almost no shift of the peak in $\sigma_{\gamma A}$, quite unlike the π-nucleus scattering situation. We consider the T matrix for forward Compton scattering, $<\vec{k}|T_{\gamma A}(\omega)|\vec{k}>$, and split it again into a Δ-resonant part and a background piece:

$$T_{\gamma A} = T_{\Delta} + T_{back} \tag{35}$$

The total photonuclear cross section is given by

$$\sigma_{\gamma A}(\omega) = - 2 \text{ Im } <\vec{k}|T_{\gamma A}(\omega)|\vec{k}>. \tag{36}$$

Following the previous developments, the Δ-resonant part T_{Δ} is written in terms of the full Δ-hole Green's function (see also Fig. 17(b)):

$$2\omega \, <\vec{k}|T_{\Delta}(\omega)|\vec{k}> = \sum_{\alpha\beta} <\vec{k}|\delta H_{\gamma N\Delta}|\alpha> \, G_{\alpha\beta}(\omega) <\beta|\delta H_{\gamma N\Delta}|\vec{k}> . \tag{37}$$

Two major differences to π-nucleus scattering are to be pointed out: first, the Δ-hole states $|\alpha>$ and $|\beta>$ now incorporate also natural parity states with $J^{\pi} = 1^{-}$, 2^{+}, ... etc., in addition to the unnatural parity states discussed previously. In fact, about one half of the resonant part of $\sigma_{\gamma A}$ comes from isobar-hole states of natural parity. Second, the transverse $\gamma N\Delta$ Hamiltonian operates now both in the in- and outgoing channel and suppresses the coherent forward propagation of π^{O} which appears in the isobar-hole Green's function G through the action of the OPE part of the full Δ-hole interaction W. This means that the strong elastic broadening and the accompanying downward shift of Δ-hole states observed in π-nucleus scattering will be nearly absent in photon scattering. The ρ exchange part of W now acts coherently in natural parity Δ-hole states, but it is of order k^{2}/m_{ρ}^{2} and therefore not of great importance. This means that resonant photon scattering is selectively sensitive to the damping

and shift resulting from Δ-self energy interactions, i.e. the coupling of the Δ(1232) to many-body excitation channels described in
section 4.2. Before we present a calculation of these, some words
must be spent on the background pieces of $T_{\gamma A}$ which emerge essentially from non-resonant parts of the charged pion photoproduction
amplitude. Their contribution to $\sigma_{\gamma A}$ is parametrized in the form
$A^\beta \, \sigma_{back}(\gamma N)$, where $\sigma_{back}(\gamma N)$ is the background part of the elementary γN cross section. Meaningful upper and lower limits are β = 1
and β = 2/3, corresponding to no damping or strong damping of back-
ground-photo-produced pions within the nucleus. This background is
then added to the resonant part calculated in the isobar-hole model.

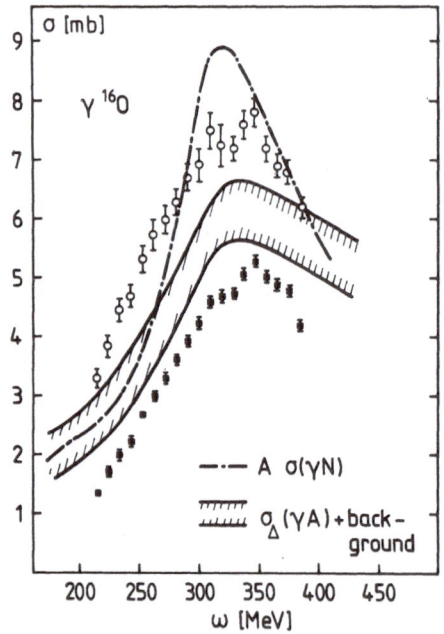

Fig. 19.   Total $\gamma\,^{16}O$ cross section in the 3.3 resonance region.
            Shown is a comparison between impulse approximation
            and a full Δ-hole model calculation, adding non-reso-
            nant background, as described in the text. The data
            correspond to total cross section for photoproduction
            of charged particles only, taken from ref.[48]. The points
            Φ are extrapolated total γA cross sections based on
            these data.

Results for ^{16}O are shown in Fig. 19. The complex isobar self-ener-
gies Σ employed in this calculation are exactly the same as those
calculated for π-nucleus scattering. The characteristic damping in
the resonance region is again evident. The absence of the shift and
broadening from elastic multiple π^0 scattering has the consequence
that the maximum of $\sigma_{\gamma A}$ is hardly moved compared to the impuls
approximation peak: repulsion from the g' part of the Δ-hole interaction, binding effects, and attractive contributions from Re Σ,
the Δ self-energy corrections, almost cancel each other. It is
instructive to associate a "damping exponent" α, defined by
$\sigma_{\Delta}(\gamma A) = A^{\alpha} \sigma_{\Delta}(\gamma N)$, with the resonant part of $\sigma_{\gamma A}$. At the peak, we
find $\alpha = 0.84$ for both ^{16}O and ^4He. This appears to be remarkably
close to the values α around 0.8 reported by Ziegler[47] for Li and
Be. For ^{16}O, a comparison is strictly possible only with the Bonn
data for total charged photoproduction[48] (included in Fig. (19));
also shown is their extrapolated total cross section, although we
should note that the procedure for obtaining these is model dependent. The main point we would like to make here is that the general
trend of the energy dependence of $\sigma_{\gamma A}$ predicted by the Δ-hole model,
especially the damping of the $\Delta(1232)$ with only a slight shift of the
maximum, seems to follow quite well the experimental observations for
total charged photoproduction. This then presents an independent test
for the $N\Delta$ -interaction mechanisms, on which the calculation of damping widths and dispersive real parts is based.

5. HIGH MOMENTUM TRANSFER PROPERTIES OF PION-LIKE NUCLEAR STATES

    We would now like to return to the question whether low-lying
nuclear states carrying pion quantum numbers show a characteristic
enhancement effect at large momentum transfers as an indication of
the proximity of a pion condensate. Such ideas have been developed
in refs.[10-13], and much of the conceptual framework is presented
in ref.[38].

5.1 Response function for pionic low-frequency modes in finite nuclei

Consider pion-like nuclear excited states with angular momentum
$J^{\pi} = 0^-$, 1^+, 2^-, ... etc. and isospin T = 1. These are formed by
particle-hole excitations correlated by the spin-isospin dependent
interaction, eq. (21), usually employing a small model space (P space)
to reproduce the long-wavelength properties of such states. The
attraction from the OPE interaction V_{π} at large q favours strong
spin-isospin polarization phenomena. Let us return to the finite-
nucleus analogue of the lowest order pion self-energy, eq. (12),
following refs.[13,39]:

$$<\vec{q}'|\,\Pi^{(o)}(\omega)\,|\vec{q}> = -\sum_{ph} \frac{<\vec{q}'|\,\delta H\,|ph><qh|\,\delta H\,|\vec{q}>}{E_p - E_h - \omega} + \quad \text{crossed term}$$

$$(38)$$

$$= \sum_{J} \frac{2J+1}{4\pi} \quad \Pi_J^{(o)}(q',q,\omega)\ P_J(\hat{q}'\cdot\hat{q}).$$

(see Fig. 20 for illustration). The partial wave self-energy $\Pi_J^{(o)}$ can be written

$$\Pi_J^{(o)}(q',q,\omega) = \sum_{LL'} a_{JL}\ [\hat{\Pi}_J(q',q,\omega)]_{LL'} a_{JL'},$$

$$(39)$$

where $a_{JL} = (J010|LO)$, and $\hat{\Pi}_J$ is a 2 x 2 matrix for the possible values of particle-hole orbital angular momenta, $L, L' = J\pm1$:

$$[\hat{\Pi}_J(q',q,\omega)]_{LL'} = -\sum_{ph} Q_{ph}^{JL}(q') \left\{ \frac{1}{E_p - E_h - \omega} + \frac{1}{E_p - E_h + \omega} \right\} Q_{ph}^{JL'}(q).$$

$$(40)$$

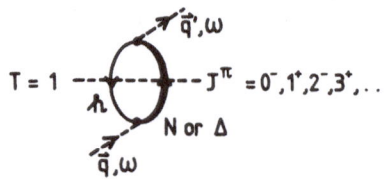

Fig. 20.   Pion self-energy in finite nuclei. The pion couples to $T = 1$ nucleon-hole and Δ-hole states with $J^{\pi} = 0^-,\ 1^+,\ 2^-,\ 3^+,\ \ldots$ etc.

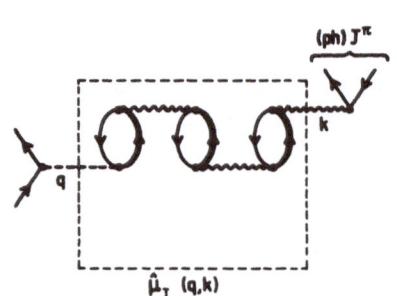

Fig. 21.   Role of the spin-isospin tensor $\hat{\mu}_J$ describing Q-space pionic polarization to all orders in a scattering process (e.g. with electrons or protons) leading to a pion-like particle-hole state in P-space. The particle lines in the particle-hole excitations building up $\hat{\mu}_J$ are nucleons or $\Delta(1232)$ isobars.

The $Q^{JL}_{ph}(q)$ are particle-hole transition form factors. Actual calculations are then performed as follows: the full particle-hole Hilbert
space is split into a model space (P space) and a large complementary
Q-space. The core polarization arising from virtual excitations in
Q-space is incorporated into the model space (P-space) calculation
as a renormalization of the relevant interactions and transition
operators, as shown in Fig. 21. In this renormalization procedure,
the sum in eq. (40) therefore extends over all Q-space ph configurations (typically a 20 - 30 $\hbar\omega_O$ Q-space is required at momentum transfers $q \sim$ 2-3 m_π). For example, an operator of the type

$$M^\lambda_{JL}(q,r) = i^L q \, [Y_L \times \sigma]_J \, j_L(qr) \, \tau_\lambda, \tag{41}$$

which excites pion-like states, is renormalized according to

$$\widetilde{M}^\lambda_{JL}(q,r) = \sum_{L'} \int_0^\infty \frac{dk \, k^2}{(2\pi)^3} \, M^\lambda_{JL'}(k,r) \, [\hat{\mu}_J(k,q)]_{L'L}. \tag{41}$$

Here $\hat{\mu}_J$ is the spin-isospin polarizability tensor, which corresponds
to the inverse of the dimesic function ε of eq. (23). In terms of
the pionic response function $\hat{R}_J$, determined by the equation

$$\hat{R}_J(k,q) = \hat{\Pi}_J(k,q) + \int_0^\infty \frac{dq'}{(2\pi)^3} \, \hat{\Pi}_J(k,q') \hat{W}_J(q') \hat{R}_J(q',q), \tag{43}$$

the $\hat{\mu}_J$ is given by

$$k^2 \, [\hat{\mu}_J(k,q)]_{L'L} = (2\pi)^3 \delta(k-q) \delta_{L'L} + [\hat{W}_J(k) \hat{R}_J(k,q)]_{L'L}. \tag{44}$$

Here $\hat{W}_J$ is related to the particle-hole interaction W by

$$[\hat{W}_J(k)]_{L'L} = [\tfrac{1}{3} V_\pi(k) + \tfrac{2}{3} V_{2\pi}(k) + g'] \delta_{L'L} +$$

$$+ \; [\tfrac{1}{3} V_\pi(k) - \tfrac{1}{3} V_{2\pi}(k) + h'] (3a_{JL'} a_{JL} - \delta_{L'L}), \tag{45}$$

where the first and second term on the right hand side corresponds
to the spin-spin and tensor part of the interaction, respectively.

Note, that as a pion-like mode of given J becomes soft, $\hat{R}_J$, and hence $\hat{\mu}_J$, tends to become very large in a certain region of momenta around the critical momentum q_c in nuclear matter. The difference between finite nuclei and nuclear matter is that $\hat{R}_J$ is non-local in momentum space for a finite system, the momentum space non-locality being inversely proportional to the nuclear radius R[39]. Thus precritical enhancement effects in small nuclei, if existent, will generally be less pronounced than in heavy nuclei. For sufficiently large nuclei, pionic soft-mode behaviour appears under similar conditions as in nuclear matter: for g' < 0.4, all pion-like modes with $J \lesssim q_c R$, where $q_c \sim$ 2-3 m_π, the critical momentum, tend to become soft. The question is then whether pionic polarization phenomena might be visible even for more realistic values of the correlation parameter, $g'(q \simeq q_c, \omega = 0) \gtrsim 0.5$.

We emphasize again that the enhancement of the pion field due to large-space core polarization is a phenomenon restricted to high momentum transfers. In the long wavelength limit, the spin-isospin dependent particle-hole interaction is entirely determined by g', with the well known consequence of quenching effects observable in transitions related to weak axial currents[26] and to the spin dependent parts of magnetic multipoles [28].

5.2 Applications

Inelastic proton scattering to unnatural parity states in heavy nuclei has been suggested[12,13] as a possible source of information on the presence or absence of precritical enhancement effects. In fact, proton nucleus scattering provides perturbations of the form $\vec{\sigma} \cdot \vec{q}\, e^{i\vec{q} \cdot \hat{r}}$, the pion source operator, at high momentum transfer. We would first like to present a schematic example by considering matrix elements of the form $S_J(\vec{\sigma} \cdot \vec{q}) = \langle J^\pi | \vec{\sigma} \cdot \vec{q}\, e^{i\vec{q} \cdot \vec{r}} | 0^+ \rangle$ from the ground state to some unnatural parity state in ^{208}Pb, mainly to illustrate how soft mode behaviour at high momentum transfer is approached in a finite system[39]. Here we consider the case of $J^\pi = 2^-$ states in ^{208}Pb at q = 2.5 m_π, close to the critical momentum for pion condensation nuclear matter. The p-space consists of all 3 $\hbar\omega_0$ excitations, while the Q-space (> 3 $\hbar\omega_0$) polarization is incorporated by the polarizability tensor $\hat{\mu}_J$, as previously described. Results for the excitation spectrum as seen by a pion source operator are shown in Fig. 22. Note the overall increase of strength with decreasing g', once the pionic polarization is turned on. Around g' = 0.4, the strength is collected into one single low-frequency mode which, upon further reduction of g', eventually becomes soft. Similar features are observed for all pion-like modes, as long as $J < q_c R$. In heavy nuclei, the precritical enhancement is localized in a narrow band of momentum transfers around $q_c \sim 2.5\ m_\pi$. No indication of such behaviour, even for g' = 0.4, can be observed in the long-wavelenght limit. Instead, there is a quenching of transitions to unnatural parity states because of the repulsion from g' at $q \simeq 0$.

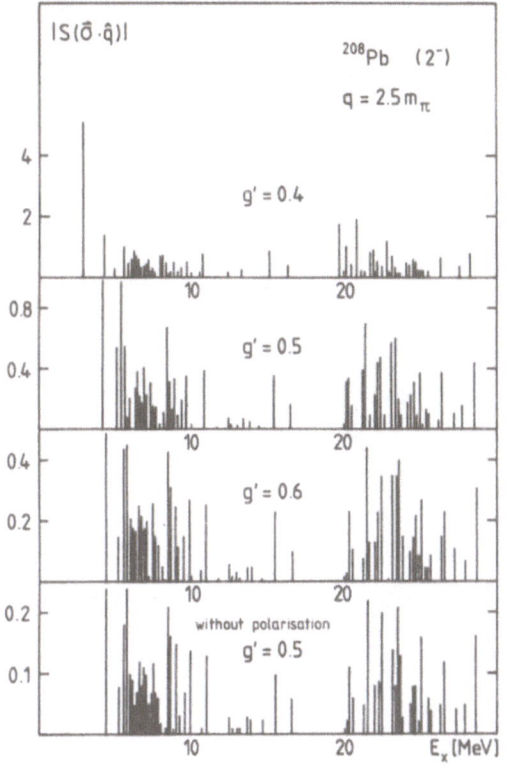

Fig. 22. $J^{\pi} = 2^-$ strength in 208_{Pb} at momentum transfer q = 2.5 m_{π} as seen by an operator $\vec{\sigma} \cdot \vec{q}$ exp $(i\,\vec{q} \cdot \vec{r})$. Bottom: results of a standard $3\hbar\omega_0$ P-space calculation using W_{ph} of eq. (7) with g'=0.5 and h'=0, omitting Q-space pionic polarization. Upper three subfigures include Q-space polarization as described in the text, and the variation with the correlation parameter g' is shown. Note different scales in various subfigures.

The situation described above is somewhat idealized, mainly because of the spreading of ph states by their coupling to more complicated (e.g. 2p2h) configurations. This spreading of strength will work against the formation of collective low-frequency, high-momentum modes. It is not obvious, however, that the spreading at high momentum transfers will be the same as the one observed at low q, because a high momentum mode does not mix easily with low lying 2p2h states. This question requires more detailed investigation.

High momentum transfer experiments involving unnatural parity excitations in heavy nuclei are extremely difficult to perform, so that present interest is directed more towards light nuclei, although the polarization effects there are generally expected to be weaker. A particularly well investigated example is the $J^{\pi} = 1^+$, T = 1 state in ^{12}C at 15.1 MeV. Its basic p-shell structure is supposed to be well known, e.g. in terms of a Cohen-Kurath wave function, which we adopt as the model space (P-space) wave function. The interest in this state is motivated by considerable difficulties to reproduce the high momentum transfer behaviour of the M1 form factor. In fact, it has been pointed in ref.[40] that spin-dependent core polarization effects are very important. The first full scale calculation of the ^{12}C M1 form factor taking into account polarization effects to all orders has been performed by Delorme et al.[41]. They come to the conclusion that the observed enhancement at large q over the Kohen-Kurath result can be interpreted as an indication of critical opalescence, in the sense that the g' required to reproduce the data is quite samll (g' $\simeq$ 0.4).

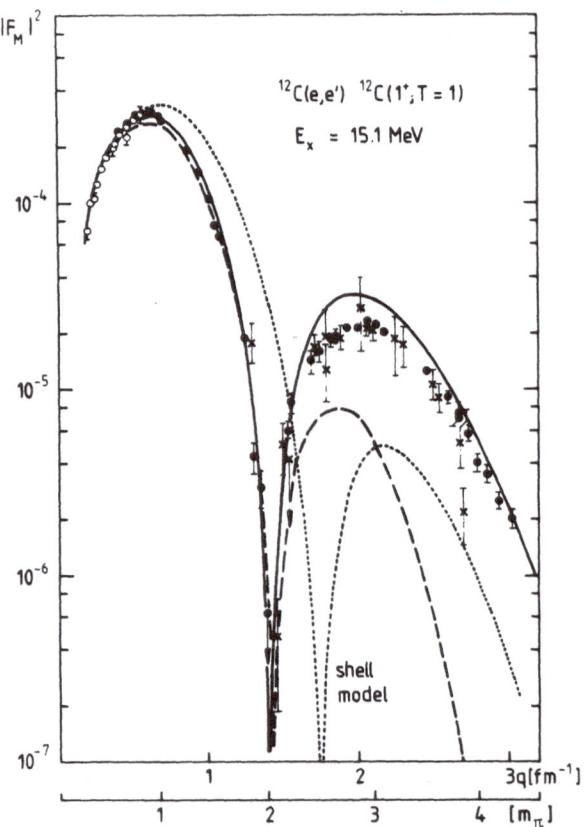

Fig. 23.   M1 form factor for ^{12}C, Short-dashed curve, standard
           Cohen-Kurath shell model; long-dashed curve: effect
           of pionic Q-space polarization, using V_π and g' = 0.55,
           but $V_{2\pi}$ = 0 (and h' = 0); Solid curve: full particle
           hole interaction W, including $V_{2\pi}$. Experimental data
           ref.[44]. See ref.[42] for details.

We have calculated[42] the M1 form factor together with $d\sigma/d\Omega$ for inelastic proton scattering at E_p = 800 MeV leading to the same state. One might expect that an M1 operator, being proportional to $\vec{\sigma} \times \vec{q}$ (i.e. not properly alligned with a pion source function $\vec{\sigma} \cdot \vec{q}$) would not be an appropriate tool for probing into pionic modes. This is true for heavy nuclei, but for light nuclei, such restrictions are not so severe, due to surface effects. The M1 form factor is shown in Fig. 23; it demonstrates the obvious failure to reproduce the high momentum transfer behaviour with the Cohen-Kurath wave function. Once large space pionic polarization is included, one observes a shift of the minimum (caused primarily by repulsive correlations proportional to g'). However, we find that the enhancement from V_π alone is only moderate, unless exceedingly small values of g' ($\lesssim$ 0.4) are introduced. It is evident that isovector two-pion exchange ($V_{2\pi}$) is of crucial importance, since this component of the particle-hole interaction is properly alligned with the $\vec{\sigma} \times \vec{q}$ nature of the M1 operator. In particular, the non-resonant part of $V_{2\pi}$ (other than ρ exchange) gives a large fraction of the enhancement required by the data. Once the full $V_{2\pi}$ is included, the pionic opalescence situation appears not to be encountered as strongly as suggested in ref.[41].

Next, let us discuss inelastic proton scattering into the same state[42]. We wish to concentrate on the data[49] available at E_p=800 MeV, where Glauber theory is a reliable tool to treat the distortions of the in- and outgoing protons. These data have recently been extended[50] to high momentum transfers covering the critical region (q $\sim$ 2-3 m_π). Results are shown in Fig. 24. The steep falloff in the interesting region of momentum transfers indicates that the elementary spin-isospin dependent two-nucleon amplitude in this region is not primarily driven by the attraction from the OPE tensor force, but is counterbalanced to a large extent by a repulsive tensor interaction of a range which falls quite naturally into the domain of the iso-vector two-pion exchange interaction[51]. Once the primary tensor interaction is reduced to such a large extent, however, the enhancement of the differential cross section around q $\sim$ 3 m_π due to pionic pola-rization effects is only very moderate. Similar conclusions can be drawn from preliminary Saclay data at E_o = 400 MeV[52], and from the Indiana experiments at 122 MeV[53].

We conclude that, so far, there is no clear evidence for pionic critical opalescence, in essential agreement with the expectation that the minimal density for the appearance of a pion condensate is certainly not lower than two or three times nuclear matter density. It is necessary, however, to extend investigations of the type just considered to a variety of different nuclei, so that one can hope to obtain a more systematic insight into the structure of nuclear spin-isospin correlations, which is of such crucial importance for the pion-nuclear many-body problem as a whole.

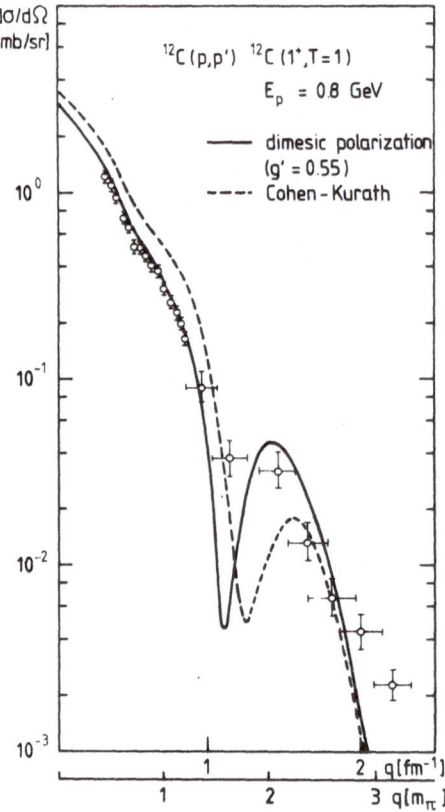

Fig. 24. Glauber calculation of 800 MeV $^{12}C(p,p')$ $^{12}C(1^+)$ using the
same model as in Fig. 23. Data are from ref.[43]. The spin
isospin dependent primary NN amplitude has been adjusted
to the low q data. See ref.[42] for further details.

I am very grateful to my friends and collaborators,
Eulogio Oset and Hiroshi Toki, whose work has been the basis of
these lectures.

REFERENCES

1.   For recent reviews see the corresponding papers in:
     Proc. 2nd Int. Conf. on Meson-Nuclear Physics,
     Houston, 1979 (AIP Conf. Proc.) Proc. 0th Int. Conf.
     on High-Energy Physics and Nuclear Structure,
     Vancouver (1979), Nucl. Phys. A 335 (1980)

2.   M. Hirata, F. Lenz and K. Yazaki, Ann. of Phys.
     108 (1977) 16;
     M. Hirata, J. Koch, F. Lenz and E.J. Moniz,
     Phys. Lett. 70 B (1977) 281 and Ann. of Phys., in print
3.   E. Oset and W. Weise, Phys. Lett. 77 B (1978) 159;
     Nucl. Phys. A 319 (1979) 477;
     Nucl. Phys. A 329 (1979) 365;
4.   W. Weise, Nucl. Phys. A 278 (1977) 402
5.   K. Klingenbeck, M. Dillig and M.G. Huber,
     Phys. Rev. Lett. 41 (1978) 387
6.   Y. Horikawa, M. Thies and F. Lenz, Nucl. Phys. (in print)
7.   A.B. Migdal, Rev. Mod. Phys. 50 (1978) 107,
     and references therein
8.   G.E. Brown and W. Weise, Phys. Reports 27 C (1976.) 1
9.   See contributions of: A.B. Migdal, R.F. Sawyer, G. Baym
     and D.K. Campbell, S.-O. Bäckman and W. Weise, in:
     Mesons in Nuclei, Vol. III, M. Rho and D.H. Wilkinson,
     eds., North-Holland (1979)
10. M. Ericson and J. Delorme, Phys. Lett. 76 B (1978) 241
11. M. Gyulassi and W. Greiner, Ann. of Phys. 109 (1977) 485
12. S.A.Fayans, E.E. Saperstein and V.E. Tolokonnikov,
     Nucl. Phys. A 326 (1979) 463
13. H. Toki and W. Weise, Phys. Rev. Lett. 42 (1979) 1034
14. F. Hachenberg and H.J. Pirner, Ann. of Phys. 112 C
15. M.K. Banerjee and J.B. Cammarata, Phys. Rev. C 17
     (1978) 1125; Phys. Rev. D 18 (1978) 4078
16. G.E. Brown and W. Weise, Phys. Reports 22 C (1975) 279
17. G.Rowe, M. Salomon and R.H. Landau, Phys. Rev. C 18
     (1978) 584
18. C. de Tar, Phys. Rev. D 17 (1978) 323, D 19 (1979) 1451
19. G.E. Brown, M. Rho and V. Vento, Phys. Lett. 84 B (1979) 383
20. R. Vinh Mau, in: Mesons in Nuclei, Vol. I (M. Rho and
     D. H. Wilkinson, eds.), North-Holland (1979)
21. O. V. Maxwell, W. Weise and M. Brack, preprint (1980);
     M. Brack, D.O. Riska and W. Weise, Nucl. Phys.
     A 287 (1977) 425
22. M. Ericson, in: Mesons in Nuclei, M. Rho and D.H. Wilkinson,
     eds., North-Holland (1979)
23. T.E.O. Ericson, in: Proc. Conf. on Common Problems in Low-
     and Medium-Energy Nuclear Physics, B. Castel, B. Goulard
     and F.C. Khanna, eds., Plenum Publ. Corp. (1979)
24. N.C. Mukhpadhyay, H. Toki and W. Weise, Phys. Lett.
     84 B (1979) 35
25. J. Speth, V. Klemt, J. Wambach and G.E. Brown, preprint (1979)
26. E. Oset and M. Rho, Phys. Rev. Lett. 42 (1979) 47
27. M. Ericson and T.E.O. Ericson, Ann. of Phys. 36 (1966) 323;
     G. Baym and G.E. Brown, Nucl. Phys. A 247 (1975) 345
28. A. Richter, Lectures presented at this school;
     A. Richter and W. Knüpfer, Proc. Int. School of Inter-
     mediate Energy Physics, Rome (1979)

29. H.M. Hofmann, Z. Physik A 289 (1979) 273
30. L.S. Kisslinger and W.L. Wang, Ann. of Phys. 99 (1976) 374
31. K. Stricker, H. McManus and J. Carr, Proc. 2nd Int. Conf.
     on Meson-Nuclear Physics, Houston (1979), AIP Conf. Proc.,
     and preprint (1980)
32. G.E. Brown, B.K. Jennings and V. Rostokin, Phys. Reports
     50 C (1979) 227
33. E. Oset, W. Weise and R. Brockmann, Phys. Lett.
     82 B (1979) 344
34. L. Tauscher, private communication;
     J. Carr, K. Stricker and H. McManus, preprint
35. E. Oset and W. Weise, Phys. Lett. 94 B (1980) 19
36. J. Koch and E.J. Moniz, Phys. Rev. C 20 (1979) 235
37. E. Oset and W. Weise, preprint (1980)
38. M. Ericson, Proc. 8th Int. Conf. on High Energy Physics
     and Nuclear Structure, Vancouver (1979),
     Nucl. Phys. A 335 (1980) 309
39. H. Toki and W. Weise, Z. Physik A 292 (1979) 389,
     Z. Physik A 295 (1980) 187
40. H. Sagawa, T. Suzuki, H. Hyuga and A. Arima, Nucl. Phys.
     A 322 (1979) 361
41. J. Delorme, M. Ericson, A. Figureau and N. Giraud,
     Phys. Lett. 89 B (1980) 327;
     J. Delorme, A. Figureau and N. Giraud,
     Phys. Lett. 91 B (1980) 328
42. H. Toki and W. Weise, Phys. Lett. 92 B (1980) 265
43. Ch. Glashausser et al., private communication, and
     Phys. Rev. Lett. to be published
44. J. Flanz et al., Phys. Rev. Lett. 43 (1979) 1922;
     R. Neuhausen et al., Mainz Linac data (1979)
45. K. Shimizu and A. Faessler, Nucl. Phys. A 333 (1980) 495
46. R. Brockmann, B.K. Jennings and R. Rockmore, preprint (1980)
47. B. Ziegler, Proc. Int. Conf. on Nuclear Physics with
     Electromagnetic Interactions, Mainz 1979,
     Lecture Notes in Physics 108 (1979) 148
48. B. Mecking, Proc. Int. Conf. on Nuclear Physics with
     Electromagnetic Interactions, Mainz 1979,
     Lecture Notes in Physics 108 (1979) 382;
     H. Rost, and B. Stanek, theses, Univ. of Bonn (1980)
49. J.M. Moss et al., Phys. Rev. Lett. 44 (1980) 1189
50. Ch. Glashausser et al., preprint (1980)
51. H. Toki and W. Weise, preprint (1980)
52. J. Deutsch and S. Austin, private communication
53. J.R. Comfort and W.G. Love, Phys. Rev. Lett. 44 (1980) 1656

NUCLEAR STRUCTURE AND HEAVY-ION REACTIONS[*+]

John P. Schiffer

Physics Division
Argonne National Laboratory
Argonne, IL 60439

INTRODUCTION

The observation of prominent structures in the excitation
functions for some heavy-ion induced reactions suggests that simple
features of nuclear structure are manifesting themselves at high
excitation energies and, presumably, high angular momentum. The
purpose of these lectures is to review the framework in which such
structures (resonances) are treated in reaction theory, a formalism
which had been formulated for light particle reactions a third of
a century ago.[1,2] Then, the somewhat more subjective criteria of when
a structure is "simple" and when it is not will be presented.
Finally, the experimental data are reviewed with particular attention
to surface-transparent nuclei which seem to show features that may
be susceptible to resonance analysis, and a case in which such a
quantitative analysis has been carried out will be discussed.

LIGHT ION REACTIONS

In the studies of light particle reactions, particularly
neutron interactions with nuclei, it quickly became evident in the
earliest stages of nuclear reaction studies that there were
enormous variations in cross sections from nucleus to nucleus and
as a function of energy. Such *resonances* were subsequently
described by expanding the scattering amplitude in terms of poles
in the scattering amplitude.[2] Space is divided into two
regions, an "outside" part ($r > R$), where the scattering wave
functions (u_ℓ) are described by appropriate hypergeometric functions
depending on the Coulomb and centrifugal field, and an inside part
which is only represented by the value of the logarithmic derivative
on the surface $r = R$

$$g \equiv \frac{r}{u} \left(\frac{du}{dr}\right)_{r = R}.$$

The logarithmic derivative has an energy dependence which can be expanded in terms of poles

$$g = (E - E_\lambda) \left(\frac{\partial g}{\partial E}\right)_{E = E_\lambda} + \ldots \quad .$$

The scattering amplitude then may be written as

$$S = (1 - \frac{i\Gamma_\lambda}{(E-E_\lambda) + \frac{i\Gamma_\lambda}{2}}) \, e^{-ikR} \quad ,$$

where

$$\Gamma_\lambda \equiv -2kR \left[\left(\frac{\partial g}{\partial E}\right)_{E = E_\lambda}\right]^{-1} .$$

All these quantities are, of course, for one partial wave, we have omitted the (ℓ,j) subscripts for simplicity. The quantity Γ_λ is the partial width characterizing a resonance λ in a particular channel and may be written as

$$\Gamma_\lambda = \overbrace{2kRP_\ell}^{\text{outside}} \overbrace{\gamma_\lambda^2}^{\text{inside}} ,$$

where the P_ℓ is the so-called penetration function and the quantity $2kRP_\ell$ depends entirely on extra nuclear boundary conditions. The nuclear structure information is contained in the "reduced width" γ_λ^2. This quantity has a limit (the so-called Wigner-Teichmann single-particle limit) which may be set from dimensional arguments as

$$\gamma_{sp}^2 = \frac{3}{2} \frac{\hbar^2}{MR^2} .$$

A resonance which involves complicated excitations will have a smaller reduced width, while one that approaches this width has a structure that is well approximated by a particle resonating within a box that is the target nucleus in its ground state. The corresponding sum rule is that

$$\sum_\lambda \gamma_\lambda^2 = \frac{3\hbar^2}{2MR^2} ,$$

where the energy interval is chosen wide enough to include all contributions.

Real nuclei tend to be complicated at the high excitation energies where resonances occur. One therefore has little hope of understanding individual resonances, but it is sensible to talk of averages. The average reduced width per unit energy is the strength function

$$\sum_{\substack{\lambda \\ \text{per unit} \\ \text{energy}}} \gamma_\lambda^2 = \frac{\langle \gamma^2 \rangle}{D},$$

where D is the average spacing of resonances. At the time the classic text of Blatt and Weisskopf[1] was written, it was assumed that the nucleus was sufficiently complicated that the reduced widths would be randomly distributed and that thus the strength function would be a constant—independent of energy.

As data became available, statistical randomness was confirmed over limited energy regions but not over a wider range. The *optical model* of Feshbach, Porter, and Weisskopf[3] gave a rather simple explanation—that the behavior of the average strength function reflected the behavior of a nucleon moving in the average potential of the nucleus. The mixing of this degree of freedom with the much more complicated states of the many-particle system was strong, but not strong enough to completely obliterate it. This mixing was introduced into the optical model by way of a damping term—an *imaginary potential*.

What we had in this picture was a model that established the connection between nucleon-induced reactions and the single-particle shell model. The shell-model assumption that the interaction of a valence nucleon with all the rest of the nucleus may be represented by an average field is modified—but retains its validity for the average strength function. The fact that at higher energies the scattering of the incident nucleon by the other nucleons may excite them from the ground state, means that there is no longer a well-defined single-particle state—but that this configuration dissolves among the more complicated configurations. The extent of this mixing is, in a sense, a measure of the validity of a model. The energy interval over which the mixing occurs is a measure of its strength, and may be thought of in terms of a mixing time

$\tau \approx \hbar/\Delta E$. The energy widths of the individual resonances
correspond to their lifetimes for decay into the outside continuum,
while the spreading width of the single-particle states into the
strength function is determined by the lifetime of the configuration
within the nucleus, or the mixing time. One may then also define a
mean-free-path (Λ) in the nucleus and

$$\Lambda \approx \frac{\hbar}{W} \sqrt{\frac{E + V}{2m}} \quad ,$$

where the optical potential is

$$U = -(V + iW).$$

E and m are the outside kinetic energy and reduced mass of the
system,   the spreading width is twice the imaginary potential (2W).
The real potential V is essentially the shell-model potential.

If one looks at the nucleus microscopically over an energy
interval small compared to the spreading width, then the individual
resonances are distributed statistically. Their average reduced
width is determined
   a)   by the average value of the strength function in that
        vicinity.
   b)   by the number of levels per unit energy.

Point b) is one about which there seems to be a fair amount
of confusion in the literature. When one starts with the model in
which the single-particle states mix into the more complicated ones,
the mixing matrix elements depend on the overlap between the
single-particle states and the fine structure resonances. The
higher the density of these fine structure resonances the weaker
the individual overlaps—in fact it is the strength function, the
total "reduced width per unit energy" that remains constant. While
W does increase slowly with increasing energy, the value of W is
not connected with the level density.

More detailed models have been developed for this mixing. One
can envision the time evolution of the mixing in terms of a
hierarchy, where the single-particle state first interacts with one
other nucleon, exciting a particle-hole state, forming a 2p-1h hole
state, and this configuration propagates further down into more
complicated states. One might even expect to see some subsidiary
"intermediate structure," on the overall envelope of the strength
function, if there were a well-defined intermediate lifetime for the
2p-1h states. Figure 1 illustrates schematically the various
stages of mixing.

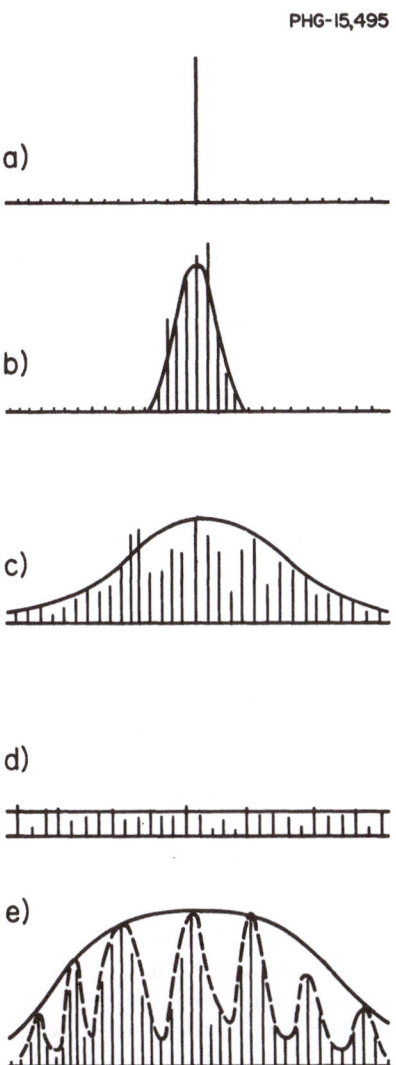

Fig. 1. Mixing modes of a degree of freedom among the states of the
nucleus. In (a) a state which corresponds completely to the
configuration described by the degree of freedom is shown,
unmixed with other states; (b) corresponds to a small amount of
mixing with nearby states; (c) to a "giant resonance"; and (d)
to complete mixing where the model degree of freedom is not a
useful one. Finally a case of "intermediate structure" is shown
in (e) by dashed lines within the giant resonance envelope.

The "giant-resonance" picture, in which a simple configuration
dissolves among the complex fine-structure states yet retains its
influence as an envelope of the strength function has quite general
validity. It applies not only to the single nucleon strength, but
also to electromagnetic excitations. The E1 dipole giant resonance
is the best known of these. The mixing width of this particle-hole
excitation is also remarkably independent of the individual nucleus
in which it occurs, again confirming that the mixing width is
independent of the fine-structure level density.

Another beautiful example of such mixing is the case of
isobaric-spin analog resonances, where mixing can only occur through
isospin violation; it is therefore relatively small, such as in
Fig. 1(b).

The concept of giant resonances is also useful outside of
nuclear physics. In a many-particle system, a giant resonance is
the sign that an approximate model, corresponding to some degree
of freedom, has a certain measure of validity. A giant resonance
is most easily observed in processes or reactions where the operator
that describes the process corresponds to the degree of freedom in
which the giant resonance appears. Thus the optical model's
single-nucleon giant resonances are seen in the scattering or
reactions of nucleons from nuclei, while the dipole giant resonance
is strongly dominant in the absorption of photons, or in electron
scattering. However, a giant resonance in some mode can also have
an influence on a reaction that is closely coupled to this mode,
but not in the exactly corresponding channel. A good example of
this is the appearance of isobaric analog resonances in excitation
functions of (d,p) reactions. Here the isospin selection rules
would forbid these resonances from being formed, yet their coupling
to the proton channel is strong enough that they definitely do
appear.

There is another aspect of reactions with light ions that is
relevant to heavy-ion reactions: the subject of *direct reactions*.
These differ from the resonant or compound nucleus processes in
the time scale on which they occur. A direct process is one which
happens on the same time scale as the transit time of the projectile
through the target, it involves a single step: the transfer of a
nucleon or group of nucleons or the inelastic excitation of a simple
degree of freedom in the target. It is a perturbation on elastic
scattering—a *quasielastic* process. The angular distributions are
generally forward peaked and the energy dependence is smooth, showing
only the giant resonances. The "distortion" of the scattered
particle by the nucleus is characterized by the appropriate optical
potentials. In contrast the compound nucleus processes show many
resonances and, on the average, angular distributions that are
symmetric about 90°. This separation of reaction into direct and
compound is somewhat arbitrary. There is a whole realm of time

scales—one talks of two-step and multistep processes that are increasingly more complex and involve not only more steps but longer times and possibly sharper energy dependences; they are closely related to the concept of intermediate structure (doorway) resonances. The status of reaction theory between the strict "direct" and the "compound" extremes is still somewhat murky.

HEAVY-ION REACTIONS

Reactions with projectile particles involving heavy ions (A >> 4) tend to differ from the light particle reactions we have just discussed in several important respects.

a)   The Coulomb barrier plays a much more important role, especially when one is concerned with the lower energies where the effects of nuclear structure are most likely to manifest themselves.

b)   The wavelength of particles is much shorter than for light projectiles, typically an order of magnitude smaller than the size of the nuclei. Thus classical approximations tend to have more validity; diffraction effects are still seen but one has to look harder to observe them.

c)   A corollary of the above two points is that one is generally dealing with higher angular momenta.

d)   Since a heavy-ion is a multinucleon projectile, the probability of a reaction taking place is increased very greatly. In the very crudest picture one would argue that the mean free path of a heavy ion of A nucleons in nuclear matter should be $\sim \frac{1}{A}$ times the mean free path of a nucleon with the same velocity.

Coulomb Effects

Let us now examine heavy-ion reactions, keeping the above considerations in mind. For classical Coulomb scattering we have a Rutherford trajectory characterized by a scattering angle θ, corresponding to an impact parameter b and a distance of closest approach D. These quantities are related by

$$\cot \frac{\theta}{2} = \frac{2b}{D_0}, \text{ where } D_0 = \frac{Z_1 Z_2 e^2}{E}$$

is the distance of closest approach for a head-on collision, and

$$D = D_0 \left(\frac{1 + \operatorname{cosec} \theta/2}{2}\right).$$

Thus if we look at elastic scattering at a fixed energy we may examine what happens to scattered particles as a function of angle and how much the scattered flux deviates from that expected for pure Rutherford scattering. Such measurements are meaningful if they are restricted to $\theta > 90°$, because at more forward angles the trajectories become appreciably modified by the attractive potential of the nucleus. Similarly, one may choose a fixed backward angle, say 180°, and look at elastic scattering as a function of bombarding energy. In both cases one is effectively mapping out elastic scattering as a function of the 'distance of closest approach' D and when data from both angular distributions and excitation functions are plotted in this way the points do follow the same smooth curve as is shown in Fig. 2. There is an interesting piece of additional information that comes from such studies with heavy

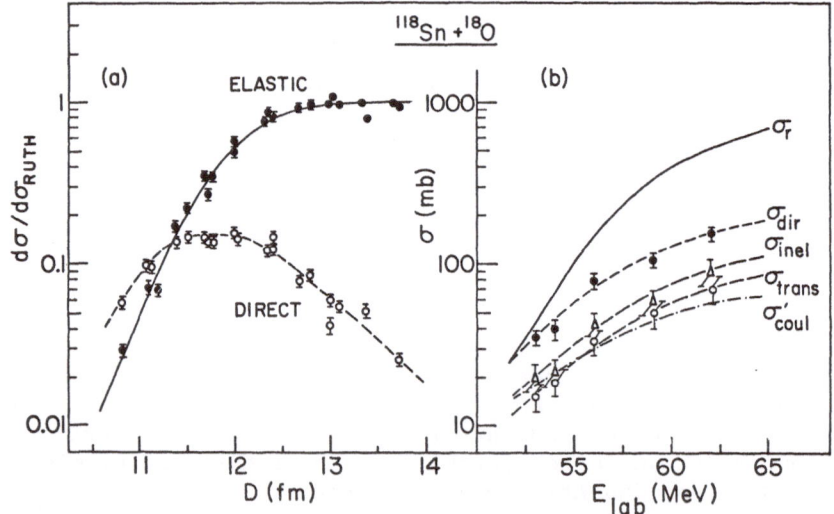

Fig. 2. (a) Elastic scattering and direct reaction of ^{18}O from ^{118}Sn plotted as a function of distance of closest approach from Ref. 4. The data are for laboratory energies between 55 and 65 MeV and several backward angles. (b) The total direct reaction cross section as a function of incident energy. The total reaction cross section is also shown.

ions. Here for the direct reactions the mass of the outgoing
particle will still be close to that of the incident heavy ion.
Normally compound nuclei decay by the emission of protons, neutrons,
and alpha particles. The evaporation of heavier particles is very
rare. Thus if one bombards a nucleus with an oxygen projectile
and sees some other isotope of oxygen, nitrogen, or carbon emerging,
one can be reasonably certain that a direct reaction has taken
place.

One may also look at the energies of the emerging particles
and see that they are consistent with the transfer of a fragment
at the distance of closest approach.[4,5,6] The yield of such particles
as a function of energy and angle may then be translated into a
transfer probability as a function of the distance of closest
approach. When this is done one sees, as was shown in Fig. 2, that
as the elastic scattering decreases from ~100% to 10% of the
Rutherford value the sum of the "quasielastic" direct yields
increases to 15% of Rutherford scattering. As one moves to even
smaller distances of closest approach both elastic and direct
yields start to decrease and absorption into a compound system
(called fusion in heavy-ion reactions) becomes dominant. In other
words, there seems to be a surface region in the interaction of
nuclei, where the approximate identity of the fragments is more-
or-less preserved, even though the flux is removed from the
'strict' entrance channel and therefore is 'absorption' in the
conventional optical-model sense. We will return to this question
of two kinds of absorption.

Another feature of the classical trajectories in the Coulomb
field is a simple expression for the reaction cross section. If
we assume that all particles that reach the top of the barrier (at
radius R_B, potential energy V_B) will interact, then

$$\sigma_{react} = \pi R_B{}^2 \left(1 - \frac{V_B}{E}\right). \tag{1}$$

This completely classical expression is independent of the details
of the shape of the attractive potential. If we plot σ_{react} as a
function of $1/E$ we get a straight line with intercepts $\pi R_B{}^2$ and
$1/V_B$. If one now separates the radius at which fusion occurs from
the barrier radius, and assumes a critical radius R_c ($<R_B$) which
has to be reached for fusion, then the above expression is valid
for lower energies (reflecting the fact that at low energies all
classical trajectories are bent toward the center of the potential)
but changes into a similar expression for higher energies, with
R_c replacing R_B and V_c (the potential energy at R_c) replacing

V_B.[4] The transition is at the intersection of the two lines—the details of rounding of the sharp corners are not of interest here, and depend on the detailed nature of the potential shape. The extrapolated intercepts of this line segment will be πR_c^2 and $1/V_c$. We will discuss such behavior later, in connection with fusion reactions and it is illustrated in Figs. 8 and 10.

There is one more classical effect that we should discuss: the behavior of elastic scattering at somewhat higher energies when the distance of closest approach is within the nuclear radius. In this case it is useful to plot the so-called deflection function, the scattering angle as a function of impact parameter. This is illustrated in Fig. 3. With a purely Coulomb field this relationship was given above, but when there is an attractive potential it is modified, and the deflection function becomes multiple-values for all but the maximum allowed values for θ. In fact, absorption eliminates all the deeper trajectories rather quickly, but near the first deviation from the 'Rutherford' deflection function there will be an increase in cross section above the Rutherford value because of the bunching of particles by the attractive field near this barely grazing angle. It is also near this angle that one expects the quasielastic direct reactions to make their largest contribution, since this is where the two nuclei are close enough for small perturbative interactions without their getting close enough in their trajectories to undergo complete absorption.

It is also in this region that we can easily envision the onset of diffraction effects that arise from the wave nature of these particles. If in this grazing region of space the two nuclei interact through a real potential but the interaction is not dominated by absorption, then we may expect to see nuclear structure effects. These may either be shape resonances of the type we discussed for nucleons or more complicated phenomena. How does it make sense to talk of shape resonances when absorption will clearly be strong over much of the nuclear volume? The answer is simple, we expect to see them in just these grazing high partial waves where the wave function is restricted to the surface region. There is a competition between the strength of the real attractive potential forming a sufficient pocket in the repulsive Coulomb-plus-centrifugal field to sustain a standing wave, and the absorptive interactions that could damp out any such effects. In the high angular momentum states we may be able to bring two nuclei together in such a way that they can only interact gently without completely disrupting each other. A few degrees of freedom may be just accessible—involving the transfer of a few particles of matter and some internal excitation. We have, in principle, an opportunity to study nuclear structure effects under very special conditions.

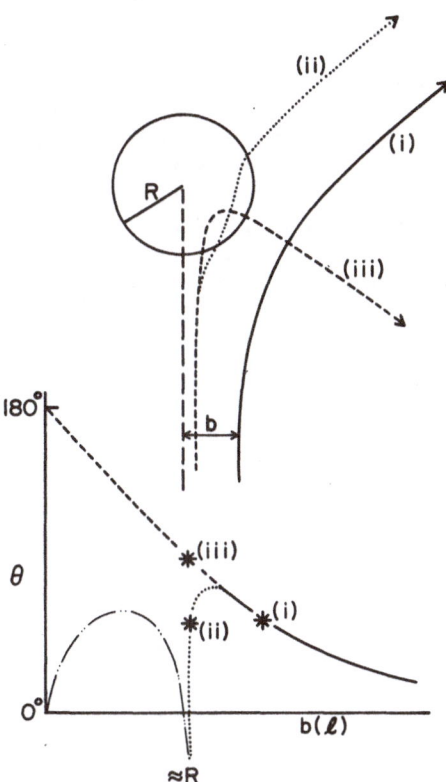

Fig. 3. Trajectories in a Coulomb and Coulomb + nuclear potential. Trajectories (i) and (ii) correspond to scattering into the same angle from different impact parameters, b. The more distant trajectory (i) is outside the range of nuclear forces while the closer trajectory (ii) goes inside the range of attractive nuclear forces (radius R). If it were not for this attraction, the point-charge Rutherford trajectory (iii) would go into a larger angle. The deflection function θ vs. b (or ℓ) is plotted below. The trajectories with small impact parameters (shown as dot-dashed line) are strongly absorbed.

Quasielastic Reactions

When a heavy nucleus, such as Sn is bombarded by oxygen nuclei near the Coulomb barrier a gradual decrease is seen in the elastic scattering as the classical distance of closest approach becomes smaller than the nuclear radius. Accompanying this decrease, as was shown in Fig. 2, is an increase in the quasielastic processes, where particles, slightly differing in mass (and/or energy) from the incident projectiles, emerge with kinematics that are essentially governed by the Coulomb field. As the distance-of-closest-approach decreases we see the quasielastic yield increase above the elastic, until eventually both fall off. Clearly this behavior is a measure of a region in which there is some interaction between target and projectile without complete absorption. We may even choose to define elastic scattering a little differently from the ordinary by including all the quasielastic yield with the elastic as is illustrated in Figs. 4 and 5. This means that the total 'elastic-plus'quasielastic' yield falls off at slightly smaller impact parameters—in other words absorption starts further in than if we had defined absorption in the conventional way as removal from the stricter elastic channel.   One is then led to some speculations: perhaps we need a reformulated optical model—in which absorption means something different from what we learned in light nuclei. Since the pseudo-absorption into quasielastic channels is highly localized on the surface, this will also imply a localization into the grazing partial waves. In this region the quasielastic process is the dominant one that bites into the elastic flux—and presumably the quasielastic channels are not only a sink but may also couple back into the elastic channel. In fact, one has a coupled channels problem that we don't know how to address. The actual number of asymptotic quasielastic channels is very large but the time scale in which the inter-coupled reactions take place is so short that it is not meaningful, from uncertainty principle arguments, to pin down energies to better than a few MeV. The system may recognize a few degrees of freedom during this interaction time and presumably these are the ones that would have to be included as the "channels" in the calculation. Calculations somewhat along these lines have been attempted by Schaeffer and Bertsch.

How do these considerations carry on to higher energies where one can no longer talk of sub-Coulomb trajectories? An example is the study of all quasielastic channels in $^{16}O + {}^{48}Ca$ at 56 MeV, about twice the Coulomb barrier, shown in Fig. 6. When one sums up the inelastic, one-, two-, three-, and four-nucleon transfers, the total of direct processes account for about 20% of the cross section. The shapes of the angular distributions suggest again that these processes are localized in a few surface partial waves—in which they cominate the reaction as shown in Fig. 7. Clearly the 'quasielastic + elastic potential' has a smaller absorptive radius than the strictly elastic potential.

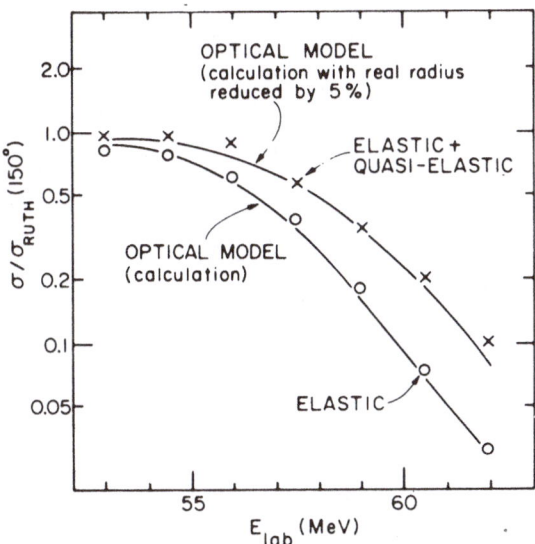

Fig. 4. Ratio of elastic scattering of ^{18}O + ^{118}Sn to Rutherford scattering at 150° from Ref. 4, plotted as a function of bombarding energy (circles) together with a simple optical-model calculation. The elastic + direct cross section (crosses) is shown in a similar form with an optical-model calculation where the radius was reduced by 5%.

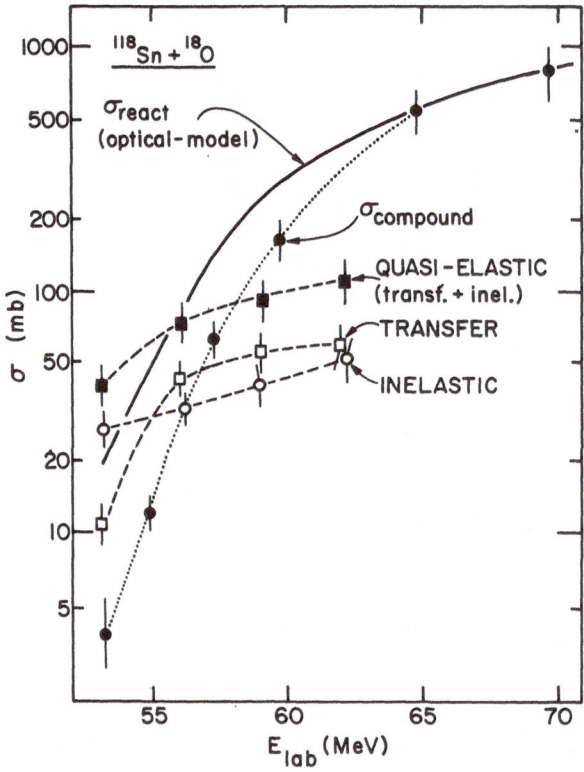

Fig. 5. The various reaction cross sections for ^{18}O + ^{118}Sn from
    Ref. 4. The solid and dotted lines correspond to the two
    optical models of Fig. 4. Note that at the lowest energies
    the reaction cross section is dominated by inelastic
    scattering (Coulomb excitation) that is not properly treated
    in the optical potential.

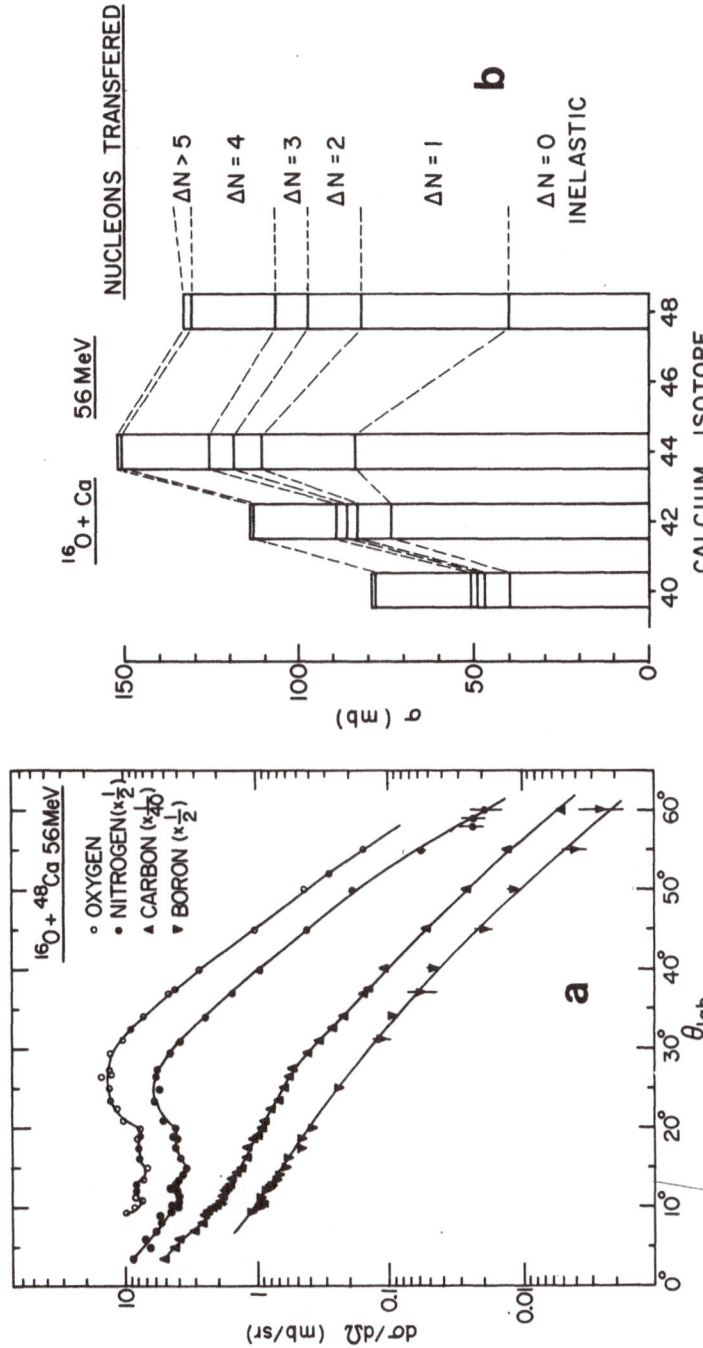

Fig. 6. (a). Angular distributions for the summed yields over all final states and all isotopes from ^{16}O from Ref. 7.
(b). The integrated direct cross sections for ^{16}O + the various Ca isotopes.

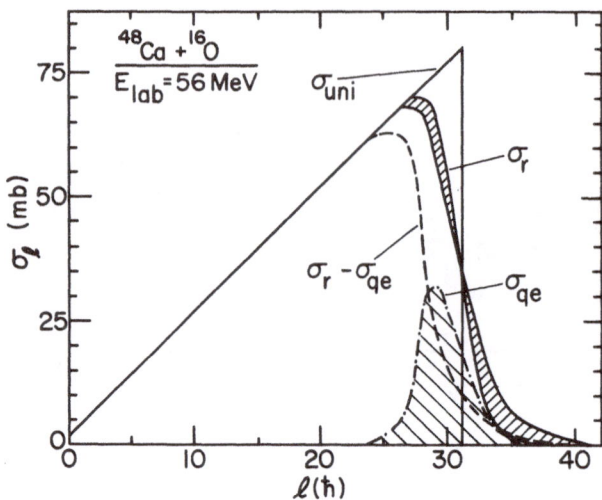

Fig. 7. Distribution of the direct cross sections from ^{48}Ca + ^{16}O
in L from Ref. 7. The triangle (σ_{uni}) represents a sharp-
cutoff model that fits the reaction cross section and exhausts
the unitarity limit; σ_r represents the distribution in ℓ of
the reaction cross section calculated from optical potentials
that fit the elastic scattering, σ_{qe} is the approximate
distribution of the quasielastic direct reactions and the
dashed curve ($\sigma_r - \sigma_{qe}$) would be the absorption in an optical
potential where elastic and quasielastic processes are not
distinguished.

Fusion Reactions

Next I would like to discuss the behavior of the fusion
process, where two nuclei interact in a way that their identity
is completely lost. Here we have more complete data for the
lighter nuclei and perhaps we will learn something from systematics.
One has to make a somewhat arbitrary separation as to what is fusion
and what quasielastic in the measured data. When fusion occurs
the fused system will travel forward with the velocity of the
center of mass. The highly excited system will decay by emitting
particles and the forward motion will be spread out as some mass
is lost by evaporation. For the data I am about to describe all
forward reaction products with masses above the mass of the
projectile were defined as fusion.   In practice the separation
between quasielastic and fusion yields is rather clear at lower

bombarding energies—ambiguities begin at higher energies when fragmentation is not easily distinguished from multiple evaporation, even in principle. There are obvious experimental problems associated with measurements at very small angles—most of these become less serious if one can use a reasonably large distance between target and detector.

The typical behavior of fusion cross sections was given in Eq. (1), it is displayed most clearly when plotted as a function of the reciprocal of the incident energy. If the critical radius R_c, which has to be reached for fusion, is distinctly smaller than the barrier radius R_B, then one expects the cross section on such a plot to follow two straight-line segments which, in classical kinematics are defined by the barrier radius height (R_B and V_B) for the lower energy segment, and the critical fusion radius and the value of the potential at this radius (R_c and V_c) for the higher energy segment. If the potential is attractive at this latter point then the intersection of the two lines defines a maximum value of the fusion cross section. Such behavior is shown for $^{18}O + ^{12}C$ and $^{19}F + ^{12}C$ in Fig. 8. The data do not go to sufficiently high energies to clearly constrain the upper line segment. However, the maximum of the fusion cross section is well defined experimentally. This quantity seems to be changing as a function of projectile atomic weight in a rather abrupt fashion. At one point it seemed that the jump in this value was completely correlated with a major shell closure, but the evidence is no longer so clear, as may be seen in Fig. 9. While the cross section seems to change by about 25% for nuclei above ^{16}O, there also seems to be a comparable difference between $^{14}N + ^{12}C$ and $^{15}N + ^{12}C$ where no new major shell is populated. On the other hand, there seems to be no difference between $^{12}C + ^{12}C$ and $^{13}C + ^{12}C$. This implies that the $^{14}N - ^{15}N$ difference is not simply a consequence of the added valence neutron.

In the model for fusion, where $R_c < R_B$, the potential is likely to be changing rather steeply in the vicinity of R_c. This implies that rather small changes in R_c could yield large changes in σ_{fus}^{max}, because the corresponding potential values differ drastically: the slope of the line segment becomes steeper very quickly as R_c is moved in by a small amount. This is illustrated schematically in Fig. 10. One may expect a difference between nuclei in the 1p shell and those in the s-d shell simply because the radius of the valence orbitals is larger ($\langle r^2 \rangle \sim (2n + \ell + 3/2)$). Since absorption is especially sensitive to the valence orbitals which contain the loosely bound and easy-to-excite nucleons—one may expect an abrupt change as one crosses a shell. There is a similar effect in atomic physics— the Ramsauer effect—anomalies are seen in the scattering of slow electrons at the noble gases. When one looks at the values of R_B and V_B (or rather $V_B/Z_1 Z_2 e^2$, to remove the obvious Coulomb dependence) as a function of the

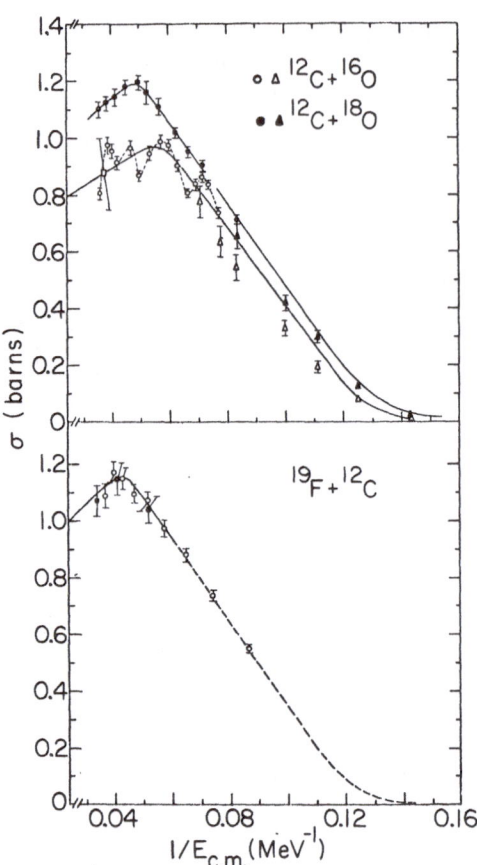

Fig. 8. The fusion cross section plotted against the reciprocal
of the center-of-mass energy shown for three systems, from
Ref. 8. For $^{12}C + ^{18}O$ and ^{19}F the cross section is described
reasonably well by two line segments given by Eq. (1). But
for $^{16}O + ^{12}C$, resonance structure seems to be present.

interacting nuclei, as is shown in Fig. 11, no obvious effects
are apparent. So the origin of the change of the maximum cross
section is not in the barrier parameters, but in parameters that
describe the fusion probability <u>inside</u>: R_c and V_c or their
equivalent.

     At the present time we do not understand the details of the A
dependence in the fusion cross section. There does seem to be a
change at the shell closure, but other effects are also apparent.

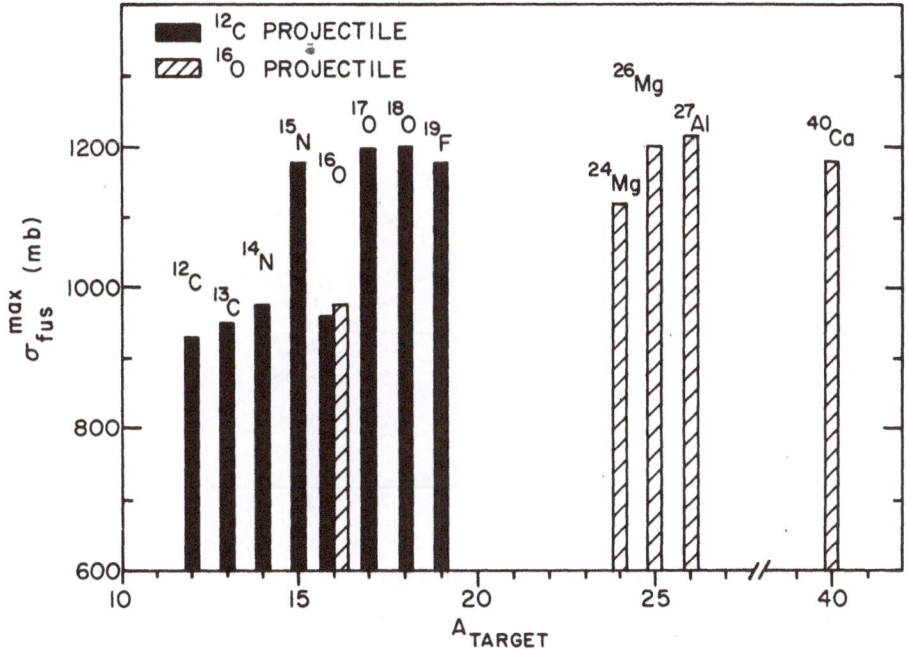

Fig. 9. Maximum values of the fusion cross sections as seen in
various systems, where ^{12}C or ^{16}O are one of the fusing nuclei,
from Ref. 8.

What is clear however, is the fact that details of nuclear structure
are important in determining this quantity and it is not a matter
of purely macroscopic, hydrodynamic parameters.

Resonances in Fusion

There is another feature of nuclear structure that appears
in the fusion cross sections. This is the presence of resonance-
like bumps in some systems, notably ^{12}C + ^{12}C, ^{12}C + ^{16}O and
^{16}O + ^{16}O, and the absence of these structures in systems that
differ by only a few nucleons; effects that we already saw in
Fig. 8. There is a considerable literature on 'molecular
resonances' for these systems but this literature usually
addresses sharper structures in specific reaction channels. The
broader maxima seen here are more nearly reminiscent of the shape
resonances with light projectiles. Indeed, if one uses the
potentials obtained by fitting elastic scattering and reduces the

FIXED POTENTIAL, VARIABLE R_c

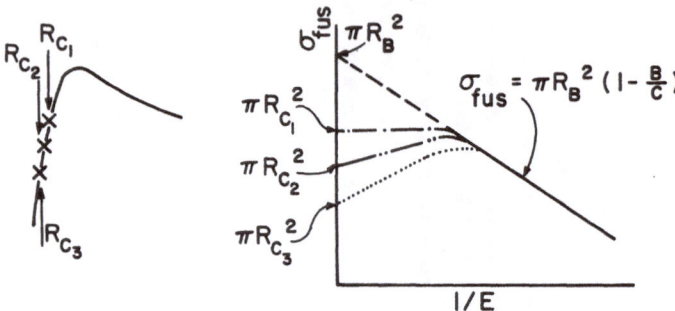

FIXED R_c, VARIABLE POTENTIAL

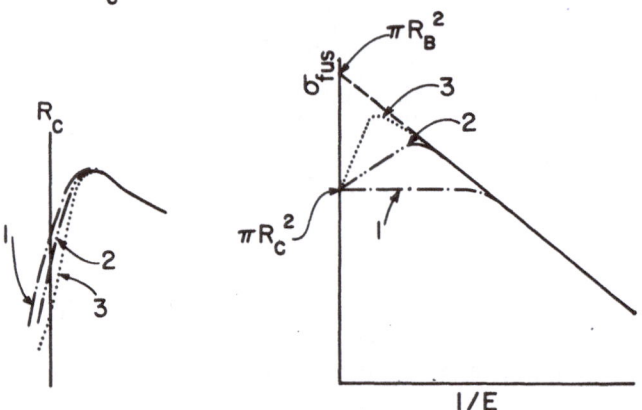

Fig. 10. The top part of the figure shows schematically the
influence of the critical radius, (inside of which everything
leads to fusion) on the fusion cross section. The bottom
part shows the influence of the value of the real potential
at a fixed critical radius.

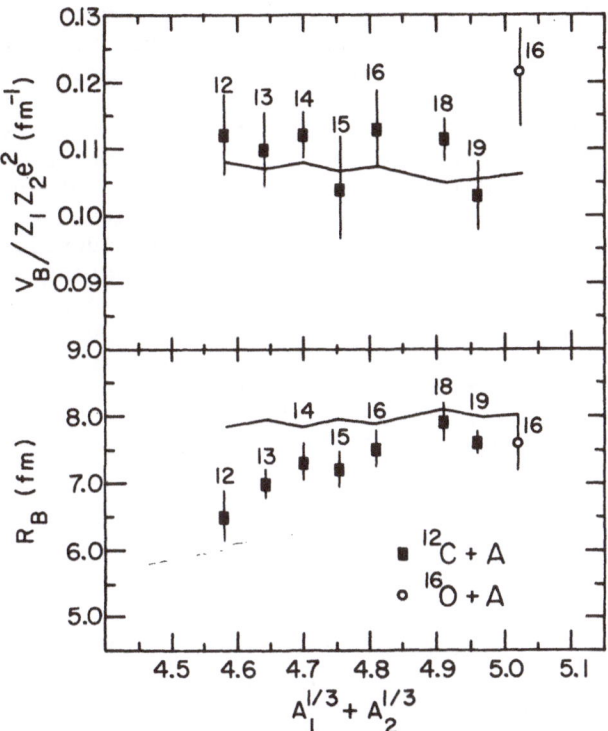

Fig. 11. Plot of the parameters that describe the line segment fitting the low-energy fusion cross sections from Ref. 8 as described in Eq. (1). The numbers next to the points refer to the atomic weight of the nucleus fusing with ^{12}C or ^{16}O. The values of the parameters are plotted against $A_1^{1/3} + A_2^{1/3}$, roughly the sum of the two nuclear radii.

imaginary radius slightly, bumps with a similar period appear in the cross section; the partial waves responsible for the resonances are the grazing ones. Unfortunately there is no clean way that can be used to determine the angular momentum responsible for the experimental bumps. And the justification of using an optical potential that has an imaginary radius reduced from the value required for elastic scattering is based only on the hand-waving arguments given earlier. So one has to take this result with some caution.

When one looks at the more detailed anatomy of these bumps another feature emerges—they seem to arise from residual nuclei which have an even number of protons, and presumably an alpha-particle structure—the odd-proton nuclei are smooth. The data for $^{12}C + ^{12}C$ are shown in Fig. 12, similar effects are seen for $^{16}O + ^{12}C$ and $^{16}O + ^{16}O$. The spacing of the bumps changes from $^{12}C + ^{12}C$ to $^{16}O + ^{16}O$ from $\sim$5.5 to $\sim$3.5 MeV, about what one would expect from shape resonances in a shallow potential whose spacing should scale roughly as $\sim R^2$. However, for the identical nuclei only the even partial waves resonate—yet the spacing of bumps observed in $^{16}O + ^{12}C$ is just about halfway between at $\sim$4.3 MeV. This may mean that the odd partial waves are suppressed and do not give rise to resonances—and thus that we may have some clue to the mechanism (e.g. the structures correspond to levels in some symmetrical object), or it may be that the spacing of shape resonances is that corresponding to a much deeper potential. In such a deeper potential an approximate coincidence between even and odd resonances may come about for essentially accidental reasons— the slope of a trajectory of resonances with the same number of nodes will not be the same as that of the line defined by the energy dependence of the grazing angular momenta.

To summarize—in the behavior of nuclei with $24 \leq A \leq 56$ there seems to be evidence that this most macroscopic of quantities, the fusion cross section, is sensitive to details of nuclear structure in the entrance channel. We do not have a good understanding of what these effects are though there is some indication that in special cases shape resonances may appear.

Elastic Scattering

The study of elastic scattering of light heavy ions has a rather long history, but interest in this subject has revived considerably in the last two years. About ten years ago there was a lot of work involving the scattering of identical particles and closely related systems. Some structures are unquestionably present in the angular distributions and excitation functions here. There has been a great deal of discussion whether these structures are 'really' resonances or not; much of the data seem to be consistent with predictions of a rather shallow optical potential with modest absorption. But not very much progress has been made in determining uniquely the angular momenta associated with the structures. In a few cases rather sharp resonances have been seen, particularly in $^{16}O + ^{12}C$, and reasonably unique angular momentum assignments were made. Some assignments have also been possible at the lowest energies near the Coulomb barrier.

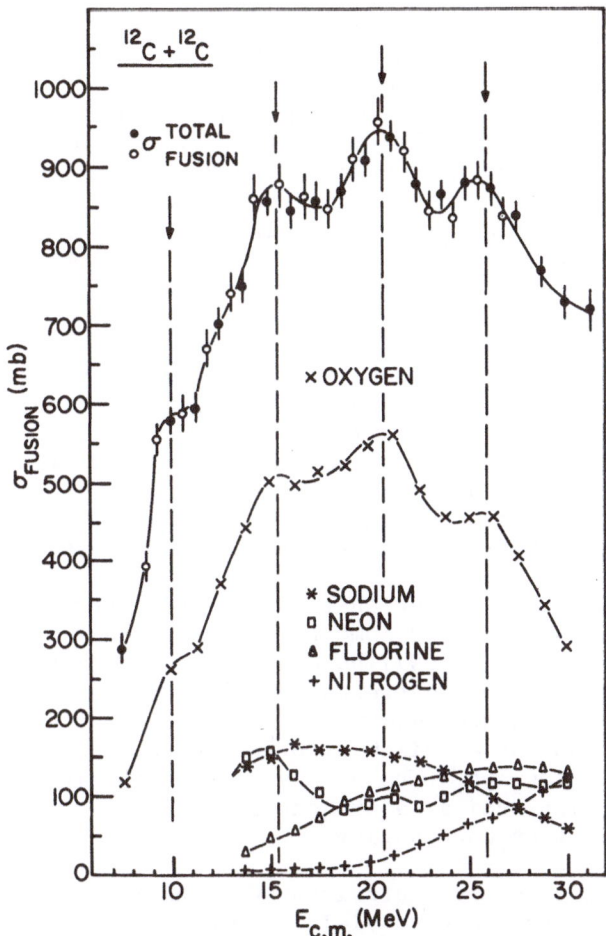

Fig. 12. Partial fusion cross sections of $^{12}C + ^{12}C$ grouped by atomic numbers from Ref. 8. Note that the bumps arise from the Ne and O yield and not that for Na, F, or N.

An interesting feature emerged recently in elastic scattering
at very far backward angles. To put this in perspective, I would
like to discuss the elastic scattering of alpha particles—which may
hold some clues on understanding these phenomena.[9] It has been known
for a long time that alpha particle elastic scattering shows
"anomalous" behavior at large angles, a behavior that changes
rather abruptly with nuclear structure. The backward (180°)
scattering is rather large for nuclei near ^{40}Ca, but when one adds
two nucleons in the f-p shell this backward cross section drops by
about an order of magnitude. A similar drop is seen in going from
^{39}K to ^{41}K or from ^{38}Ar to ^{40}Ar—in other words it is the occupation
of the f-p shell that seems to matter. The enhanced backward yield
is characteristic of the surface transparency of optical potentials,
by increasing the absorption in the surface region the backward-
angle yield in optical-model calculations is strongly reduced.
Classically one may talk of orbiting but since the back angle
yield is sharply oscillatory it is perhaps more sensible to think
of it in a quantum-mechanical picture. In a few, surface-localized
partial waves the potential is sufficiently transparent that the
scattering amplitudes do not follow the simple destructive
interference pattern predicted by strong absorption—these partial
waves combine to give a maximum in the yield at 180°—often it may
be a single ℓ-value that dominates and the angular distribution
exhibits a Legendre polynomial-like behavior. The effect of these
partial waves at forward angles is somewhat difficult to untangle
since here the absorbed amplitudes also add constructively.

For heavy ions rather sharp oscillatory structure at forward
angles has been known for some time in certain s-d shell targets
but not in others. In 1977 however, sharp, back-angle structure
was discovered in the scattering of ^{16}O and ^{12}C from ^{28}Si using
an ingenious technique—beams of Si were accelerated onto ^{12}C or
^{16}O targets and the recoil C or O ions were detected with a
magnetic spectrograph.[10] The data are shown in Fig. 13. Similar
experiments with ^{30}Si as a projectile or ^{18}O as a target showed
much less yield at back angles.

What was even more surprising was the fact that the energy
dependence of the backward yield showed rather rapid changes—both
a broader structure on the MeV scale and a narrower fine (or
intermediate) structure as is shown in Fig. 14. This structure
appears in the backward yield of inelastic particles with a rather
close correlation between the elastic and inelastic channels.
Furthermore the angular distributions at back angles appeared to
be close to those expected from a single Legendre polynomial
squared and thus some tentative assignments of angular momenta
were made to the observed bumps. These angular momenta were close
to the maximum allowed (or grazing) L-values, but there is no
obvious regular sequence.

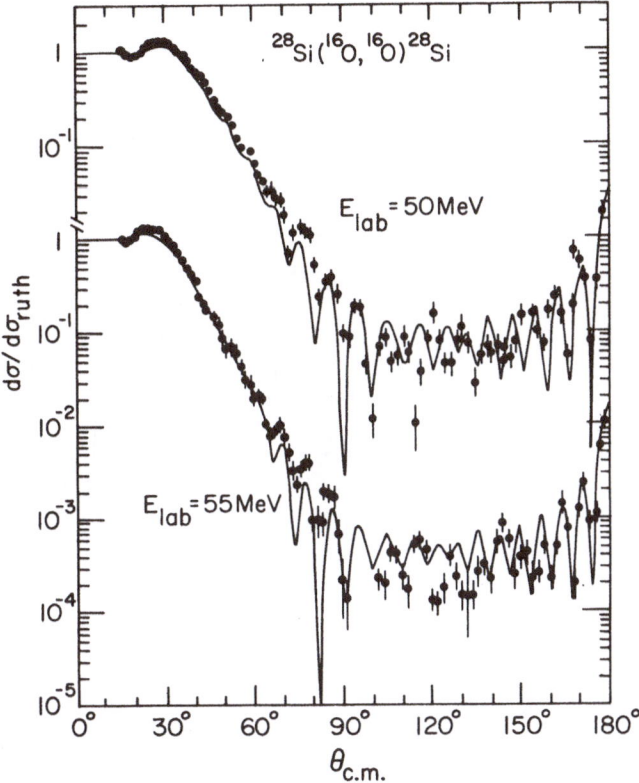

Fig. 13. Elastic scattering of ^{16}O from ^{28}Si from Ref. 10. The
lines represent optical-model calculations.

Other studies have been carried out on the Ca isotopes for
the scattering of ^{12}C.[12] Here it is clear that oscillatory
structure in the angular distributions over a rather wide range
of angles is qualitatively sharper for ^{40}Ca than the other
isotopes. The Woods-Saxon potentials that reproduce these data
definitely have stronger absorption in the surface region for
the heavier Ca isotopes. The fact that the valence nucleons are
more easily excited and therefore must contribute more heavily to
absorption than the tightly bound ones ought to be the source of
an explanation. The fact that above ^{40}Ca valence nucleons are

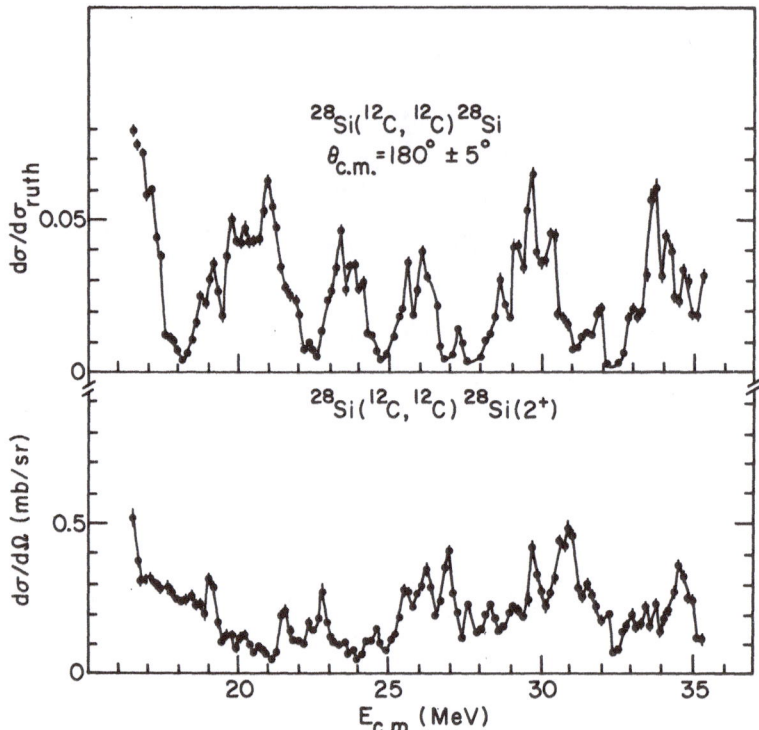

Fig. 14. Elastic and inelastic excitation functions at θ ≈ 180°
   for ^{12}C + ^{28}Si from Ref. 11.

fed into a new (larger) oscillator orbit could have a large effect
on surface absorption.

     The effect of absorption in the surface region on calculated
shapes of elastic scattering angular distributions is illustrated
in Figs. 15 and 16. Two potentials were used that differed only
in an incremental absorptive term in the critical region. As is
seen in the figure this increased absorption has a dramatic effect
on the forward-angle and backward-angle oscillations, as well as
on the absolute magnitude of the back-angle scattering cross section.

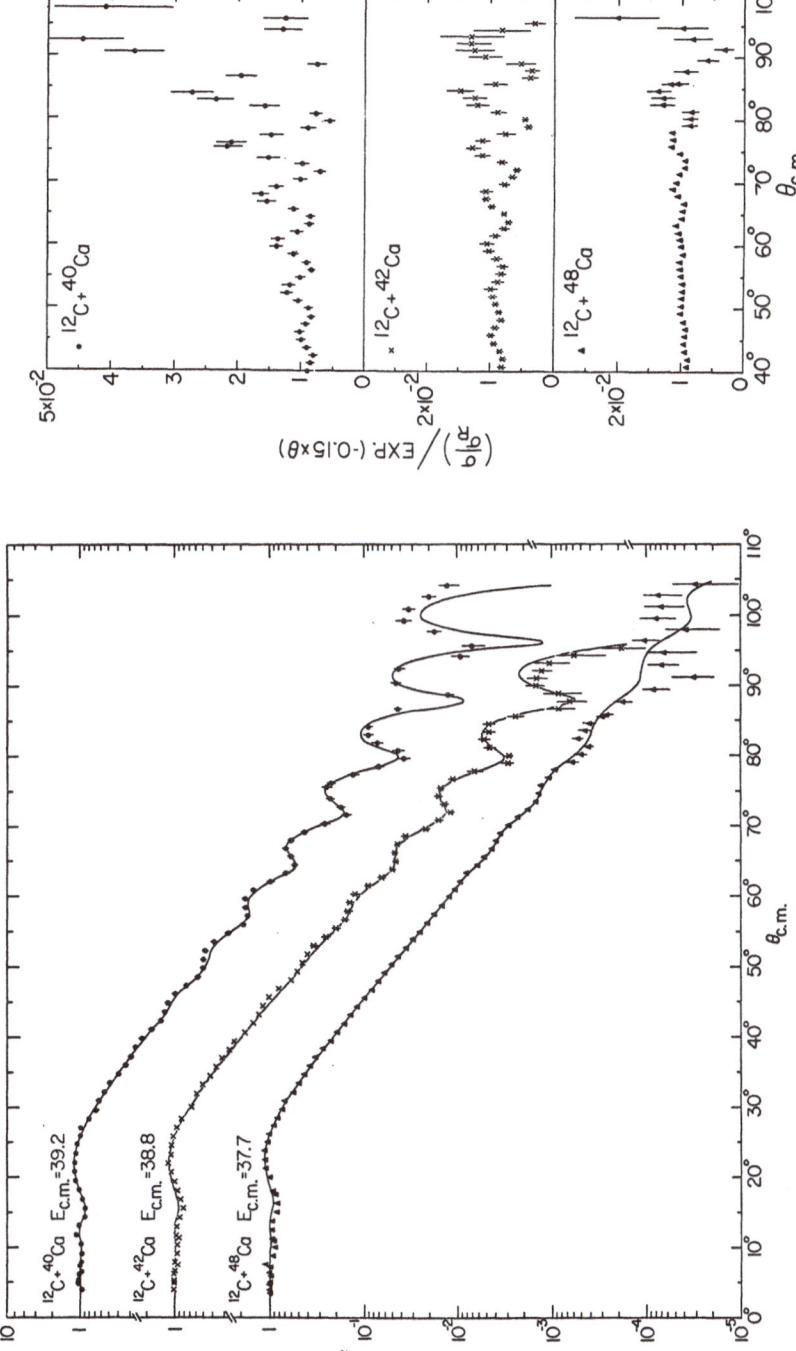

Fig. 15. Elastic scattering of ^{12}C from ^{40}Ca, ^{42}Ca, and ^{48}Ca, from Ref. 12. The left-hand curve shows some calculated optical-model curves with the data. The right-hand plots show the same data, with the exponential dependence of the cross sections on angle removed, in order to emphasize the difference in the oscillatory structure.

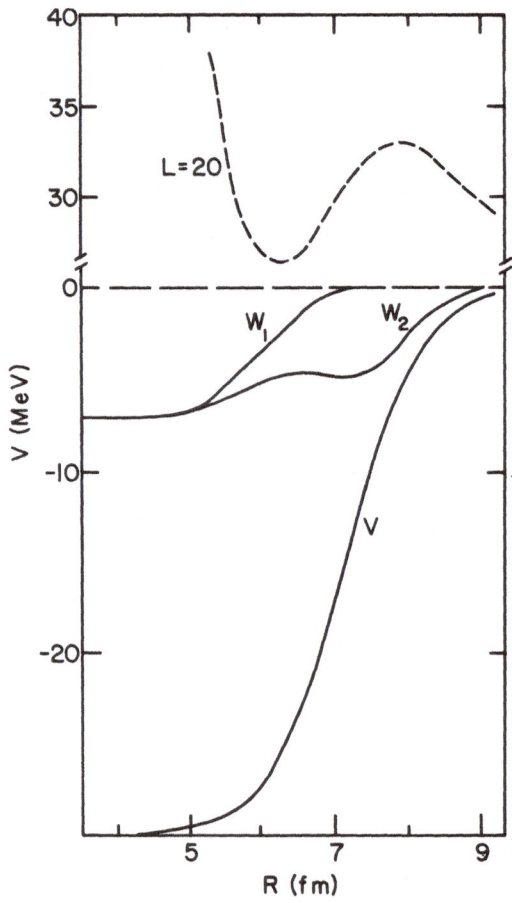

Fig. 16. (a) Calculated elastic scattering for $^{16}O + {}^{24}Mg$ from two
potentials at $E_{lab} = 52$ MeV. Shown above are the two potentials:
potential 1 has $V = 30$ MeV, $R = 7.1$ fm, $a = 0.49$, $W_{vol} = 7$ MeV,
$R_W = 5.9$ fm, and $a_W = 0.30$ fm. Potential 2 has an additional
term of surface-derivation absorption, $W_{surf} = 5$ MeV,
$R_{W'} = 7.1$ fm, $a_{W'} = 0.49$ fm. The top part of (a) shows the
total (nuclear + Coulomb + centrifugal) potential for the
$L = 20$ grazing partial wave; the center-of-mass energy is
31.2 MeV.

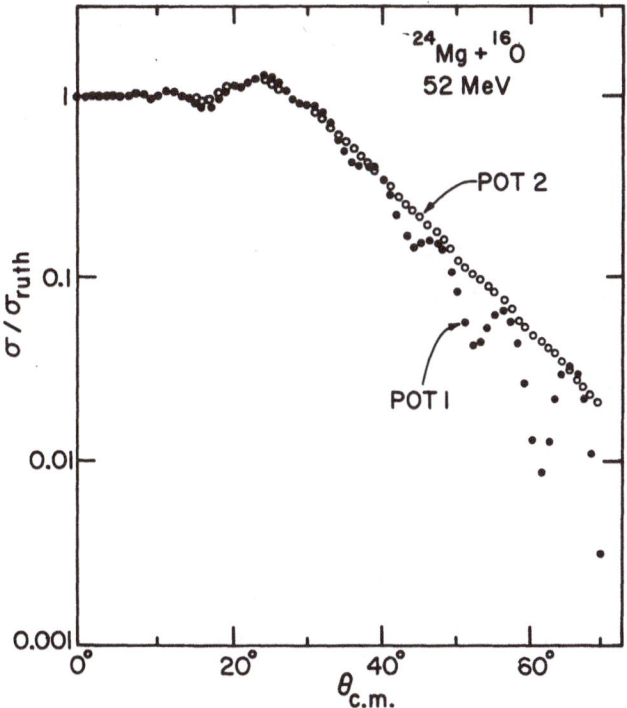

Fig. 16(b). Calculated angular distributions at forward angles.

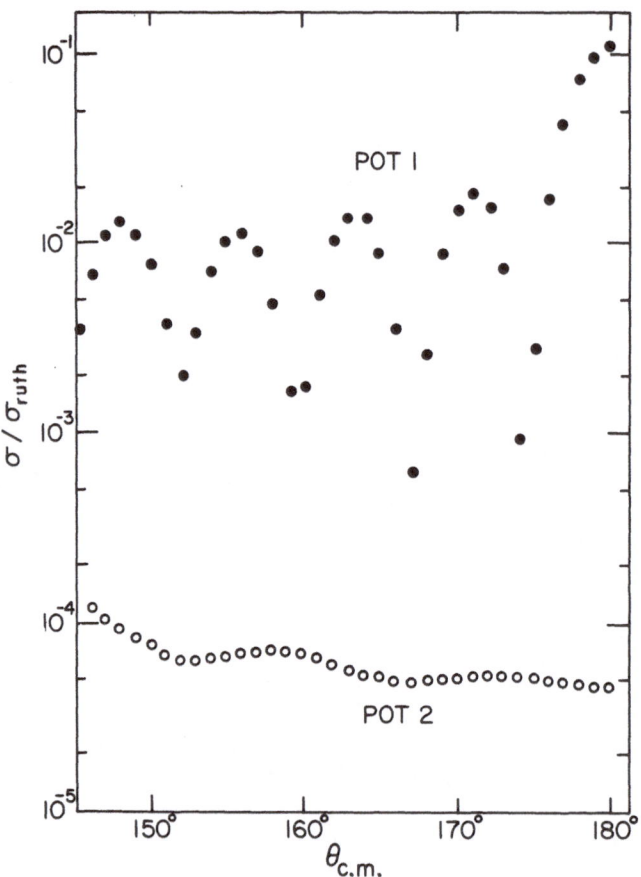

Fig. 16(c).　Calculated angular distributions at backward angles.

Resonances in Heavy Ion Reactions

The behavior of elastic scattering, the apparent surface transparency, and some sort of resonance structure, are now well established experimentally, but understanding these phenomena has not progressed very far. Structure has now been seen even in systems as heavy as $^{40}Ca + {}^{16}O$ and approximate angular momenta were assigned. But the underlying physics has been very elusive. Are these shape resonances in some potential model? If so, how can we tie them to nuclear structure? There is talk of rotational bands—but the language is vague and unclear.

It is tempting to try to find other handles on these phenomena. In elastic scattering, a large number of partial waves contribute. Inelastic scattering to low-lying states generally has a Coulomb excitation component and again a large number of partial waves contribute to the reaction. On the other hand if there are reaction channels which are well matched, the grazing angular momenta are the same in the entrance and exit channels, then the reaction, even if it is direct, will take place in a relatively few grazing partial waves.

The $^{24}Mg(^{16}O,^{12}C)^{28}Si$ reaction had been studied some years ago and was found to have a strongly oscillatory angular distribution characteristic of a reaction with sharp L-localization. When excitation functions were measured for this reaction, resonance-like structures were indeed seen at forward as well as back angles. The analysis of the angular distributions, however, has turned up considerable complexity in the interpretation of the structure.[13]

I would like to discuss the details of the analysis a little bit further at this point. From Fig. 17 we note that the yield at 0° is always higher than at 180° by about an order of magnitude. Thus clearly, several partial waves contribute to the reaction with a phase relationship that causes coherent addition between odd and even partial waves, overall the interference is constructive at forward angles and destructive backwards. But the resonant structures are not simply correlated between 0°, 90° and 180°. For the sake of this analysis, we assume that there is a clean separation between a direct component to the reaction that is smooth in its energy dependence and has negligibly small contributions at 180°, and several resonating amplitudes. At 0° we would see coherent interference between the direct and resonant amplitudes, while at 180° only the resonances can contribute. If the resonances were isolated and not overlapping with each other, then the back-angle angular distributions would tell the whole story. If they do overlap, however, the story becomes more complicated. The 90° excitation function is valuable because here only even partial waves can contribute, whether there is interference or not. In addition, of course, the detailed shapes of the angular

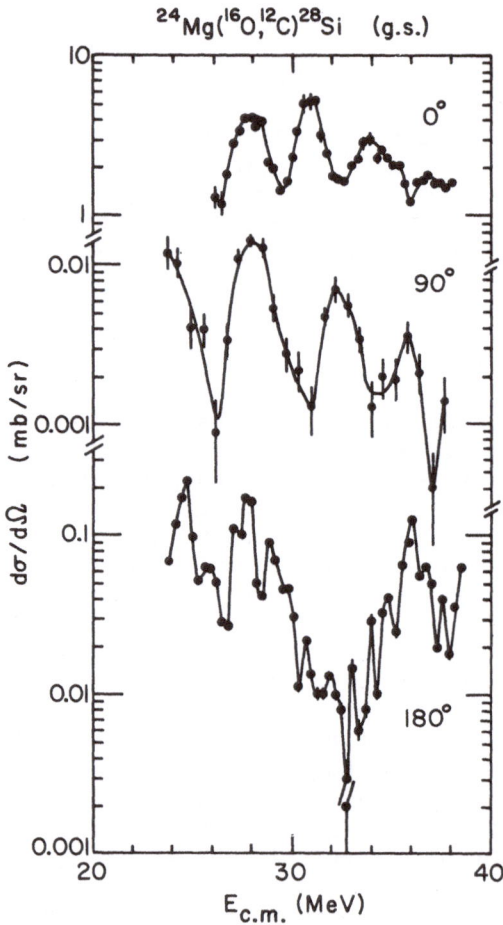

Fig. 17.   Excitation functions for the ^{24}Mg(^{16}O,^{12}C)^{28}Si reaction
   at 0°, 90° and 180°, from Ref. 13.

distribution have to be analyzed. The data shown in Fig. 18 for
E_{cm} between 29 and 31 MeV were first thought. to require three
resonances. The bump in Fig. 17 at 29 MeV seems to be consistent
with an L = 20 assignment. However, interference effects are clearly
important for the 31-MeV resonance. The fact that the 180° yield
shows a minimum at the same energy where the maximum in the 0° yield
occurs, suggests interference between odd and even partial waves.
But the L = 20 resonance at 29 MeV contributes sufficiently at
31 MeV, because of the energy dependence of the penetration function,
to cause such interference. The upper resonance therefore must have
odd parity—both L = 21 or 23 are possible but L = 23 is preferred.
One may actually do the equivalent of a phase shift analysis and
fit the angular distributions as a function of energy with a direct
term, that is assumed to have a smooth energy variation predicted
by DWBA calculations, and partial wave amplitudes that are allowed
to vary, either freely, or in accordance with a Breit Wigner
resonance. The results of such analysis are shown in Fig. 19. The
principal uncertainty in this process is the choice of distorting
potentials for the direct calculations.

If we look at states at higher excitation energies in ^{28}Si,
many of them show evidence of the 29 and 31-MeV bumps—though with
varying strengths. The L-matching for ^{28}Si in the 3^-, 4^+, and 5^-
states is still rather good. These are generally the states that
are strongly coupled to the ^{28}Si ground state—the states that
appear to be strong in the α-transfer reaction seem to be the same
as the ones that are prominent in inelastic excitations.

The three resonances identified in this reaction are plotted
against the grazing L-values in Fig. 20. They do fall very close
to these lines suggesting that kinematic selection rules form an
important constraint on whether a simple underlying structure will
be observable or not. The widths of these resonances seem to be a
few percent in entrance and exit channels—if the widths are
comparable in the closely coupled states of ^{28}Si, ^{24}Mg, etc. then
perhaps we can account for most of it as some sort of a shared
resonance between structurally similar channels. The spin
sequence (J = 20, 23, 26) suggests that the observed resonances
perhaps belong to different families (bands or trajectories) whose
slopes may be steeper than those of the grazing L-values that
provide us with the surface-transparent window.

There appears, at this time, no simple correlation between the
bumps seen at back angles in the $^{24}Mg + {}^{16}O$ and $^{28}Si + {}^{12}C$ elastic
scattering and the bumps seen in the reactions connecting the two.
Clearly any resonance effects in the reaction must be present in
the elastic scattering as well. But what the data indicate is that
only some of the structures (and apparently not the most prominent
ones) in the elastic channels are connected by the α-transfer
reaction.

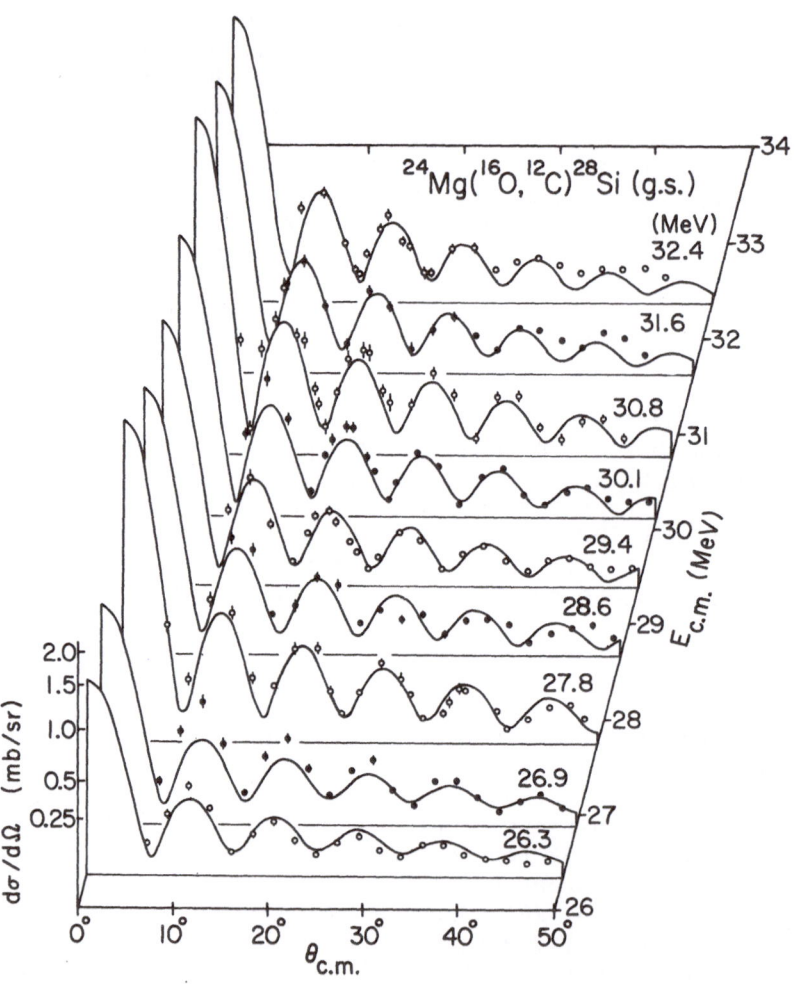

Fig. 18. Angular distributions for the ^{24}Mg(^{16}O, ^{12}C)^{28}Si reaction from Ref. 13. The solid lines represent one of the preliminary attempts to fit the data.

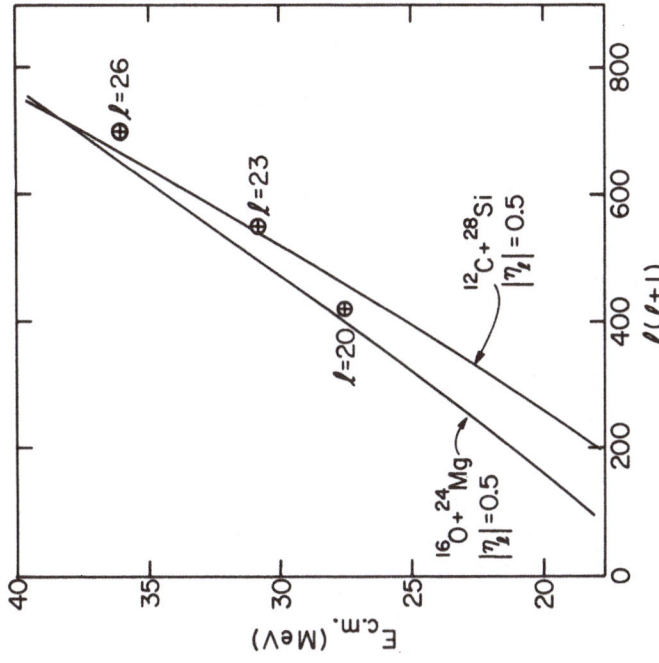

Fig. 20. Energy against $L(L+1)$ for the resonance of Ref. 13 in $^{24}Mg(^{16}O, ^{12}C)^{28}Si$. The lines indicate the grazing partial waves in the entrance and exit channels.

Fig. 19. Plot of the $L = 20$ complex amplitude over the resonance near 27.6 MeV from Ref.13. The shaded area shows the uncertainty in the amplitude from the analysis of the data. The line describes the Argand plot for an $L = 20$ Breit-Wigner resonance amplitude added to a direct amplitude.

It is certainly possible to choose potentials for the entrance
and exit channels that produce shape resonances. Such potentials
would also give rise to resonances in the DWBA calculations. But
this is not very sensible physically. Resonances in the $^{16}O + ^{24}Mg$
channel and the $^{12}C + ^{28}Si$ channel are not orthogonal: one might
expect substantial overlap between them. But unless one does a
coupled-channels calculation, this non-orthogonality is not included.
Then, the 2+ states (and even higher ones) also seem to show the
resonances, and thus the coupling to these also has to be taken
into account—then perhaps the coupling to the $^{20}Ne + ^{20}Ne$ channel
(not yet established experimentally). Clearly one will be
approaching the problem of the structure of ^{40}Ca rather than just
a simple one-channel reaction problem.

Some experiments have been done to look for resonances in
excitation functions of other systems. No effects were seen in
$^{26}Mg(^{16}O,^{12}C)$ or $^{26}Mg(^{16}O,^{14}C)$ nor on $^{48}Ca(^{16}O,^{14}C)$. All these
reactions have angular distributions that appear to be sharply
oscillatory [or as well localized in L as the $^{24}Mg(^{16}O,^{12}C)$ reaction
was]. Some bumps have been reported, however, in the $^{24}Mg(^{18}O,^{14}C)$
reaction though both the absolute yield and the relative size of the
bumps is down. In $^{28}Si(^{16}O,^{12}C)^{32}S$ resonant effects were present
at forward angles but they seem to be a much smaller percentage of
the yield (20% oscillations rather than a factor of three). It is
likewise the case for ^{12}C elastic scattering from ^{32}S that the
backward yield is down by an order of magnitude over that from ^{28}Si.

The spacing of the observed structures in $^{24}Mg(^{16}O,^{12}C)^{28}Si$ is
much greater than the estimated level densities in ^{40}Ca for these
angular momenta, by at least three orders of magnitude. These
structures therefore are 'simple' in some sense but we do not have
a reasonable idea as to what model-Hamiltonian could explain it.
Whenever we have seen such phenomena they were the signature of a
simple symmetry in some degree of freedom. At this time we do not
know enough to be able to say what degree of freedom may be
responsible. The surface transparency in certain classes of
heavy-ion reactions seems to open a narrow window through which we
are getting glimpses of new features of the nucleus. All the
indications are that this class of phenomena is a rich field for
nuclear structure investigations, but it will take years of careful
and detailed work to untangle the data sufficiently so that we may
see the simple physics that at present is still hidden.

Bibliography and References

No attempt has been made in this paper to give references
beyond some pedagogical books and articles and the sources of
specific data.

FOOTNOTES AND REFERENCES

*Work performed under the auspices of the Department of Energy.

+These notes are a slightly modified version of lecture notes to be published in the Proceedings of the 1979 Enrico Fermi Summer School on Nuclear Structure, Varenna, Italy, 9-16 July 1979.

[1] J. M. Blatt and V. F. Weisskopf, Theo. Nucl. Phys. Wiley (1952).

[2] A. M. Lane and R. G. Thomas, Rev. Mod. Phys. 29, 191 (1957).

[3] H. Feshbach, C. E. Porter, and V. F. Weisskopf, Phys. Rev. 96, 448 (1954).

[4] W. Henning et al. Phys. Letts. 58B, 129 (1975).

[5] J. P. Schiffer et al., Phys. Lett. 44B, 47 (1973).

[6] W. Henning et al., Phys. Rev. C 17, 2245 (1978).

[7] S. Vigdor, Proc. of Symposium on Macroscopic Features of Heavy-Ion Collisions, Argonne Informal Report, ANL/PHY-76-2 (Apirl 1976).

[8] D. G. Kovar et al., Phys. Rev. C 20, 1305 (1979).

[9] G. Gaul et al., Nucl. Phys. A137, 177 (1969).

[10] P. Braun-Munzinger et al., Phys. Rev. Lett. 38, 944 (1979).

[11] J. Barrette et al., Phys. Rev. Lett. 40, 445 (1978).

[12] T. Renner, Phys. Rev. 196, 765 (1979).

[13] S. Sanders et al., Phys. Rev. C 21, 1810 (1980).

GIANT RESONANCES

A. van der Woude

Kernfysisch Versneller Instituut, University of

Groningen, Groningen, The Netherlands

I. INTRODUCTION

Giant resonances are highly collective normal modes of nuclear excitation. A very schematic picture based on the liquid drop
model is given in figure 1 for the lowest multipoles. Here the
nuclear matter is thought to consider out of 4 components: protons, neutrons, particles with spin-up and particles with spin-
down. Oscillations in which protons and neutrons move in phase
are the isoscalar resonances, if they move out of phase we have
the isovector resonances. In the electric resonances particles
with spin up and down move in phase while in the magnetic resonances they move out of phase. In these lecture notes the emphasis
will be on electric isoscalar resonances but a short summary of
the isovector resonances will be included. The magnetic resonances
will not be discussed at all, although a lot of interesting work,
experimentally as well as theoretically, has recently been done
on this subject |1|.

The isoscalar giant dipole resonance is not a real nuclear
excitation which is the reason that it is not shown in fig. 1.
Probably the most interesting mode is the isoscalar giant monopole
resonance. As is clear from the figure it is the only mode in
which the density of the nuclear matter oscillates. Its excitation
energy is directly related to the nuclear compressibility . In
fact it turns out that the only way to determine this fundamental
quantity of nuclear matter is from the giant monopole excitation
energy.

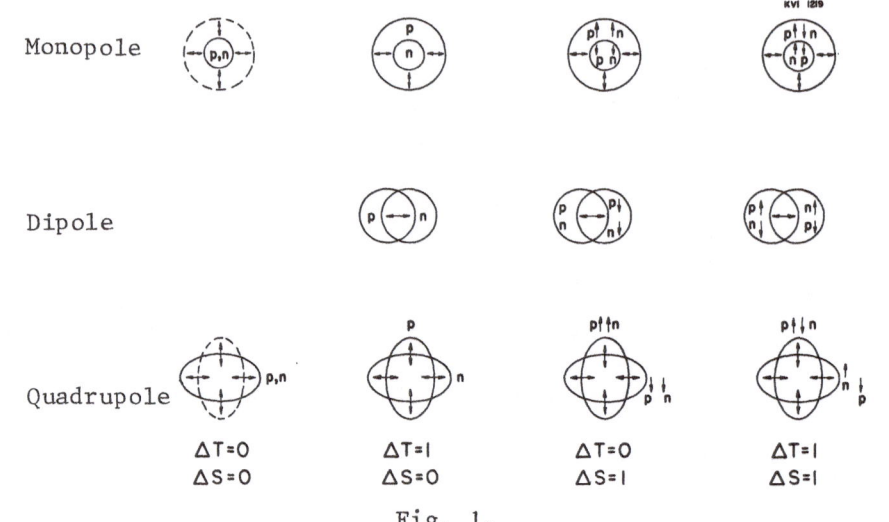

Fig. 1.

In the shell model picture giant resonances are described as strongly coherent particle-hole excitations resulting from applying the multipole operator:

$$Q_{LM} = \sum_{i=1} r_i^L \, Y_{LM}(\theta_i, \phi_i)$$

to the ground state of the nucleus. A schematic picture is given in figure 2. The successive shells are denoted by N, N+1,.... and have alternating parity. Their energy distance is given by $\hbar\omega \sim 41A^{-1/3}$MeV. In the ground state all levels up to the Fermi level located somewhere in the shell N, are filled. The operator $r^L Y_{LM}$ can promote a particle at most L shells. Taking into

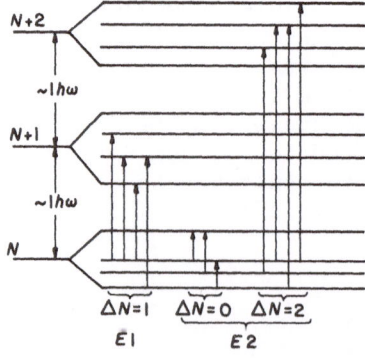

Fig. 2.

account the parity change $(-1)^1$ we get the following scheme:

Dipole   L = 1 $1\hbar\omega$
Quadrupole   L = 2 $o\hbar\omega$ $2\hbar\omega$
Octupole   L = 3 $1\hbar\omega$ $3\hbar\omega$
Hexadecupole   L = 4 $o\hbar\omega$ $2\hbar\omega$ $4\hbar\omega$

The transitions with $o\hbar\omega$ correspond to the low-lying intershell transitions. Because of the spin orbit interaction all pure particle-hole transitions with $L\hbar\omega$ are spread out in excitation energy. They are mixed through the residual particle-hole interaction and the transition strength is concentrated in one or in a few levels located within a narrow energy interval. The envelope of the strength function is the collective vibrational state. The excitation energy of this collective state is moved up or down with respect to the original value of $L\hbar\omega$ depending on whether it is an isovector (breaking-up of neutron-proton pairs) or an isoscalar vibration. The results of a real calculation using harmonic oscillator wave functions are summarized in table 1 |2|.
Here the strength is given as a fraction S of the energy-weighted sumrule (EWSR) which is defined as $\sum_x E_x^i \times B^i(E_L)$. It is a nearly model independent quantity depending only on ground-state parameters.

Table 1.   Theoretical predictions for excitation energy E_x and strength S (expressed as a percentage of the appropriate sum rule strength) for the T=0 and T=1 components of different multipoles.

		T=0		T=1	
L	$\hbar\omega$	$E_x(A^{-1/3}MeV)$	S(%)	$E_x(A^{-1/3}MeV)$	S(%)
2	2	58	100	135	100
3	1	25	28	53	2
	3	107	72	197	98
4	2	62	51	107	3
	4	152	49	275	97

The main point to note is that for L>2 the multipole strength is divided over different shells and that for instance $2\hbar\omega$, T=0, L=4 strength will coincide with the $2\hbar\omega$, L=2 strength. More sophisticated calculations, using RPA wave functions and taking into account the spreading of the 1p-1h states into 2p-2h components, confirm these predictions |3|.

A giant resonance is thus characterized by three properties:

(i)   it is a general property of many nuclei
(ii)   its excitation energy E_x and width Γ depend smoothly on A
(iii)   it exhausts an appreciable fraction S of an appropriate
         sumrule.
Experimentally, a giant resonance of which S is large and Γ is
small, will be the easiest one to observe.

The best known example of such a vibrational resonance is the
giant dipole resonance (GDR) which is a collective vibration of
neutrons vs protons. It is selectively excited in photo-absorption
experiments. It was discovered in 1947 by Baldwin and Klaiber |4|.
In heavy nuclei it is located around $E_x = 78\ A^{-1/3}$ MeV, in medium
and light nuclei at slightly lower energies. It has all the
properties characteristic for a giant resonance − see for instnace
the review papers |5, 6| and section III.

Since about 1972 another collective vibration is known, the
isoscalar giant quadrupole resonance (GQR) in which protons and
neutrons oscillate in phase. The main features of this resonance
are well known |14,15|. It has been located as a single resonance
in many nuclei with A>36 at an excitation energy of approximately
$E_x = 63\ A^{-1/3}$ MeV; it exhausts 50 to 100% of the energy weighted
sumrule (EWSR); its width varies from about 6 MeV for A ~ 60 to
about 3 MeV for A ~ 200. It is interesting to note that this resonance was predicted at an excitation energy $E_x = 58\ A^{-1/3}$ MeV,
very close to its actual value, already several years before it
was actually experimentally observed.

Very recently strong evidence for yet another resonance has
been found, the isoscalar giant monopole resonance (GMR). The
available data seem to indicate that it is located at about
$E_x = 80\ A^{-1/3}$ MeV and that it also exhausts an appreciable amount
of the EWSR |15|.

Also in many nuclei a concentration of $\Delta T=0$, 3^--strength has
been observed, corresponding to the $1\hbar\omega$, octupole strength (Low
Energy Octupole Resonance = LEOR). In addition there is now some
evidence for the existence of the $3\hbar\omega$, 3^- resonance (HEOR) and
for the $\Delta T=1$, GQR.

II. TOOLS FOR STUDYING GIANT RESONANCES

Table 2 gives a survey of the main techniques used in the
study of giant resonances. The electromagnetic and hadronic probes
are in a sense complementary: because of the relatively weak interaction the cross sections for excitation of giant resonances
through e.m. probes is small, but the initial state is much better
known than for hadronic probes while for hadronic probes the cross
sections are much larger, but the initial state is often not as
well known.

Table 2. Summary of the main techniques used in the study of electric giant resonances.

	E0 T=0	E0 T=1	E1 T=1	≥E2 T=0	≥E2 T=1
electromagnetic interaction					
γ-absorption: (γ,n), (γ,f),...			+	(+)	(+)
radiative capture (p,γ)			+		+
(α,γ)			+	+	
inelastic electron scattering (e,e')	+	+	+	+	+
strong interaction					
inelastic hadron scattering (p,p')	+	(+)	(+)	+	(+)
(d,d')	+			+	
(³He,³He')	+	((+))	((+))	+	((+))
(⁴He,⁴He')	+			+	
(heavy ion)				+	
decay experiments (α,α',c)	+			+	

The bulk of our present knowledge about the GDR and its properties are obtained from the photo-absorption process using mono-energetic photons |5,6|. The dominance of E-1 excitation in photo-absorption is due to the low momentum transferred, $\Delta p = \hbar\omega/c$. The E-2 resonance is also excited but much weaker. Since the GDR in all nuclei is located above the particle-emission threshold and since in medium and heavy nuclei the height of the Coulomb-barrier prevents any charged particle decay, the photo-absorption cross section can be accurately determined from a measurement of the neutron-yield as a function of γ-energy, that is by measuring the excitation function of the (γ,n)-reaction. If $E_\gamma > E_{2n}$, the threshold for two-neutron emission, also the (γ,2n)-cross section has to be added. Likewise for the actinide nuclei one has to add the (γ,f)-cross section. The total γ-absorption cross section is then obtained from $\sigma(\gamma) = \sigma(\gamma,n) + \sigma(\gamma,2n) + \sigma(\gamma,f)$. Figure 3 shows such a measurement for ^{208}Pb.

Radiative capture is, by detailed balance, the inverse of a photo-nuclear process in which a particle like a proton or alpha particle is ejected and the residual nucleus is left in its ground state. It has the advantage that precise excitation functions can be measured, the disadvantage that only part of the total photo-absorption cross section is measured. It has been used extensively for the study of giant resonances in light nuclei (see for instance |7|. In (p,γ)-reactions the E-1 emission is dominating. However, in (α,γ)-reactions, especially on light self-conjugate T=0 nuclei, E-1 emission can only occur through T=0 admixtures into the T=1 giant dipole resonance, which amounts to only a few percent. So in this case also E-1 emission is hindered and the ΔT=1 and ΔT=0

resonances are about equally strongly excited. By carefully measuring
the angular distribution of the emitted γ-rays, it is possible to
disentangle E-2 contributions from the still dominating E-1 con-
tributions |8|.

Another feature of capture reactions like the (p,γ)-reaction
is that by measuring the decay to excited states one can study
giant resonances built on excited states instead of the ground
state. An extreme case of this is the recent suggestion made from
the analyses of a $^{11}B(p,\gamma)^{12}C$ reaction at E_p=40, 60 and 80 MeV,
that one observes a 1 ℏω resonance built upon the GDR in ^{12}C,
which in itself is a 1 ℏω resonance |9,10|.

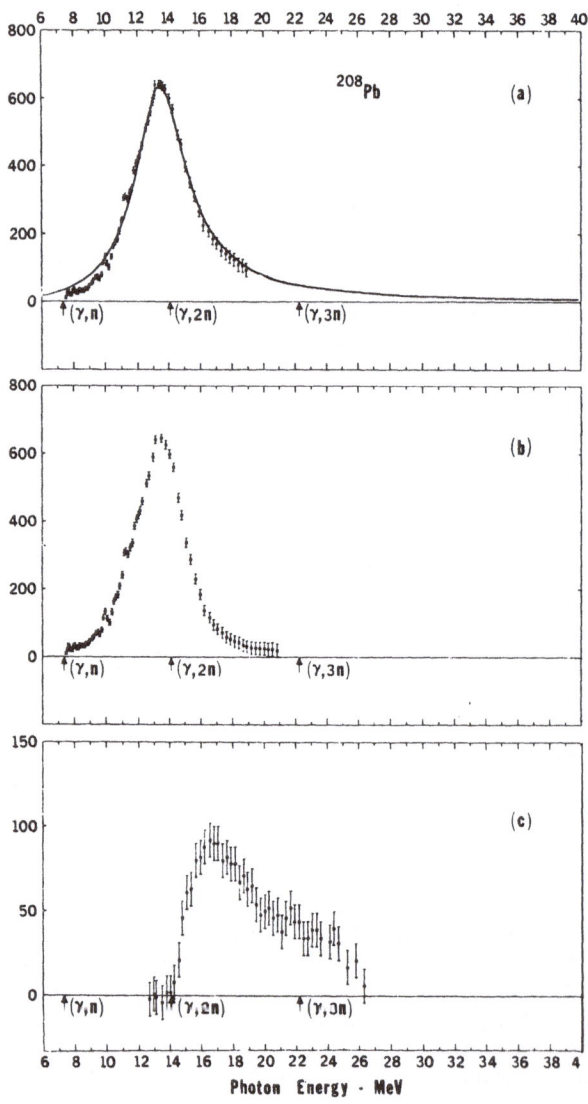

Figure 3.

The photon-absorption
curve for ^{208}Pb. Fig.
4a shows the total
cross section $\sigma(\gamma)$
which is the sum of
$\sigma(\gamma,n)$ and $\sigma(\gamma,2n)$
shown in figures 4b
and c, respectively.

In inelastic electron scattering all possible multipolarities with T=0 and T=1 are excited. By measuring the cross section for a resonance as function of momentum transfer, for instance by measuring it as a function of bombarding energy or scattering angle, the different multipolarities can in principle be disentangled. However, there is no way to distinguish between T=0 and T=1 excitations and this information has to come from other experiments. Another disadvantage is the large continuum underlying the resonances. The main contribution to this continuum is due to the radiation tail, which in principle can be calculated and subtracted. But since the resonance cross section is often more than an order of magnitude smaller than the one for the underlying continuum, a small error in the continuum subtraction can have a large effect on the deduced resonance cross sections. Figure 4 shows an (e,e')-spectrum for ^{140}Ce. A possible decomposition in continuum and ΔT=0 and ΔT=1 multipoles is also indicated |11|. It is clear that in this way rather large errors can be introduced. Nevertheless many important contributions to our knowledge about giant resonances have been obtained by (e,e') scattering, including the first evidence for the existence of the GQR |12|, one of the

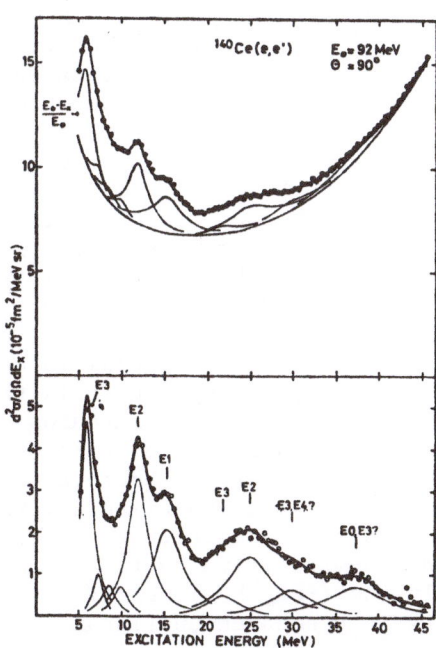

Fig. 4. An (e,e')-spectrum for ^{140}Ce.
A possible decomposition in different multipoles is also shown |11|.

first suggestions for the T=0 monopole resonance |13| and the
evidence for the existence of a ΔT=1 GQR |15,11|.

Most information on electric isoscalar giant resonances has
been obtained through inelastic hadron scattering and especially
inelastic alpha scattering. Inelastic alpha-particle scattering in
particular has several advantages: T=0 resonances are selectively
excited which is for instance not necessarily true in (p,p') scattering, and the angular distributions are often pronounced and
rather indicative for the transferred L-value. But most important,
good quality α-beams in the desired energy range from say 100 to
150 MeV are nowadays readily available at a number of cyclotron
facilities throughout the world, while standard detection techniques with solid state counter telescopes can be used for the
detection of the scattered particles. In another section the
analysis of and problems involved in the (α,α') experiments will
be discussed in more detail.

Inelastic proton scattering at about 60 MeV has played an
important role in the discovery of the giant quadrupole resonance
|92|. Higher energy protons have not often been used, mainly
because they were not readily available and because it is hard
to detect them. Nowadays, such beams (E_p ≥ 100 MeV) are available
at a number of new facilities like the meson factories LAMPF,
TRIUMPH and SIN and especially at the Indiana University Cyclotron
Facility. At several of these facilities one has a magnetic
spectrograph available for the detection of the scattered particles, so that (p,p') experiments at these energies now become
feasible. The incensive for these experiments is shown in figure
5; by a suitable choice of the scattering angle one should be able
to distinguish between different multipole strengths (from |16|).
Note, however, that the L=0 and L=2 angular distributions only
differ at very forward angles.

Recently there has been also a special interest in exciting
nuclear collective modes using heavy-ion inelastic scattering,
primarily because the angular momentum-matching conditions favor
large L-transfer and thus higher multipole resonances may show up
stronger. Also, since it is generally believed that part of the
continuum observed in scattering experiments like (p,p') or
(α,α') is due to quasi-elastic scattering one might hope that the
magnitude of the continuum relative to the resonances would be
decreased. Especially the (^{6}Li,^{6}Li')-reaction looks promising in
the sense that the ratio resonance-continuum is very favorable,
but the extraction of multipolarity and strength from the data
seems to be somewhat ambiguous |17|. Also in inelastic ^{12}C, ^{14}N
and ^{16}O scattering the GQR is excited but the spectrum is rather
complicated and it is not yet clear what one can learn from these
experiments |18,19|.

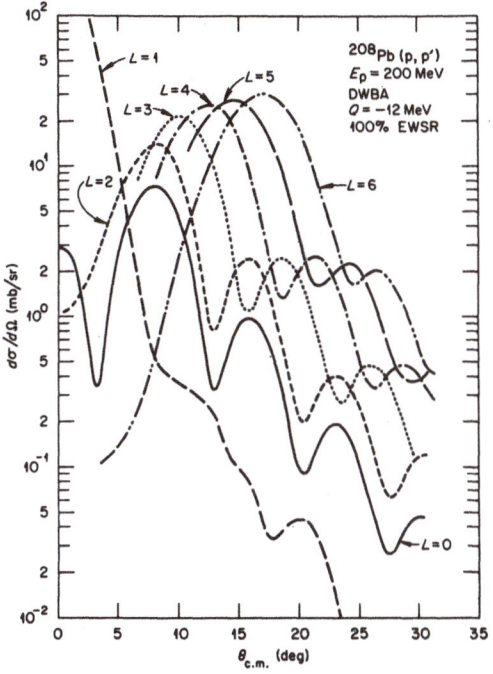

Fig. 5. DWBA calculation for the angular distribution of the various multipoles (all T=0 except for the L=1 multipole) for the reaction ^{208}Pb(p,p') at E_p = 200 MeV. (From reference |16|.

III. THE ISOVECTOR RESONANCES

Isovector resonances are collective modes of excitation of the nucleus in which the protons oscillate against the neutrons according to a multipole pattern. The best known example is the Giant Dipole Resonance (GDR) which was already discovered in 1947 |4|. Since then it has been the subject of numerous experimental and theoretical studies, the results of which are summarized in a number of recent review papers or conference proceedings |5,6,15|. Recently there is also mounting evidence for the existence of an isovector giant quadrupole mode located at about E_x = 130 $A^{-1/3}$ MeV. Other isovector modes like the isovector giant monopole resonance have not yet been established although some evidence for its existence has been presented |11,20|.

The giant dipole resonance is selectively excited by the photon-absorption process. In a semiclassical theory of the interaction of photons with non-deformed nuclei, the resonance shape can be written as a Lorentz-curve (see also |21|);

$$\sigma(E) = \frac{\sigma_m(E)}{1 + \left[(E^2 - E_m^2)^2 / E^2 \Gamma^2 \right]}$$

In a rotationally symmetric statically deformed nucleus the
GDR is split in two parts, one corresponding to a vibration along
the symmetry axis and the other one, with about double the intensity, to a vibration along the axes perpendicular to the symmetry axis. All photon-absorption data for $50 \leq A \leq 238$ can be very
well fitted by one or two Lorentz curves and there is an impressive amount of systematics available with respect to the resonance
energy E_x, width Γ and total absorption cross section σ.

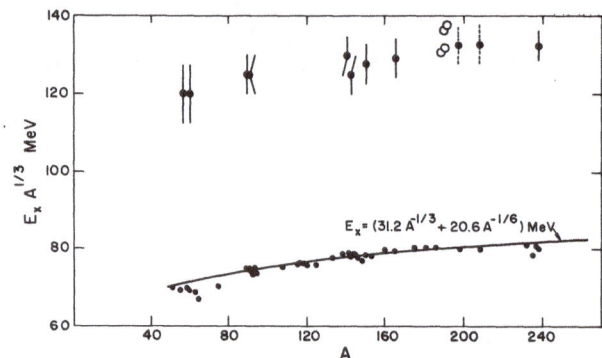

Figure 6.   The excitation energy $E_x A^{-1/3}$ vs A for the
            $\Delta T=1$ GDR (lower part) and $\Delta T=1$ GQR (upper part).

The data for E_x are given in figure 6. For prolate deformed
nuclei E_x was determined from $E_x = (E_{m1} + 2E_{m2})/3$ $|5|$.

Historically two hydrodynamical models have been used for the
giant dipole. In the GT mode the dipole oscillation is thought to
be an oscillation of a proton sphere against a neutron one $|22|$ –
it has an energy dependence $E_x \sim A^{-1/6}$. In the Steinwedel-Jensen
model the neutrons first pile up on one side and then on the
other while the protons move in opposite direction so that the
total density is constant. This model $|23|$ predicts $E_x \sim A^{-1/3}$.
The actual data, as shown in figure 6, indicate that none of the
models gives a correct description of the data over the mass range
$50 < A < 238$. Recently in a droplet model of the GDR, Myers et al.
$|24|$ have shown that the GDR should be considered as a linear
superposition of the GT and SJ-mode. In their solution the GT-mode
tends to dominate for light nuclei and it contributes more than
the SJ-mode for all mass-numbers up to $A \sim 250$. The amount of SJ-
mode increases with A and almost reaches parity for the heaviest
nuclei. The droplet model predicts to a good approximation
$E_x = c_1 A^{-1/3} + c_2 A^{-1/6}$ and it has been found experimentally that with
$c_1 = 31.2$ MeV and $c_2 = 20.6$ MeV the E_x-data are reasonably well des-
cribed $|5|$, as is shown by the full curve in figure 6.

The splitting of the giant dipole in statically deformed
nuclei is illustrated in figure 7 for ^{238}U. In a hydrodynamical
model it is directly related to the ratio of the long axis a and
the short axis b of the ellipsoidal nucleus |25|:

$$E_b/E_a = 0.911 \ a/b + 0.089$$

and thus also the intrinsic quadrupole moment Q_o. Table 3 shows
the impressive agreement obtained for Q_o through measurements
of the B(E2) values and from the shape of the giant dipole ab-
sorption curve |6|.

Table 3.   Intrinsic quadrupole moments Q_o' and Q_o (in b) of
           statically deformed nuclei, from B(E2) measurements
           and from the splitting of the GDR |6|, respectively.

	Sb	^{127}I	^{143}Rh	^{150}Nd	^{152}Sm	^{154}Sm
Q_o	−1.8±0.4	−2.3±0.4	1.7±0.2	6 ±1	5.9±0.4	6.6 ±0.4
Q_o'				5.1	5.9	6.65

	Er	Lu	W	Re	^{232}Th	^{238}Np	U
Q_o	6.96±0.4	6.93±0.3	6 ±0.5	6 ±0.5	10.2 ±1	11.3±1	11 ±1
Q_o'	7.6	7.2	6.2	5.9	9.66	10.9	11.3

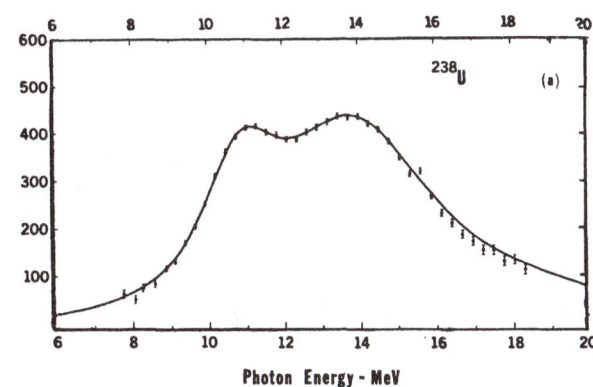

Fig. 7 . The giant dipole resonance in ^{238}U. The
         splitting of the dipole strength in two
         components is clearly demonstrated |5|.

The total dipole absorption cross section $\Sigma^o = \int_{0}^{\infty} \sigma(E)$ can be
compared with the classical Thomas-Reiche-Kuhn sumrûle:
$\Sigma^o = \frac{60NZ}{A}$ (1+K) MeV mb, where K is an enhancement factor due to
neutron-proton exchange contributions which is characteristic for
all ΔT=1 resonances. Recently it has been found that if one in-

cludes all photon-absorption cross section up to the meson-produc-
tion threshold $E_\gamma = 140$ MeV, that $K \sim 0.75$ gives a fairly good
description of the data |6|.

From this short summary it should be clear that in medium and
heavy nuclei the GDR has all the properties of an ideal giant resonance: it occurs systematically in all nuclei, its excitation energy E_x and width Γ vary smoothly as function of A and it exhausts a
large part of the appropriate sumrule. In light nuclei like ^{16}O or
^{12}C the GDR-strength becomes fragmented: it cannot be described as
one Lorentzian curve. This phenomenon, which seems to be a common
feature of all giant resonances, will be described in more detail
for the $\Delta T=0$ GQR.

The hydrodynamic model relates directly the excitation energy
E_x of the $\Delta T=1$, GQR to the excitation energy of the GDR:
$E_x(GQR)=1.6E_x(GDR)\approx130$ $A^{-1/3}$ MeV. Also in a microscopical model
one can calculate that $E_x(GQR)\approx130$ $A^{-1/3}$ MeV (see table 1).

In a number of (e,e')-experiments such a bump at the excitation energy predicted has been found indeed (see figure 8). Recently resonances at the expected energy have also been found to be
present in photon-absorption experiments on the Os-isotopes |26|.
The resonance energy deduced from a Lorentz curve fit to the excess
strength (excess in the sense of non-GDR strength) are indicated
by o in figure 6. Since these $\Delta T=1$ GQR resonances are rather weakly
excited, broad and possibly overlapping with other resonances,
their characteristic properties are by far not as well determined
as in the case for the GDR. The available data indicate that the
strength, corresponding to the $\Delta T=1$ GQR is concentrated only in nuclei
with $A\gtrsim60$. In lighter nuclei its distribution becomes very broad,
as deduced from radiative capture reactions. Also the available
data seem to indicate that it exhausts an appreciable fraction
(40 to 80%) of the appropriate sumrule (without exchange term) and
that its width Γ ranges from 6 to 10 MeV.

IV. THE ISOSCALAR GIANT RESONANCES

1. Introduction

The presently available experimental information on isoscalar
resonances is nicely demonstrated in figure 8, which shows for
^{120}Sn α-spectra taken at $E_\alpha = 153$ MeV |27|. At low excitation energies a number of bound states are strongly excited. In the ex-
citation energy range $8 \lesssim E_x \lesssim 30$ MeV one observes small broad bumps
superposed on a rather flat and featureless continuum. These bumps
are due to the excitation of giant resonances. At still higher excitation energy around $E_x \sim 35$ MeV there is a broad peak which is
due to elastic scattering of the alpha-particles from hydrogen

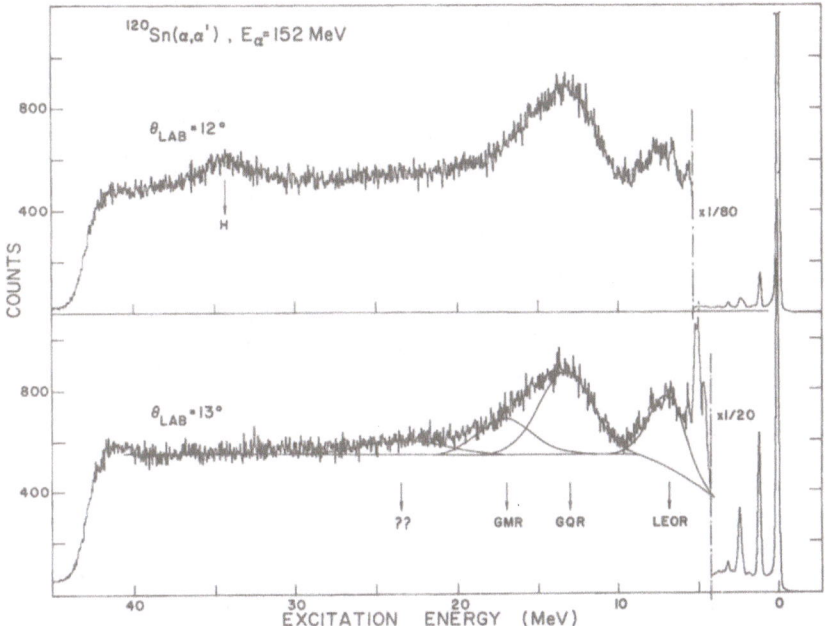

Fig. 8. ^{120}Sn(α,α')-spectra at $E_\alpha = 152$ MeV. A possible
decomposition into a continuum and in different
multipoles is indicated for the 13º-spectrum |27|.

present in the target. In order to analyse such a spectrum one has
to assume a certain shape and magnitude for the continuum, for
instance as indicated in figure 8. If one further assumes that the
observed structures consist of a number of different resonances
each with a Gaussian shape, then one can decompose this spectrum
as shown in figure 8. Besides the large bump around $E_x \sim 14$ MeV,
which is mainly due to GQR and GMR excitation, and the one around
$E_x \sim 7$ MeV, due to the excitation of the low energy $1\hbar\omega$ octupole
resonance (LEOR), there is some evidence for a broad bump around
$E_x \sim 22$ MeV which may indicate the presence of still different multipole strength, like the $3\hbar\omega$ T=0 octupole resonance (HEOR). As will
be also clear from figure 8 though, the magnitude of the bump
around $E_x \sim 23$ MeV is very sensitive to the assumed shape of the
continuum. In fact it is possible to draw the continuum in figure
8 in such a way that the bump at $E_x \sim 23$ MeV nearly completely disappears.

There is no a priori justification for assuming a Gaussian
(or a Lorentzian or Breit-Wigner) shape for each individual resonance: the resonance in figure 8 might well be a single one with
an asymmetric shape. The justification stems from the fact that

the shape of the structure taken as a whole changes as a function of scattering angle as will be discussed later on.

By far most data to be discussed in this section have been obtained from inelastic scattering of protons, deuterons, ^{3}He- and α-particles. Especially α-particle scattering has been frequently used, since being itself a T=0 probe, it will selectively excite ΔT=0 states in nuclei. The excitation of for instance the ΔT=1 giant dipole can only occur through Coulomb-excitation and this is known to be weak |28|. Also in (^{3}He,^{3}He') scattering, ΔT=1 resonances are only weakly excited and quite a lot of data has been obtained in this way, especially for the GMR.

2. Analysis and problems

In order that one can judge properly the information obtained from inelastic scattering data a short survey of the way these data are usually analyzed and of the problems involved will be given here. In inelastic hadron-scattering the usual procedure involves several steps:

1. Decomposition of the spectrum into a resonance bump and a continuum. The problems involved here are illustrated in figure 9 which shows a ^{209}Bi-spectrum taken at E$_\alpha$=120 MeV. Two possible backgrounds are shown: background I is probably an upper limit, background II a lower one. The corresponding resonance cross section (see figure 10) can differ by as much as 50%, so that due to this fact alone resonance cross sections may have an uncertainty of ±25%.

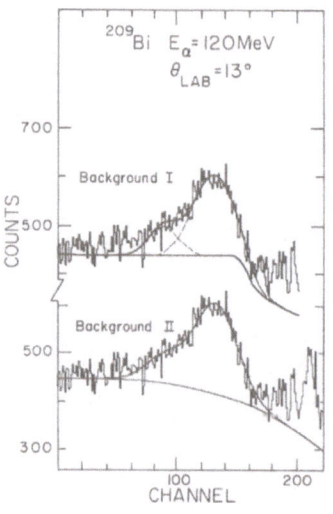

Fig. 9. ^{209}Bi(α,α')-spectra at E$_\alpha$=120 MeV. Two possible backgrounds and the decomposition of the rest spectrum into two bumps are indicated. The bump located around channel 90 is presumably the GMR located at E$_x$~14 MeV and the one around channel 135 the GQR located at E$_x$~11 MeV |29|.

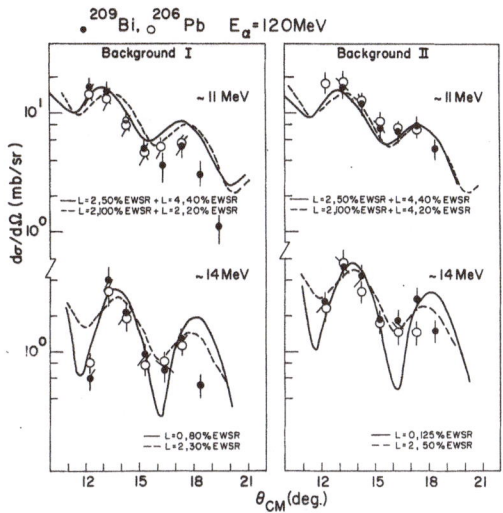

Fig. 10.
Angular distributions for the
two bumps in the $^{209}(\alpha,\alpha')$-
spectrum of figure 9, corresponding to the two choices
for the background, The fits
are DWBA calculations using
different mixtures of multipoles as indicated. The data
for ^{206}Pb behave in a simular
way |29|.

        In the following we will discuss the characteristics of the
GQR and the GMR separately. However, it should be realised that
in most experiments both are excited simultaneously and that the
two resonances overlap to some extent, as is illustrated in
figure 8. Thus in order to analyse such data in terms of two resonances, one has to make certain assumptions on the shape of
these resonances. Nearly all available data have been analysed
assuming a symmetric, Gaussian or Breit-Wigner-like shape for the
resonances. It is not yet clear how well such an assumption is experimentally justified. Probably for special nuclei like ^{208}Pb,
^{142}Nd and ^{90}Zr it is a good assumption, but for deformed nuclei
like ^{150}Nd the shape of the GQR is certainly more complex. Also,
in a recent analysis of (d,d')-data taken at very forward angles,
where the GMR cross section is small, it was found that the GQR
shape in certain nuclei is slightly asymmetric |30|.

2. The cross section angular distributions are analyzed in the
   framework of the vibrational model, using DWBA calculated
   angular distributions for comparison, to determine the multipolarity and strength of the resonance. For low-lying collective states this procedure is known to give reliable multipolarity assignments, but it should be realized that for high-
   lying excited states it has been tested only in a few specific
   cases where it was possible to determine the multipolarity of
   the resonances independently |31,32|.   Problems arise
   whenever there is a suspicion that various multipolarities are
   involved: the inherent uncertainty with which the cross sections
   can be measured often does not allow a determination of whether
   and to what extent this is the case. This is illustrated in

figure 10 where an attempt was made to describe the data as an
L=2 and L=4 mixture - several combinations fit the data equally
well.

3. The resonance strength is determined from the mass deformation
parameter β_L^m according to:

$$S_L(IS) = (\beta_L^m)^2 \, E_x / \tilde{S}_L \tag{1}$$

where for L>0 the sumrule limit $\tilde{S}_L$ is given for a uniform mass
distribution by:

$$\tilde{S}_L = (\hbar^2/2mR^2)(4\pi/3A)L(2L+1) \tag{2}$$

Appropriate correction factors for a more realistic Fermi mass
distribution have been tabulated |33|. For a discussion of
L=0 transitions see section IV.5.

Here a well-known problem is that for instance for low-lying
collective states in light nuclei the β_L^m -values derived from
(α,α')-scattering in the way described above, are considerably
smaller than the ones one would derive from the corresponding B(EL)
values as obtained from life-time measurements or (e,e')-scattering
data. This discrepancy can be largely resolved by using a folding
model analysis |34| or by applying an L-dependent normalization
factor |35|.

A major assumption in the procedure outlined above is that
the form factor of the vibrational model gives a good description
of the real one. Theoretical studies using microscopic wave functions indicate that this is not necessarily the case - see for
instance reference |28|. Also, the sumrule limit S_L in formula 2
has been derived for the electromagnetic transition operator
$r^L Y_L^m$, but the real transition operator may have a substantially
different r-dependence. Especially for inelastic proton scattering
appreciable errors may occur, possibly as large as 50% for 156 MeV
protons on ^{208}Pb |90|.

Similar problems occur in the analysis of (e,e')-data. The
first step here is to subtract the huge but in principle calcu-
lable radiation tail. Then, in order to determine from the resulting spectrum the isoscalar structures, one has to subtract first
the isovector components like the GDR, which are strongly excited.
This subtraction strongly depends on the model used for the $\Delta T=1$
excitation. For instance for ^{208}Pb it was shown |36| that for
$7.4 \leqslant E_x \leqslant 12.5$ MeV the T=0, E-2 strength could vary from $(92 \pm^{14}_{8})$% of
the T=0. EWSR to $(52 \pm^{12}_{5})$%, depending on whether a Goldhaber-Teller
or Jensen-Steinwedel model for the GDR was assumed.

It is then clear that rather large uncertainties are involved
in extracting the resonance parameters E_x, Γ and S from the experimental data. These uncertainties are the largest for S, while

the value of E_x is not so much affected. It is reassuring though
to find that the data obtained from very different techniques like
(e,e') and (α,α') scattering, often give very similar results, as
will be shown in the next sections.

3. The Giant Quadrupole Resonance in nuclei with A > 40

Selected data for the excitation energy E_x, width Γ and fraction of sumrule strength S, which characterize the GQR for nuclei
with A ≳ 40, are shown in figure 11. Representative errors have been
indicated. For this summary three groups of data have been used:
(1) Inelastic hadron scattering data in which the total resonance
bump was decomposed into a GQR and a (GMR+GDR) contribution |29,
37,38,91|, assuming a symmetric shape for the two resonances. These
data are shown as full dots. (2) Inelastic hadron scattering data
for which it might be reasonably assumed that the effect of the
(GDR+GMR)-excitation is small, as for instance the data of refer-
ence |39|, - open dots. (3) Inelastic electron scattering data from
various groups |11,20,36,40,41|, given by the open squares.

The data from the three groups overlap to a large extent
- there is perhaps a tendency for the electron scattering data to
give a somewhat lower value for E_x. It is clear that the main conclusions on the systematics, as given in the 1976 review-paper by
Bertrand |14|, describe the new data also quite well:
$E_x \sim 63 A^{-1/3}$ MeV, Γ decrease from around 6 MeV at A=40 to 3 MeV
at A=208 and S is about (50-100)% of the EWSR.

However, the systematics for E_x seem to favor a somewhat
higher value than $63 A^{-1/3}$ MeV. In fact, the line for $E_x = 64.7 A^{-1/3}$
MeV, which results from a recent macroscopic calculation with no
free parameters |42| gives quite a good description of the avail-
able data for A ≥ 90. For A ≤ 90, E_x decreases slightly, a tendency
which is continued in the s-d shell.

The data for Γ, as shown in figure 11, also show some systematic behavior, at least for the spherical nuclei around A ∼ 90, 120,
142 and 208. The data indicate a smooth $\Gamma = cA^{-n}$ dependence with
$1/3 \lesssim n \lesssim 2/3$. A $\Gamma \sim A^{-2/3}$ dependence is predicted in a macroscopic
model |42| in which the damping is assumed to be due to two-body
viscosity. The value for the damping coefficient derived from fission, a large scale oscillation mode, would give rise to a
$\Gamma = (58\pm20)A^{-2/3}$ MeV dependence, which is somewhat below the "best"
value $\Gamma \sim 85 A^{-2/3}$ MeV. A $\Gamma \sim A^{-1/3}$ dependence would result from a
one-body dissipation mode |42|.

Significant deviations from this smooth behavior occur in the
deformed region around A=150 and around A=100 |39|. The "low" point
at A=40 is from ^{40}Ca and seems to fall far outside the systematic
trend. However, the 2^+ strength in this nucleus (and also in other
fp shell nuclei) is divided among two bumps, one around $E_x \sim 60 A^{-1/3}$
MeV and one at lower excitation energy, so that defining a value
of Γ in this case becomes somewhat ambiguous.

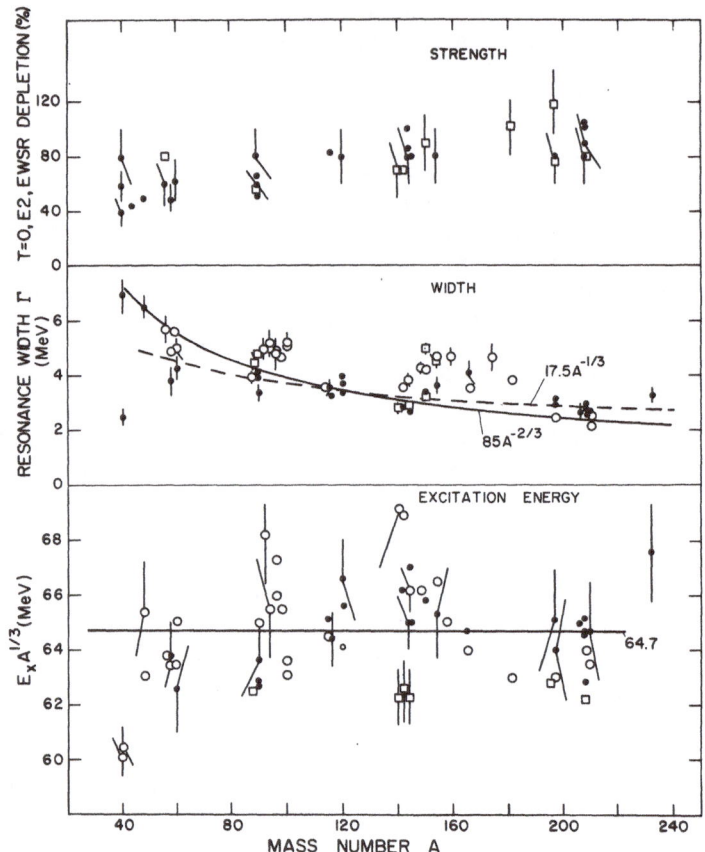

Fig. 11. Systematics for the excitation energy E_x,
        width Γ and fraction S of the T=0, E-2
        energy weighted sumrule strength of the GQR
        as observed in nuclei with A>40. The different
        symbols refer to different data sets - see
        text.

    Notwithstanding the reservations made above with respect to
the strength determination, the results of different experiments
seem to agree pretty well within the (large) uncertainties indicated. No systematic difference between (e,e') and inelastic hadron
scattering results is observed. For A>90 (80-100)% of the EWSR is
exhausted, which means that, taking into account the ~10% exhausted in the low-lying levels, the GQR exhausts completely the
available strength. For A<90, there seems to be a tendency for S
to decrease slightly, to about 60% of the EWSR for ^{40}Ca, a trend
which again continues in the s-d shell. On the other hand, the
amount of 2^+ strength in low-lying levels increases, so that for
^{40}Ca still nearly the full 100% of the EWSR has been found |35,43|.

4. The GQR in s-d shell nuclei

The situation with respect to a GQR in s-d shell nuclei was confusing for many years – see |14| and references therein. Radiative capture experiments, like the one of reference |8|, indicated that the E-2 strength is very fragmented |44|, an observation which was supported by the 96 MeV (α,α')-data of the Texas A&M group |45|, where only weakly excited structures in the giant resonance region were observed. However, data taken at higher bombarding energies and with a resolution of 200 to 400 keV |47,48|, showed a clear resonance-like bump with some fine structure superposed. But in the 120 MeV (α,α')-data taken with a resolution of about 120 keV |43| the whole structure was found to be fragmented as is illustrated in figure 12: the broad structures around $E_x \sim 18$ MeV and 14 MeV in ^{40}Ca are dissolved into many probably partially over-lapping fragments in the lighter nuclei. In fact, recent higher resolution experiments on ^{40}Ca |35,38| show that also for ^{40}Ca a lot of fine structure is superimposed on the broad bumps. This suggests that also for these nuclei one might consider the GQR as being composed of many partially overlapping fragments. Most of these fragments, but not all, have an L=2 multipolarity |35|.

The various (α,α')-data can be reconciled with each other by realizing that at higher bombarding energies the 2^+ strength will be relatively more strongly excited than the continuum |48| and that the strongly fragmented structure as observed in the 120 MeV (α,α')-data will be smeared out because of resolution.

The T=0, E-2 strength as observed in the nuclei $16 \leq A \leq 40$ is shown in figure 13: the strength for $E_x \leq 12$ MeV (full squares) and $E_x \geq 12$ MeV (full dots) are shown separately. In general there is fair agreement between the different experiments. The only exception is ^{20}Ne where an (e,e')-experiment located about 90% (open square) |49| but an (α,α')-experiment only ~55% of the EWSR. For some nuclei like ^{16}O, ^{20}Ne, ^{24}Mg and ^{40}Ca, (70-100)% of the T=0, E-2 EWSR is observed, while for instance for ^{28}Si and probably also ^{32}S only a total of (40-50%) is observed. In these nuclei a large part of the E-2 strength must be located at high excitation energies and is probably very fragmented. This is in qualitative agreement with recent theoretical calculations |50,51|.

If one calculates from the data shown in figure 12 the centroid $\bar{E}_x$ of the E-2 strength with $E_x \geq 14$ MeV one finds for ^{24,26}Mg, ^{28}Si and ^{40}Ca the values $E_x = 16.9$, 18.9, 18.3 and 17.0 MeV, respectively. Only for ^{40}Ca is this value in line with the systematics observed for $A \geq 40$ nuclei. However, if the missing E-2 strength of about 30%, 50% and 60% in ^{24,26}Mg and ^{28}Si, presumably located at higher excitation energies is taken into account, the E_x-values would also be pushed up.

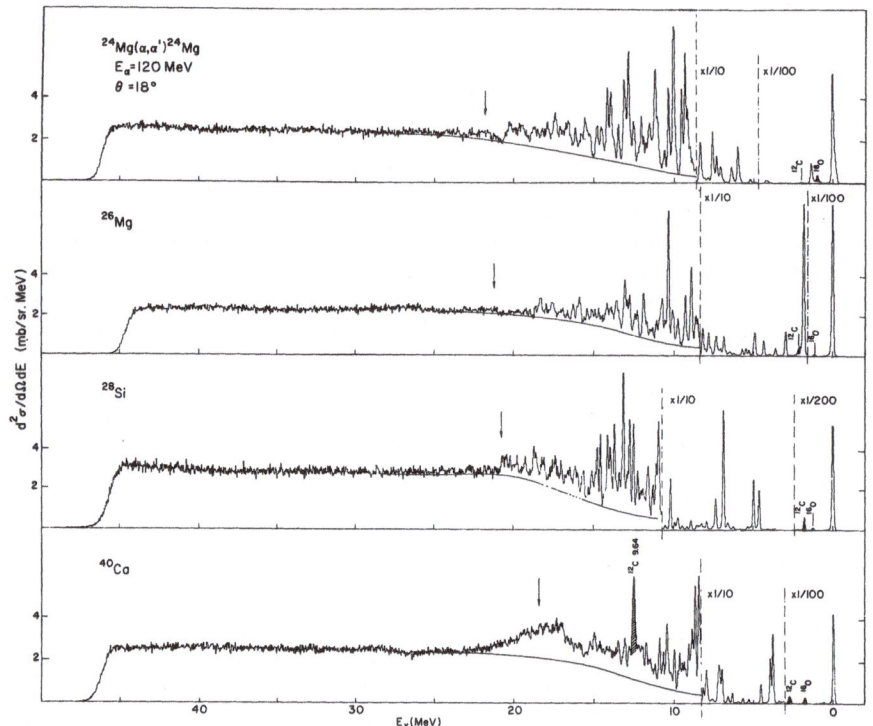

Fig. 12. (α,α')-spectrum at $E_\alpha = 120$ MeV with a resolution
         of about 120 keV for different s-d shell nuclei.
         The arrows indicate the value of $E_x = 63\ A^{-1/3}$ MeV.

A qualitative explanation for this missing strength in light
nuclei can be given using figure 14, where for ^{16}O the levels involved in a $2\hbar\omega$, (1p-1h) excitation are shown. The main components
are the (f 5/2 - p 3/2,1/2) ones. From (d,p)-experiments on ^{16}O it
is known that the f 5/2-level, which is located at high excitation
energies and thus particle-unstable, is in itself already very
broad and not well defined. Thus it is no surprise that also the
(1p-1h)-component including this level will be broad and thus hard
to detect in (α,α')-scattering experiments. Recently a number of
shell model calculations including the $2\hbar\omega$ space |52| and even
some special $4\hbar\omega$ configurations |53| have been performed, which
quantitatively explain the increase in spreading and fragmentation
going from ^{40}Ca to ^{16}O.

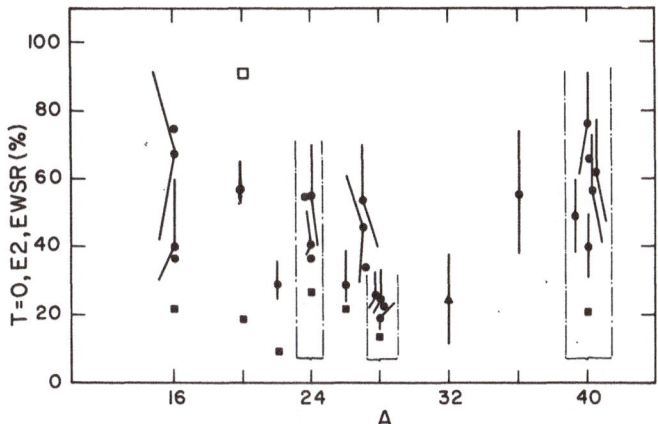

Fig. 13. T=0, E-2 strength expressed as a fraction of the
          EWSR observed in nuclei with $16 \leq A \leq 40$. The full
          dots are for $E_x > 12$ MeV, the full squares for
          $E_x < 12$ MeV, the open square refers to an (e,e')-
          experiment.

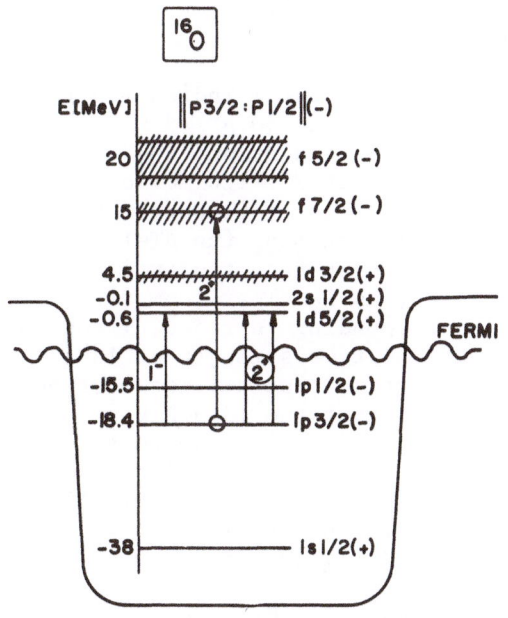

Fig. 14.

It is interesting that the behavior of the GQR in nuclei with
$16 < A < 40$ is quite similar to what is found for the GDR $|5,6|$.
Here also no compact resonance is observed, the centroid energy
is lower than is observed for $A > 40$ nuclei and only part of the
total strength has been located. Finally, it should be noted that
in still lighter nuclei like ^{12}C only a small amount of the GQR-
strength has been located, less than 16% $|54|$, which probably
signifies the disappearance of this collective phenomenon in very
light nuclei $|55|$.

5. The giant monopole resonance

The first suggestion for the existence of a $\Delta T=0$, 0^+ giant
resonance was made in 1975 by Marty et al. $|93|$. They compared
their (d,d')-spectra on ^{40}Ca, ^{90}Zr and ^{208}Pb taken at $E_d = 80$ MeV
with previously published (α,α')-data from the Texas A&M group at
$E_\alpha = 96$ MeV in which, because of the relatively low bombarding
energy, the monopole would be only weakly excited. From such a
comparison they concluded that in their (d,d')-spectra there was
some additional cross section at about $80\ A^{-1/3}$ MeV, which they
attributed to $\Delta T=0$, E-0 excitation. Although their conclusions
were met with much scepticism, it stimulated many groups throughout the world to search for possible signs of monopole excitation.
Evidence for a giant monopole resonance in ^{90}Zr $|40|$ and ^{208}Pb
$|36|$ was also reported in 1976-1977 by the Sendai-group from an
analysis of their (e,e')-data.

The first clear experimental indication that there is isoscalar strength located around $80\ A^{-1/3}$ MeV with a multipolarity
different from the GQR located around $65\ A^{-1/3}$ MeV came in 1977
from an (α,α')-experiment at $E_\alpha = 120$ MeV on a number of nuclei in
the lead-region $|29,57|$. This is illustrated in figure 15 for
^{208}Pb. Since in (α,α')-scattering the $\Delta T=1$ is only weakly excited
$|28|$, there is no doubt that the structure around $80\ A^{-1/3}$ MeV
is isoscalar. Moreover, as is clear from figure 15 the structure
around 10.9 MeV (corresponding to the GQR) and 13.9 MeV are excited with different relative strength at 12° and 14°, which can
only mean that different multipolarities must be involved. Although it was suggested that the 13.9 MeV structure might be due
to giant monopole excitation, $\Delta L=2$ and to a lesser extent also
$\Delta L=4$ transfer could not be excluded, since the $\Delta L=0$ and the $\Delta L=2$
angular distributions do not differ drastically for scattering
angles larger than $\sim 10^O$. Thus a firm proof for the existence of
the giant monopole was still lacking.

As is shown in figure 5 a drastically different behavior of
the $\Delta L=0$ versus $\Delta L=2$ angular distributions occurs at very forward
angles, where the $\Delta L=0$ distribution has a sharp minimum. Figure 16
shows the result of such a measurement, performed in 1977 at Texas

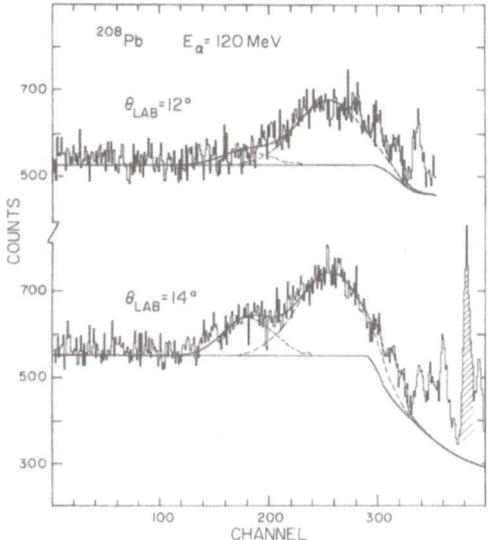

Fig. 15.

(α,α')-spectra at $E_\alpha = 120$ MeV for ^{208}Pb. The relative contribution of the strengths located at 80 $A^{-1/3}$ MeV and 65 $A^{-1/3}$ MeV is angle dependent, clearly indicating that more than one multipolarity is present.

A&M, with 96 MeV alpha-particles on ^{208}Pb and ^{144}Sm, in which it was shown for the first time that the bump located at about 80 $A^{-1/3}$ MeV is "the" giant monopole resonance indeed |58|. Since then many more small-angle scattering data have been obtained notably at Texas A&M |58,59,60,61| and Grenoble |37,38|, but more recently also at Orsay |30| and at Osaka |62|. Some typical results of the Grenoble-group, where they have concentrated on small-angle ^{3}He-scattering at $E_{3_{He}} = 108$ MeV, are shown in figure 17 |38|. These data clearly show, similar to the (α,α')-data discussed above, the minimum around $\theta \sim 5°$ characteristic for monopole excitation.

Having thus established that an isoscalar giant monopole resonance does exist, we can now summarize the available data on its characteristics – excitation energy E, width Γ and strength S. This has been done in figure 18 for E_x.
For this summary we not only used the small angle scattering data mentioned above, but also the data taken at larger scattering angles which often can be quite well analysed in terms of two different, symmetric Gaussian-shaped structures. Especially (α,α')-data taken at $E_\alpha > 150$ MeV turn out to be very useful – the ratio bump/continuum is relatively large and the underlying continuum is well defined as illustrated in figure 8 |27|.

As indicated in figure 18 giant-resonance-like $\Delta T=0$, 0^+ strength has been located in many nuclei with $A \geq 58$. The strength decreases gradually from about 100% of the EWSR in nuclei

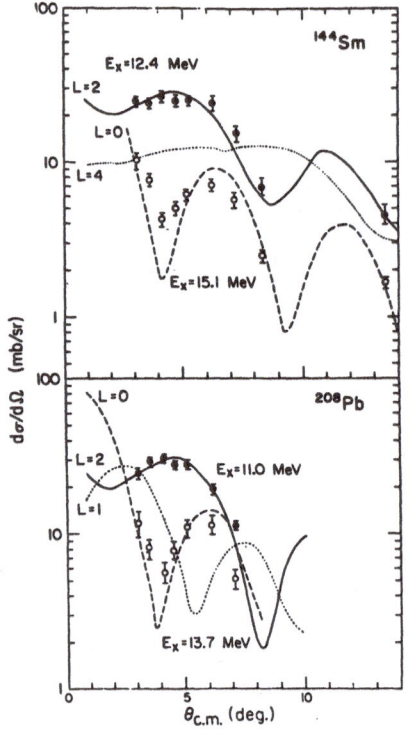

Fig. 16.

The angular distributions obtained for the $E_x = 65$ $A^{-1/3}$ MeV bump (full dots) and $E_x \sim 80$ $A^{-1/3}$ MeV bump (open circles). These data for the first time unambiguously identified the 80 $A^{-1/3}$ MeV bump as the isoscalar giant monopole resonance |58|.

with $A \geq 120$ to about 40% in ^{58}Ni. No definite evidence for an appreciable amount of monopole strength in nuclei with $A \leq 58$ has yet been found although many experiments were designed to look for it, especially in ^{40}Ca |30,60,62|.

The data for E_x as shown in figure 18, indicate that for $A \geq 120$ the excitation energy $E_x \sim 80$ $A^{-1/3}$ MeV, while for $A \leq 120$ there is a definite tendency towards lower values. This could be an experimental problem: since the monopole is relative to the quadrupole weaker excited in lighter nuclei, it will be harder to differentiate it from the GQR and from the underlying conti-nuum. This may result in an apparent "shift" towards lower ex-citation energies. As shown for instance by Blaizot et al. |63| the excitation energy of the GMR is given by:

$$E_x = \frac{1}{\eta_o} \sqrt{\frac{\hbar^2}{m} K_A}$$

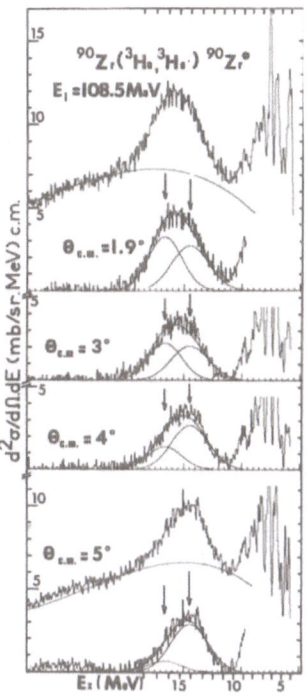

Figure 17.   Small-angle (^{3}He,^{3}He') inelastic scattering data at
             E_x(^{3}He) =108 MeV for ^{90}Zr. The E_x 16 MeV bump has a
             minimum around θ=5°, which identifies it as a ΔL=0
             transition | 38 |.

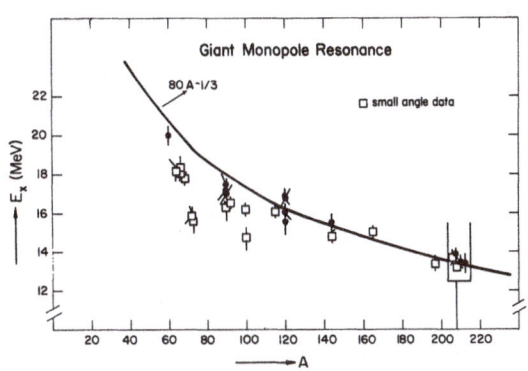

Figure 18.   The excitation energy E_x of the T=0 GMR as a function of A.

where η_o is the root-mean-square radius: $\eta_o = <r^2>^{1/2}_{g.s.} \cdot \alpha A^{-1/3}$ and K_A the compression modulus of the nucleus A:

$$K_A = \eta^2 \left. \frac{\delta^2 \ (E/A)}{\delta\eta^2} \right|_{\eta=\eta_o}$$

The quantity K_A can be related to K_{nm}, the compression modulus of nuclear matter by means of the semi-empirical mass formula |63|:

$$K_A = K_{nm} + K_{surf} \ A^{-1/3} + K_{symm} \ (\frac{N-Z}{A})^2 + K_{coul}$$

The coefficients K_{nm}, K_{surf}, K_{symm} and K_{coul} can be viewed as the second derivatives with respect to η of the corresponding co-efficients in the mass equation. Clearly the quantity K_A is A-dependent and thus one might expect that the dependence of E_x on A is more complicated than a simple $A^{-1/3}$-relation. The small-angle scattering data as shown in figure 18 have been used to determine the values K_{nm}, K_{surf} and K_{symm} from a least squares fit |61|. The results are shown in table 4 together with the values obtained from a microscopic calculation within the Random Phase Approxima-tion using different types of interaction |63|. The theoretical and experimental values for K_{nm} are for the Brink and Boeker (B1) and for the Gogny interaction D1 in reasonable agreement.

Table 4. Coefficients of the incompressibility (MeV).

	I[a]	II[b]	B1[c]	D1[c]	Ska[c]
K_{nm}	222± 32	274± 40	190	228	263
K_{surf}	−403±130	−651±171	−300	−315	−394
K_{symm}	−224±265	−378±274	−500	−500	−610

a Calculation included experimental K_A values for ^{64}Zn, ^{66}Zn, ^{90}Zr, ^{115}In, ^{116}Sn, ^{118}Sn, ^{120}Sn, ^{124}Sn, ^{144}Sm, ^{148}Sm, ^{197}Au, and ^{208}Pb.
b Calculation without experimental values of ^{64}Zn and ^{66}Zn.
c Theoretical values using the indicated interactions |63|.

In judging these results one should realise that because of the weak A-dependence of $(K_{nm} - K_A)$ the results of the least squares fit are weighted heavily by the E_x-values obtained for the light-est nuclei ^{64} ^{66}Zn shown in figure 18. In these nuclei though only (30±15)% of the total $\Delta T=0$, 0^+, EWSR has been found in the struc-tures associated with the monopole resonance |61|. It is quite likely that in these nuclei more monopole strength is present but too spread out to be detectable in single scattering experiments. Thus the center of gravity of the monopole strength might well be located at a different and presumably higher excitation energy than the "resonance" energy plotted in figure 18 and that would result in quite different values for K_{nm}, K_{surf} and K_{symm}.

   As remarked before, for nuclei with A$\gtrsim$120 nearly 100% of the $\Delta T=0$, 0^+, EWSR has been found but for smaller A the observed strength gradually decreases and only little or no E-0 strength has been found for nuclei with A<50. These values are obtained by fitting the data with a DWBA-calculation using a standard deformed Woods-Saxon potential model |64|.

$$U_{tr}(r) = \alpha[\,3U(r)+\frac{rdU(r)}{dr}]\quad\text{for }\Delta L=0$$

The deformation parameter α which would exhaust the total monopole sumrule S_o is given by |27|:

$$E\alpha^2 = \hbar^2/(2m<r^2>_{g.s.} A)$$

It should be pointed out that as yet there has been no experiment in which it was possible to check in an independent way the procedure outlined above for determining the strength S of the monopole resonances. Thus it cannot be excluded that in reality the values for S should be very different.

   In principle a more fundamental approach to calculate the transition potential $U_{tr}(r)$ is to use a folding model in which the transition density $\rho_{tr}(r)$ is folded with the nucleon projectile interaction or with a nucleon-nucleon interaction averaged over the projectile nucleons (|27|,|65| and references therein). The effective nucleon-projectile interactions which can be taken complex, can be determined from a fit to the elastic scattering data |27| or to low-lying excited states |65|. For the transition density $\rho_{tr}(r)$ one can use the Tassie-density, which for the monopole resonance is |36|.

$$\rho_{tr}(r)=\alpha_o [\,3\rho(r)+rd\rho(r)/dr]$$

Also for some nuclei like ^{208}Pb microscopic densities calculated in the Randomn Phase Approximation are available. This density is very close to the Tassie-density especially in the important surface region which is for ^{208}Pb (10±2) fm. Recently it has been shown that whether one uses a Woods-Saxon deformed potential or a folded potential makes a large difference for the giant monopole. This is demonstrated in table 5 which is obtained from an analysis of 152 MeV (α,α')-data |27|.

Table 5. EWSR depletion for giant resonances (from 27).

Target	E_x (MeV)	% EWSR Folded Potential	% EWSR Deformed Potential
		Isoscalar Quadrupole	
^{46}Ti	17.6	70 ± 14	50 ± 10
^{58}Ni	16.0	43 ± 10	45 ± 10
^{90}Zr	14.1	75 ± 15	65 ± 13
^{120}Sn	13.3	70 ± 15	70 ± 15
^{208}Pb	10.9	73 ± 15	77 ± 15
		Isoscalar Monopole	
^{46}Ti	not observed		
^{58}Ni	20.0	120 ± 20	40 ± 10
^{90}Zr	17.0	180 ± 40	80 ± 20
^{120}Sn	16.9	150 ± 20	100 ± 20
^{208}Pb	13.9	140 ± 30	100 ± 20

6. Isoscalar multipole strength with L > 2.

6.1. Isoscalar octupole strength. As discussed before, two kinds
of octupole resonances are predicted, the $1\hbar\omega$ (LEOR) at $E_x \sim 25$ A$^{-1/3}$
MeV, exhausting about 30%, and the $3\hbar\omega$ (HEOR) at $E_x \sim 110$A$^{-1/3}$ MeV,
exhausting about 70% of the T=0, E-3, EWSR. Evidence for a HEOR
comes for instance from (e,e')-data, but the analysis of such data
is quite difficult. The continuum underlying the 3⁻strength is
quite large, the 3⁻ strength is much more spread out than the 2⁺-
strength and the T=0, E-3 strength is bracketed between the T=1,
GDR and T=0 GMR at around 80 A$^{-1/3}$MeV and possibly the T=1, E-2
strength at $E_x \sim 130$ A$^{-1/3}$ MeV. As an example, a model dependent
analysis of the total structure observed in electron scattering
above the continuum in ^{208}Pb has been made by Sasao et al. |36|.
They find E-3 strength at $E_x \sim 17$ MeV, with a width $\Gamma \sim 6$ MeV and exhausting 50% to 180% of the EWSR, depending on whether the Steinwedel-Jensen or Goldhaber-Teller model was used for calculating
the contributions from the T=1 resonances. Similar results are
obtained by Pitthan |11|, for instance for ^{140}Ce |66|. In the
(e,e')-experiments the T=0 nature is inferred from the approximate
correspondence with the calculated value of the excitation energy.
Also some recent inelastic hadron scattering experiments can be
analysed in a way which is not inconsistent with the presence of
a HEOR. Yamagata et al. |62| from ^{3}He scattering on ^{118,120}Sn at
$E_{^3He} = 120$ MeV, claim evidence for L=3 strength at $E_x \sim 23$ MeV with a
width Γ of 7 MeV, but this result seems to depend strongly on the

way the continuum is subtracted. It should be noted that the (α,α')-spectrum of figure 8 shows a similar bump at roughly the same excitation energy and with the same width. Similar remarks can be made with respect to [208]Pb where a recent analysis of 172 MeV α-scattering data again indicated some evidence for structures around $E_x = 17.5\pm0.8$ MeV and 21.3 ± 0.8 MeV, respectively. The angular distribution of the 17.5 MeV structure is consistent with L=3 excitation |67|. Also in the (α,α')-spectrum for [208]Pb at $E_\alpha = 150$ MeV |27| there is some indication for a bump around $E_x \sim 18$ MeV which could indicate the presence of L=3 strength. Summarizing, one might say that although there is evidence for T=0, E-3 strength around $E_x \sim 110\ A^{-1/3}$, it is not yet very well established, mainly because of the uncertainties in continuum subtraction.

For the $1\hbar\omega$ 3⁻ strength (LEOR) the situation is different though. Moss et al. |68| have shown that it is systematically present for many nuclei with 66 < A < 200. In these nuclei it is seen as a 1 to 2 MeV broad bump, located at about $30\ A^{-1/3}$ MeV and exhausting from 7 to 23% of the EWSR. In addition detailed studies of [208]Pb |29| and a number of s-d shell nuclei |35,43| have shown that for these nuclei also about 15% of the 3⁻ strength is present around and below $30\ A^{-1/3}$ MeV, but in these nuclei the strength is located in well separated states or clusters of states. Thus the occurrence of $1\hbar\omega$ E-3 strength with roughly the predicted strength is a general phenomenon for nuclei with 16 < A < 208, but whether it is concentrated in one bump or fragmented depends on the specific nuclear properties.

6.2. <u>Isoscalar hexadecapole and higher multipole strength</u>. No evidence has been found for the $4\hbar\omega$, E-4 strength, predicted to occur around $E_x \sim 150\ A^{-1/3}$ MeV. With respect to the $2\hbar\omega$, 4⁺ strength, which is predicted to coincide with the T=0, E-2 strength and/or E-0 strength, some evidence for it has been reported for nuclei in the [208]Pb-region. In the analysis of the 120 MeV (α,α')-data on [206,208]Pb and [209]Bi the best fit to the data was obtained |29,57| by including 20 to 40% of the EWSR 4⁺-strength, as shown in figure 10. Also, in recent analysis of the 150 MeV (p,p')-data on [208]Pb, it was found that a better fit to the data was obtained by including 20% E-4 strength |69|. A similar conclusion was reached in the analysis of 172 MeV α-scattering data on [208]Pb |65|, where it was also indicated that, if one assumes that in addition to L=2 strength centered at 10.9 MeV excitation there is also some L=3, 4 and/or 6-strength present, a better fit to the measured angular distribution is obtained.

It should be remarked though that all the evidence presented depends on only a few datapoints in the angular distribution, which are in the minima of the angular distribution. Thus their magnitude is very sensitive to the way the continuum is subtracted.

7. The isoscalar resonances in deformed nuclei

Perhaps the most spectacular property of the giant dipole resonance is that it splits in two components in statically deformed nuclei |5|. For the GQR, the effect of deformation is not as spectacular but it is there. The calculations of Suzuki and Rowe |13| show that in a deformed nucleus with a deformation parameter δ (which is approximately equal to the relative difference between the long and short axis of the ground state of an axially symmetric deformed nucleus) the energy of the K=0, 1 and 2 components of the GQR are given approximately by:
$E(K=0)=\sqrt{2}\hbar\omega(1-\delta/3)$, $E(K=1)=\sqrt{2}\hbar\omega(1-\delta/6)$ and $E(K=2)=\sqrt{2}\hbar\omega(1+\delta/3)$ respectively. So in a deformed nucleus with $\delta\sim0.3$ and $E_x\sim12$ MeV, like the Sm- or Nd-isotopes, the splitting is only 1 or 2 MeV, much smaller than for the GDR. It is too small to be directly observable as a splitting in different components, but it can manifest itself as a broadening of the resonance. For light nuclei like ^{20}Ne, where the bulk of the GQR-strength is located around $E_x\sim18$ MeV and where $\delta\sim0.5$ |21|, one expects a larger effect. The ^{20}Ne(α,α')-spectrum taken at $E_\alpha=155$ MeV shows a clear splitting in two components indeed.

As expected, in heavy nuclei the effects of deformation show up as a broadening of the giant resonance bump at the low energy side and/or as a shift of E_x towards lower excitation energies, much faster than predicted from systematics. This trend is already obvious in figure 11: the Γ-values around A=150 are about 1 MeV too high and the E_x-values around A~100 and A~150 drop much too fast with increasing A. Several experiments have been performed to study the effect directly by comparing the spectra for spherical nuclei (^{142}Nd and ^{144}Sm) and deformed ones (^{150}Nd and ^{154}Sm) |46,70,71,72|. Typical (α,α')-spectra for ^{142,150}Nd are shown in figure 19: the difference between the two spectra is a slight broadening in the ^{150}Nd-spectrum at the low excitation energy side |72|. The same effect is observed for ^{144,150,154}Sm as the background-subtracted spectra in figure 28 show |73|.

V. THE WIDTH AND DECAY-PROPERTIES OF GIANT RESONANCES

1. Introduction

Giant resonances are located at excitation energies well above the particle emission threshold so that they mainly decay by particle emission. Macroscopically the decay properties of giant resonances are reflected in the width Γ, as determined from an analysis of single scattering data. More information can be obtained though from studying the decay properties of these resonances, that is from studying the energy spectrum and angular distribution of the emitted particles. Such experiments require measuring the emitted particle and the inelastically scattered particle in

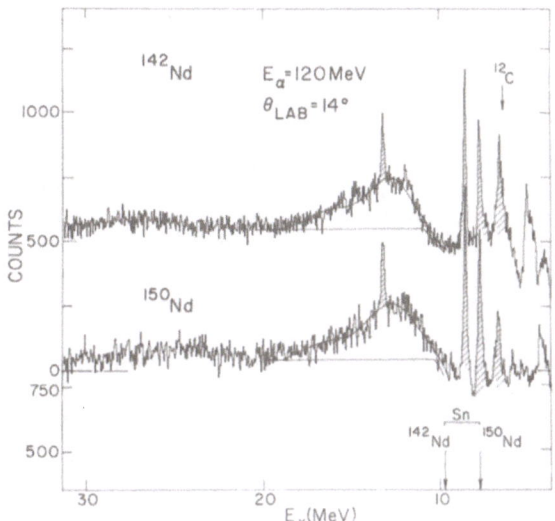

Fig. 19. ^{142,150}Nd(α,α')-spectra at E_α=120 MeV.
The hatched peaks are contaminants.
The neutron binding energies are indicated
by S_n. Note the broadening on the low-
energy side of the GQR in ^{150}Nd with
respect to ^{142}Nd.

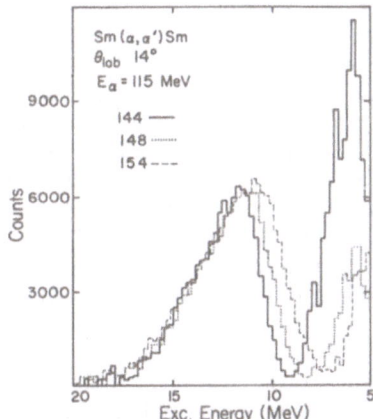

Fig. 20. Continuum subtracted spectra for
^{144,150,154}Sm at E_x=115 MeV, clearly
demonstrating the broadening and shift
to lower energies of the resonance
bump to deformation |73|.

coincidence. They are more difficult and time consuming than
single-scattering experiments but they are worthwhile for two
reasons. First of all they can give insight into the microscopic
properties of the resonances. For instance the answer to questions
like "to what extent is the (1p-1h) collective state mixed with
more complicated states" or "are there decay channels which are
specially favoured or suppressed", are fundamental for our understanding of the microscopic structure of giant resonances. But
there is also a practical reason for studying the decay properties.
As has been pointed out many times above, the main experimental
problem in giant resonance research is the high continuum underlying the resonances. If it would be possible to find a decay
mode in which the resonance is favoured over the continuum, the
"signal/noise"-ratio would increase and more information could be
obtained. As will be shown below, this is the case indeed with
respect to the alpha-decay channel in some light nuclei like ^{16}O
and ^{24}Mg.

2. The width of giant resonances

In general the total width Γ of a giant resonance can be
written as the sum of three different contributions:

$$\Gamma = \Gamma\uparrow + \Gamma\downarrow + \Delta\Gamma .$$

Here $\Delta\Gamma$ is due to the fact that the intrinsic collective (1p-1h)-
state, which is the doorway state for the giant resonance, can be
split into several components in specific nuclei. A well-known
example is the splitting in deformed nuclei as described above
in section IV 7. The component $\Gamma\uparrow$ is due to the coupling of the
(1p-1h)-state to the continuum, while the $\Gamma\downarrow$ component is due to
the fact that the simple (1p-1h) doorway state mixes with many
more complicated states of the nucleus. This effect can be calculated macroscopically by introducing a nuclear viscosity through
which the collective nuclear vibration is damped. Such an approach
has been recently applied by Nix et al. |42| for the isoscalar
resonances. They find that assuming a 2-body viscosity the width
of resonances with $\Delta L>2$ is given by $\Gamma_L^{\downarrow}=c(L) A^{-2/3}$. Here c(L)
depends on the multipolarity of the resonance and can be calculated
from the information on the energy dissipation in the fission
process (see section IV 3).

Microscopically $\Gamma\downarrow$ can be calculated from taking into account
the coupling of the (1p-1h)-state to the much more numerous
(2p-2h)-states which are present in the vicinity of the resonance:

$$\Gamma\downarrow = 2\pi < \overline{2p-2h|V|1p-1h} >^2 \rho(E_x)$$

where V is the coupling matrix element between the relevant
(1p-1h)-state and the (2p-2h)-states and $\rho(E_x)$ is the level
density of the (2p-2h)-states at E_x. One assumption often made is
that the (2p-2h)-states couple again to (3p-3h) ....(np-nh)-states

so that the excitation energy is statistically equilibrated. If the spreading mechanism would be the main effect, the decay characteristics of a resonance would be that of a compound nucleus of the same excitation energy i.e. the energy spectrum of the emitted particles resembles that of an evaporation spectrum. If on the other hand the direct process ($\Gamma\uparrow$) would be dominant, the particle decay will mainly occur to well defined hole-states in the residual nucleus resulting in a very different particle spectrum. Thus by a measurement of the particle decay spectrum one can decide on the relative magnitude of $\Gamma\uparrow$ and $\Gamma\downarrow$.

A few measurements indicate that for the GDR in heavy nuclei the spreading width $\Gamma\downarrow$ is dominating, while for lighter nuclei like ^{16}O the coupling to the continuum ($\Gamma\uparrow$) is the dominating decay process (see for instance |6|). For the heavy nuclei this conclusion is based on a direct measurement of the neutron-spectrum in the $^{209}Bi(\gamma,n)$-reaction which indicates that about 90% of the emitted neutrons are resulting from a statistical decay process while 10-15% could be due to the direct coupling to the continuum. Also the fact that the fission probability of the GDR in actinide nuclei is about equal to the one measured for compound nuclei is in agreement with this picture |74,75|.

For the GQR and GMR in heavy nuclei no experimental information has yet been published. Theoretically it is predicted |76| that in heavy nuclei like ^{208}Pb also for the GQR the spreading width is the main component. Some very recent and preliminary data from Osaka, in which the neutron decay of the GQR excited by (α,α')-scattering at E_α=109 MeV on ^{119}Sn was measured, seems to indicate that the GQR does decay by fast neutrons (that is $E_n > 4$ MeV) and that $\Gamma_{(fast\ neutrons)}/\Gamma_{total} \simeq$ 30%. That would imply that $\Gamma\uparrow/\Gamma \sim 30\%$ which is not as small as mostly assumed |77|. The information on the fission probability is contradictory |78,79,10,80,81,82|. On the other hand for light nuclei like ^{16}O and ^{24}Mg it is known experimentally that a large non-statistical component is present.

In figure 21 the width of the GQR is plotted as a function of excitation energy. The data are obtained from a number of non-deformed nuclei so that $\Delta\Gamma$ may be assumed to be small. Since we also have reason to believe that $\Gamma\uparrow$ is small, as discussed above, the main contribution to the experimentally observed Γ comes from $\Gamma\downarrow$, the spreading width. The curve which is seen to give a reasonable fit to the data is obtained from a least-square fit to similar data for the giant dipole resonance GDR |6|. Thus from this comparison it is suggested that the spreading mechanism for the GDR and the GQR are rather similar, at least within the accuracy of the available data. This does not seem to be true for the isoscalar GMR as also shown in figure 21; the width of the

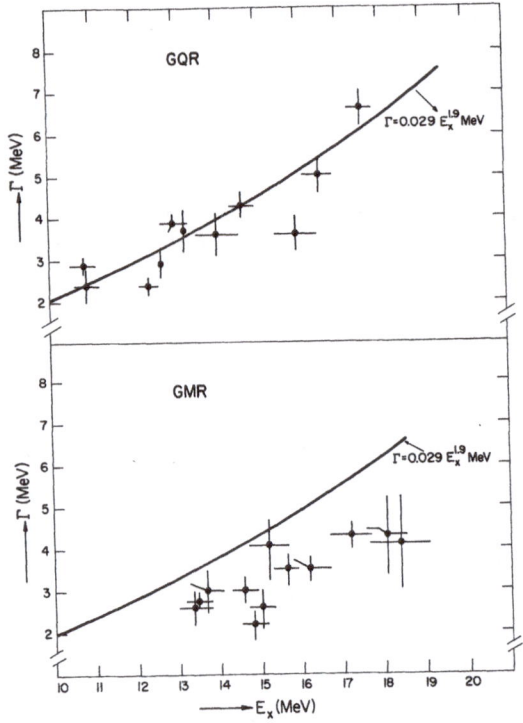

Fig. 21.

The width of the $\Delta T=0$ GQR and GMR as a function of excitation energy for specially selected nuclei (see text).

GMR is about 1 to 1.5 MeV smaller than the one of the GQR or the GDR at the same excitation energy. This effect is in qualitative agreement with a recent microscopic calculation for ^{208}Pb in which an even stronger suppression of the GMR spreading width with respect to the GQR one was obtained |76|.

3. The decay of the isoscalar giant quadrupole in light nuclei

In the last few years the decay of the fragmented giant quadrupole in the nuclei ^{16}O |32|, ^{20}Ne |83|, ^{24}Mg |84,85|, ^{28}Si |86,85| and ^{40}Ca |87,88| has been extensively studied. For these nuclei the charged particle and especially the alpha-decay branches are large if not dominant and this makes such experiments attractive from an experimental point of view.

Figure 22 illustrates schematically the different decay channels for ^{24}Mg. As indicated, the neutron decay treshold is much higher than the charged particle decay treshold, even if one takes for the latter the effective treshold in which the Coulomb barrier is taken into account |83|. This means that even if the decay would be purely statistical, charged particle decay would be favoured.

Especially interesting is the α_0-decay branch, that is the α-decay branch to the ground state of the residual nucleus. The angular distribution of these α_0-particles with respect to the symmetry axis, from a resonance with multipolarity L which is excited by inelastic scattering at a forward angle where the differential cross section is at a maximum, is simply given by $W_L(\theta) \alpha \hat{L} P_L^2(\cos \theta)$, with $\hat{L} = (2L+1)$. Here θ is the angle with respect to the symmetry axis. In a Plane Wave Born Approximation (PWBA) the symmetry axis is the axis along which the excited nucleus will recoil as a result from the excitation process. It turns out from Distorted Wave calculations (DWBA) $|81|$ and from experiments $|81,32|$ that this is a good approximation. In case of overlapping resonances with different multipolarities L and L' the decay pattern is more complicated:

$$W_{LL'}(\theta) \alpha |A|^2 \hat{L} P_L^2(\cos\theta) + |B^2| \hat{L}' P_{L'}^2(\cos\theta) + 2|A||B|\sqrt{\hat{L}\hat{L}'} \, P_L(\cos\theta)$$

$$P_{L'}(\cos\theta) \cos \delta,$$

where $|A|^2 + |B|^2 = 1$ and δ a phase factor. Because of the interference term the decay pattern of a resonance can be drastically changed. For instance a large forward ($\theta \sim 0^\circ$) – backward ($\theta_\pi = 180^\circ$) asymmetry can be introduced in a positive parity decay pattern of a resonance J $= L^+$ by a small mixture ($B^2/A^2 < 0.1$) of negative parities, $J^\pi = L'^-$ $|32,83|$.

As indicated in figure 22 the α_0 and p_0 decay channels can be considered as the inverse of the (α,γ) and (p,γ) capture reactions respectively. The (p,γ) reactions will take place mainly through the dipole channels, but in (α,γ) capture reactions, especially on T = 0 target nuclei like ^{20}Ne or ^{24}Mg, the dipole channel is suppressed because of isospin conservation, so that dipole and quadrupole excitation are of the same order of magnitude $|8,7|$.

The main results of the decay work on ^{16}O, ^{20}Ne, ^{24}Mg and ^{28}Si can be summarised as follows.
1. The ratio α-decay strength/proton-decay strength is about 5 for ^{16}O and ^{20}Ne, decreases from about 3 in ^{24}Mg to about 1 in ^{28}Si, and is about 0.3 for ^{40}Ca and still smaller for ^{58}Ni $|83|$. This trend can be roughly explained by considering the effective particle-decay thresholds.
2. The (α,α')-spectrum taken in coincidence with the α_0-decay particles (the $(\alpha,\alpha', \overline{\alpha}_0)$-spectra) show a strong relative surpression of the continuum. This is illustrated in figure 23 for ^{16}O $|32|$ and in figure 24 for ^{24}Mg $|85|$. Both sets of data clearly show the large improvement in "signal/noise" ratio that can be obtained by these techniques for these specific nuclei.
3. The α-decay branch for ^{16}O, ^{20}Ne and ^{24}Mg has a large, non-statistical component. This was already clear for ^{24}Mg from

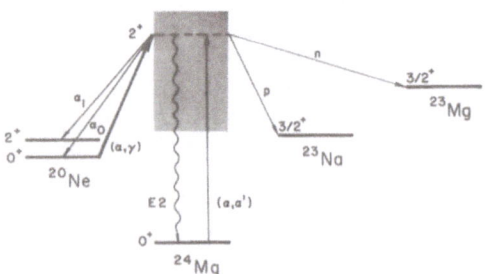

Fig. 22.

Schematic representation of
the excitation of the GQR
in ^{24}Mg through (α,α') and
radiative α-capture reactions. Some particle and
gamma-decay branches are
indicated.

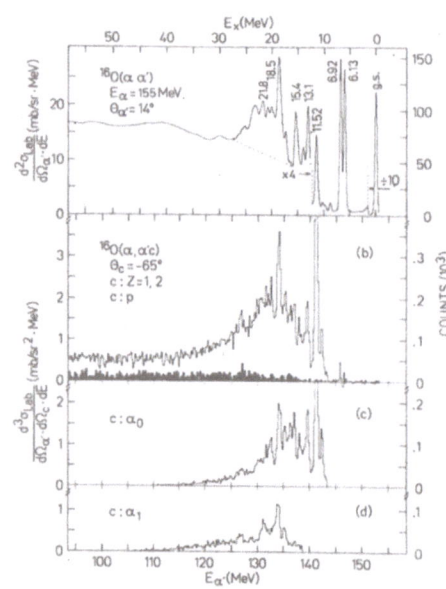

Fig. 23.

^{16}O(α,α')-spectra at E_α=155 MeV.
The upper spectrum is a single
spectrum, the lower ones are
taken in coincidence with all
charged decay particles (b),
with only α_0 (c) and only α_1 (d)
respectively. Note the strong
suppression of the continuum
especially in (c) and (d) |83|.

the E-2 strength observed in the (α,γ) radiative capture experiment
| 8, 35, 31 |. For ^{16}O the effect is illustrated in figure 25, which
is taken from | 83 |. As shown the α-decay brances are about twice
as strong as expected from such a simple statistical decay. This
leads to the at first sight somewhat surprising conclusion that
there must exist a large overlap between the basic (1p-1h)
collective giant resonance structure and the (α-cluster   residual
state) structure. It has been showed though | 89 | that in SU(3) such
effects are to be expected and in fact a detailed calculation for
^{16}O and ^{40}Ca has been able to explain the observed ratios quite
well | 53 |.

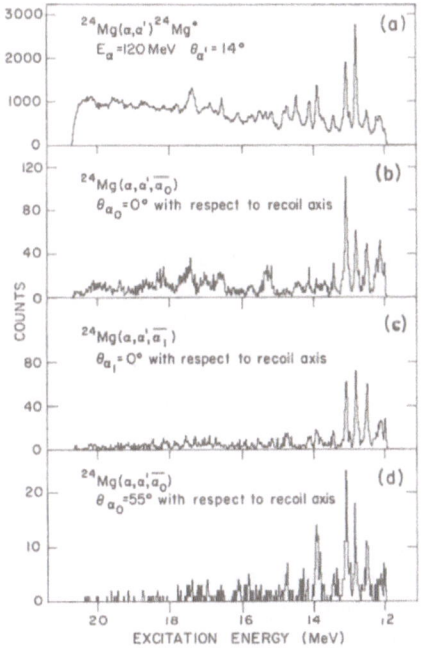

Fig. 24.

The ^{24}Mg(α,α')-spectra at $E_\alpha=$ 120 MeV taken with an energy resolution $\Delta E \sim 70$ MeV |85|. The upper part (a) shows a singles spectrum. Figures b, c and d are (α,α')-spectra taken in coincidence with the α_o-decay particles along the symmetry axis, the α_1-decay particles along the symmetry axis and with the α_o-decay particles at an angle of 55° in the c.m.-system with respect to the symmetry axis respectively.

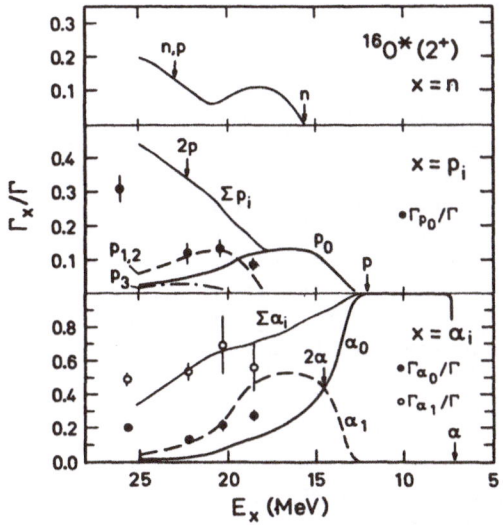

Fig. 25.

Calculated and measured branching ratios Γ_x/Γ as a function of excitation energy for the 2^+-resonance in ^{16}O. The lines are predictions assuming statistical decay |83|.

4. The angular distribution of the α_0-particles confirm the dominant L = 2 multipole character of the excited resonances. In addition, in all nuclei studied, a strong forward-backward asymmetry with respect to the symmetry axis is observed. This effect which was observed for the first time in ^{40}Ca by Moalem et al. |87| indicates that the dominant $J^\pi = 2^+$ resonances overlap with a negative parity state. The nature of these negative parity structures, which only have to be weakly excited in order to explain already the large observed anisotropy is not yet clear. But from (α,α')-experiments in several of these nuclei it is already known that some 3^--strength and also possibly $3\hbar\omega, \Delta T = 0$, 1^--strength might be present in the resonance region.

5. The high resolution $(\alpha,\alpha',\overline{c})$ work on ^{24}Mg |85| and ^{28}Si |83,85| has shown that the ratios Γ_x/Γ_{total} for the resonances, where $x = \alpha_0$, α_1 or p_0, can fluctuate drastically as a function of excitation energy. In figure 24 if one compares the (α,α')-spectrum with the $(\alpha,\alpha',\overline{\alpha}_0)$ taken at the recoil axis one can see that even for the large single peak at $E_{exc} \sim 17.6$ MeV the ratio $\Gamma_{\alpha_0}/\Gamma_{total}$ is not constant. Similar effects are observed for ^{28}Si |83|. A detailed explanation of this effect has not yet been given but it is likely that such information will lead to a new insight into the microscopic structure of the resonances.

Finally one should point out the potentiality of this kind of measurements for detecting weak non-2^+-strength like isoscalar monopole strength. This is illustrated for ^{24}Mg in figure 24. This is an $(\alpha,\alpha',\overline{\alpha}_0)$ - spectrum taken at $\theta \sim 55°$ with respect to the symmetry axis, that is at an angle where the α_0-decay pattern for a L = 2 resonance is at a minimum |85|. All 2^+-strength should be absent in such a spectrum and figure 24 shows that this is largely the case. The states which do show up relatively strong must have a different multipolarity. Of special interest is the one at ~ 13.7 MeV for which it was already suggested from the angular distribution of the singles (α,α')-experiment that it might have $J^\pi = 0^+$ |35|. The fact that it is seen strongly in figure 24, seems to confirm this assignment. It is clear that with such high resolution decay work one has a sensitive tool for detecting weakly excited multipolarities which might be present but cannot be observed in single scattering experiments because of the overwhelming strong 2^+-resonance. This might be especially important for detecting the 0^+-strength which has not yet been located in these light nuclei.

References

|1| D.E. Bcinum, J. Rapaport, C.D. Goodman, D.H. Horen, C.C. Foster, M.B. Greenfield and C.A. Goldding, Phys. Rev. Lett. 44:1751 (1980).

|2| J. Hamamoto, Physica Scripta, Vol. 6:266 (1972).

|3| J. Wambach, V.A. Madsen, G.A. Rinker and J. Speth, Phys. Rev. Lett. 39:1443 (1977).

|4| G.C. Baldwin and G.S. Klaiber, Phys. Rev. 13:1156 (1948).

|5| B.L. Berman and S.C. Fultz, Rev. Mod. Phys. 47: 113 (1975).

|6| R. Bergère, Lecture Notes in Physics 61:1.

|7| S.S. Hanna, Lecture Notes in Physics, 108:288.

|8| E. Kuhlmann, E. Ventura, J.C. Calerco, D.G. Mavis and S.S. Hanna, Phys. Rev. C11:1525 (1975).

|9| M.A. Kovash, S.L. Blatt, R.N. Boyd, T.R. Donoshue, H.J. Hausman and A.D. Bacher, Phys. Rev. Lett. 42:700 (1979).

|10| L.G. Arnold, Phys. Rev. Lett. 42:1253 (1979).

|11| R. Pitthan, Nuckleonika, 24:449 (1979).

|12| R. Pitthan and Th. Walcher, Phys. Lett. 36B:563 (1971).

|13| T. Suzuki and D.J. Rowe, Nucl. Phys. A289:461 (1977) and A292:93 (1977).

|14| F.E. Bertrand, Ann. Rev. Nucl. Sci. 26:457 (1976).

|15| Proceedings of Giant Multipole Resonance Topical Conference Oak Ridge Tenn. 1979; F.E. Bertrand editor.

|16| F.E. Bertrand, Varenna Summer School on Nuclear Structure and Heavy Ion Collisions, Varenna, Italy 1979.

|17| H.J. Gils, H. Rebel. J. Buschmann and H. Kleve Nebenius, Phys. Lett. 68B:427 (1977).

|18| R. Kamermans, J. van Driel, H.P. Morsch, J. Wildzynski and A. van der Woude, Phys. Lett. 82B:221 (1979).

|19| U. Garg, P. Bogucki, J.D. Bronson, Y.W. Lui, K. Nagatani, E. Takada, N. Takahashi, T. Tamya, D.H. Youngblood, Phys. Lett. 93B:31 (1980).

|20| R. Pitthan, F.R. Buskirk, E.B. Dalby, J.M. Deyer and X.X. Marvyama, Phys. Rev. Lett. 33:849 (1974).

|21| A. Bohr and B.R. Mottelson, Nuclear Structure, Vol. II (Benjamin, Reading, 1975).

|22| M. Goldhaber and E. Teller, Phys. Rev. 74:1046 (1948).

|23| J. Steinwedel and J.H. Jensen, Z. Naturforsch. 52:413 (1950).

|24| W.D. Meyers, W.J. Swiatecki, T. Kodama, L.J. El-Jaick and E.R. Hilf, Phys. Rev. C15:2032 (1977).

|25| M. Danos, Nucl. Phys. 5:23 (1958).

|26| B.L. Berman, D.D. Faul, R.A. Alvarez, P. Meyer, D.L. Olsen, Phys. Rev. C19:1205 (1979).

|27| F.E. Bertrand, G.R. Satchler, D.J. Horen, J.R. Wu, A.D. Bacher, G.T. Emery, W.P. Jones, D.W. Miller and A. van der Woude, Phys. Rev., to be published.

|28| E.C. Halbert, J.B. MacGrory, G.R. Satchler and J. Speth, Nucl. Phys. A245:189 (1975).

|29| M.N. Harakeh, B. van Heyst, K. van der Borg and A. van der Woude, Nucl. Phys. A327:373 (1979).

|30| A. Willis, M. Morlet, N. Marty, R. Frascaria, C. Sjalali, V. Comparat and P. Kitching, Nucl. Phys., to be published.

|31| F.E. Bertrand, K. van der Borg, A.G. Drentje, M.N. Harakeh, J. van der Plicht and A. van der Woude, Phys. Rev. Lett. 40:635 (1978).

|32| K.T. Knöpfle, G.J. Wagner, P. Paul, H. Breuer, C. Mayer-
     Böricke, M. Rogge and P. Turek, Phys. Lett. 74B:191 (1978).
|33| A.M. Berstein, Adv. Nucl. Phys. 3:325 (1969).
|34| H. Rebel, Z. Physik A277:35 (1976).
|35| K. van der Borg, Thesis University of Groningen, 1979.
|36| M. Sasao and Y. Torizuka, Phys. Rev. C15:217 (1977).
|37| M. Buenerd, C. Bonhomme, D. Lebrun, P. Martin, J. Chauvin,
     G. Duhamel, G. Perrin, P. de Saintignon, Phys. Lett. 84B:305
     (1979).
|38| D. Lebrun, M. Buenerd, P. Martin, G. Perrin and T. de
     Saintignon, private communication and in ref. 15.
|39| A. Moalem, Y. Gaillard, A.M. Bernolli, M. Buenard, J. Chauvin,
     G. Duhamel, D. Lebrun, P. Martin, G. Perrin, P. de Saintignon,
     Phys. Rev. C20:1593 (1979),
|40| S. Fukuda and Y. Torizuka, Phys. Lett. 62B:146 (1976).
|41| H. Miura and Y. Torizuka, Phys. Rev. C16:1688 (1976).
|42| J.R. Nix and A.J. Sierk, Phys. Rev. C21:396 (1980).
|43| K. van der Borg, M.N. Harakeh, A. van der Woude and
     F.E. Bertrand, Nucl. Phys. A341:219 (1980).
|44| S.S. Hanna, Proc. Int. Conf. on Nucl. Struct. and Spectr.,
     Amsterdam, 1974, Vol. 2:249 (Scholar's Press 1974).
|45| J.M. Moss, C.M. Rozsa, D.H. Youngblood, J.D. Bronson and
     A.D. Bacher, Phys. Rev. Lett. 34:478 (1975).
|46| T. Kishimoto, J.D. Moss, D.H. Youngblood, J.D. Bronson and
     A.D. Bacher, Phys. Rev. Lett. 35:552 (1975).
|47| K.T. Knöpfle, G.T. Wagner, A. Kiss, M. Rogge, C. Mayer-
     Böricke and Th. Bacher, Phys. Lett. 64B:263 (1976).
|48| D.H. Youngblood, C.M. Rosza, J.M. Moss, D.R. Brown and
     J.D. Bronson, Phys. Rev. C15:1644 (1978).
|49| Z.M. Szalata, K. Ituh, G.A. Peterson, J. Flanz,
     S.P. Fivozinsky, F.J. Kline, J.W. Lightbody, X.K. Maruyama
     and S. Penner, Phys. Rev. C17:435 (1978).
|50| Y. Abgrall, B. Morands, E. Cautier and B. Grammaticos,
     Phys. Rev. Lett. 34:922 (1977).
|51| W.F. Knüpfer in ref. 15.
|52| T. Hoshino and A. Arima, Phys. Rev. Lett. 37:266 (1976).
|53| A. Faessler, D.J. Millener, P. Paul and D. Strottman,
     Nucl. Phys. A330:333 (1979).
|54| H. Riedesel, K.T. Knöpfle, H. Bruer, P. Boll, G. Mairle and
     G.J. Wagner, Phys. Rev. Lett. 41:377 (1978).
|55| G.J. Wagner, Lecture Notes in Physics 92 (1979), Springer
     Verlag.
|56| N. Marty, M. Morlet, A. Willis, V. Comparat and R. Frascati,
     Nucl. Phys. A238:93 (1975).
|57| M.N. Harakeh, K. van der Borg, T. Ishimatsu, H.P. Morsch,
     A. van der Woude and F.E. Bertrand, Phys. Rev. Lett. 38:676
     (1977).
|58| D.H. Youngblood, C.M. Rosza, J.M. Moss, D.R. Brown and
     J.D. Bronson, Phys. Rev. Lett. 39:1188 (1977).

|59| C.M. Rosza, D.H. Youngblood, J.D. Bronson, Y-W. Lui and
     U. Garg, Phys. Rev. C21:1252 (1980).
|60| D.H. Youngblood in ref. 15.
|61| Y-W. Lui, P. Bogucki, J.C. Bronson, U. Garg, C.M. Rozsa and
     D.H. Youngblood, Phys. Lett. 93B:31 (1980).
|62| T. Yamagata, K. Iwamoto, S. Kishimoto, B. Sacki, K. Yasa,
     T. Fukuda, K. Obada, I. Miura, M. Inove and H. Ogata,
     Phys. Rev. Lett. 40:1628 (1978) and in Proc. 79.
|63| J.P. Blaisot, D. Gogny and B. Grammaticos, Nucl. Phys.
     A265:315 (1976).
|64| G.R. Satchler, in Elementary Modes of Excitation in Nuclei
     (ed. A. Bohr and R.A. Broglia, North Holland Publishing Co.,
     Amsterdam, 1977),
|65| H.P. Morsch, C. Sköza, M. Rogge, P. Turek, H. Machner and
     C. Mayer-Böricke, to be published.
|66| R. Pitthan, D.H. Meyer, F.R. Buskirk, J.N. Dreyer, Phys.
     Rev. C19:1251 (1979).
|67| H.P. Morsch, M. Rogge, P. Turek and C. Mayer-Böricke, Phys.
     Rev. Lett. 45:337 (1980).
|68| J.M. Moss, D.H. Youngblood, C.M. Rozsa and D.J. Bronson,
     Phys. Rev. C18:741 (1978).
|69| J. Wambach, F. Osterfeld, J. Speth and V.A. Madsen, Nucl.
     Phys. A324:77 (1979).
|70| A. Schwierczinski, R. Frey, E. Spamer, H. Theissen and
     Th. Wolcher, Phys. Lett. 55B:171 (1975).
|71| F.E. Bertrand, G.R. Satchler, D.J. Horen and A. van der Woude,
     Phys. Rev. C18:2788 (1978).
|72| A. van der Woude et al., unpublished data KVI.
|73| D.H. Youngblood, J.M. Moss, C.M. Rosza, J.D. Bronson,
     A.D. Bacher and D.R. Brown, Phys. Rev. C13:994 (1976).
|74| A. Veyssiere, H. Beil, R. Bergere, P. Carlos, A. Lepretre
     and K. Kernbath, Nucl. Phys. A199:45 (1973).
|75| J.T. Caldwell, E.J. Dowdy, B.L. Berman, R.A. Alvarez and
     P. Meyer, Phys. Rev. 21:1215 (1980).
|76| G.F. Bertsch, P.F. Bortignon, R.A. Broglia and C.H. Dasso,
     Phys. Lett. 80B:161 (1979).
|77| H. Ejiri, private communication.
|78| J.D.T. Arunda Neto, S.B. Herdade, B.S. Bhandari and
     I.C. Nascimento, Phys. Rev. C18:863 (1978).
|79| J.D.T. Arunda Neto, S.B. Herdade and I.C. Nascimento, Nucl.
     Phys. A334:297 (1980).
|80| A.C. Shotter, C.K. Gelbke, T.A. Ames, B.B. Back, J. Mahoney,
     T.J.M. Symons and D.H. Scott, Phys. Rev. Lett. 43:569 (1979).
|81| J. van der Plicht, M.N. Harakeh, A. van der Woude, P. David
     and J. Debrus, Phys. Rev. Lett. 42:112 (1979).
|82| F.E. Bertrand, J.R. Beene, C.E. Bemis, E.E. Gross, D.J. Horen,
     J.R. Wu and W.P. Jones, submitted to Phys. Rev. Lett. and
     private communication, 1980.

|83| K.T. Knöpfle, Habilitationsschrift Heidelberg 1979 and Lecture Notes in Physics 108:311.

|84| A. Djaloeis, H. Machner, M. Manko, C. Mayer-Böricke, M. Rogge and P. Turek, Proceedings Int. Conf. Nucl. Phys. with Electromagnetic Probes, Mainz, June 1979.

|85| F. Zwarts, K. van der Borg, A.G. Drentje, M.N. Harakeh and A. van der Woude, private communication and KVI Annual Report 1979.

|86| K.T. Knöpfle, H. Riedesch, K. Schindler, G.J. Wagner, C. Mayer-Böricke, W. Ockert, M. Rogge and P. Turek, Lecture Notes in Physics, 92:944 (1979).

|87| A. Moalem, W. Benenson, G.M. Crawley and T.L. Khoo, Phys. Lett. 61B:167 (1976) and Nucl. Phys. A281:461 (1977).

|88| D.H. Youngblood, A.D. Bacher, D.R. Brown, J.D. Bronson, J.M. Moss and C.M. Rosza, Phys. Rev. C15:246 (1977).

|89| K.T. Hecht and D. Braunschweig, Nucl. Phys. A295:24 (1978).

|90| J. Aschenbach, R. Haag and H. Krieger, Z. Phys. A292:285 (1979).

|91| F.E. Bertrand et al, Phys. Lett. 80B: 198 (1979).

|92| M.B. Lewis and F.E. Bertrand, Nucl. Phys. A196: 337 (1972).

|93| N. Marty et al, Int. Symp. on Highly Excited States in Nuclei, Jülich 1975.

MULTIPLE COULOMB EXCITATION OF HIGH SPIN STATES

D. Schwalm

Gesellschaft für Schwerionenforschung
Darmstadt, W. Germany

1. INTRODUCTION

Nuclear Coulomb excitation, that is the excitation of a nucleus via the electromagnetic field produced by another, swiftly passing nucleus, has been a very important tool in nuclear spectroscopy ever since the first Coulomb excitation experiment had been performed by T. Huus and Č. Zupančič in the early fifties[1]. The fruitfulness of Coulomb excitation for the study of the properties of excited nuclear states is mainly due to two reasons: (i) As long as the two colliding nuclei remain well outside the range of the nuclear forces, the interaction between the two nuclei can be assumed to be purely electromagnetic. Thus the excitation process itself is theoretically well understood, in contrast to most production processes involving nuclear forces. Consequently, the cross-sections observed in Coulomb excitation measurements can be solely used to determine electromagnetic properties of the nuclear states involved in the excitation process. (ii) Coulomb excitation, on the other hand, provides also a very clean and well determined way for producing nuclei in excited states. Thus the application of standard in-beam γ-spectroscopic techniques as e.g. the Recoil-Distance, the Doppler-Shift-Attenuation and the Perturbed-Angular-Correlation methods developed to study specific properties of excited states, is very often simplified significantly by using Coulomb excitation to prepare the ensemble of excited nuclei.

The exploitation of the possibilities offered by Coulomb excitation for nuclear structure studies was intimately connected with the development of suitable heavy ion accelerators; as the excitation is caused mainly by the Coulomb force, projectiles with large nuclear charges Z are required to maximize the excitation

cross-sections. Thus a marked increase of the activity in this
field took place in the sixties due to the advent of the Tandem-
van-de-Graaff accelerators, the largest of which can accelerate
nuclei with $Z \leq 25$ up to energies close to the Coulomb barrier.
Many important contributions to our understanding of the structure
of the atomic nucleus stem from Coulomb excitation measurements
done at these machines; in particular, the so-called reorientation
effect[2] could be exploited to determine static quadrupole moments of low lying nuclear states. It was only recently, however,
that the ultimate dream of all votaries of Coulomb excitation
became true; in 1976, the UNILAC accelerator built by the Gesellschaft für Schwerionenforschung (GSI) at Darmstadt went into operation, which is capable of producing intense ^{208}Pb beams with energies high enough to reach the Coulomb barrier for all elements
occuring in nature. For many Coulomb excitation studies ^{208}Pb con-
stitutes the ideal projectile not only because of its large nuclear
charge ($Z = 82$) but also because of its doubly-magic structure.
The first excited state of ^{208}Pb, the well-known $I^{\pi} = 3^{-}$ state
at an excitation energy of $\Delta E = 2.6$ MeV, is only weakly excited
in Coulomb excitation experiments. This is not only important from
an experimental point of view, it also leads to a considerable
simplification in the theoretical description of the excitation
process as the ^{208}Pb nucleus can be viewed as a monopole charge
not possessing any internal structure.

In the present lecture I will restrict myself to the dissussion
of Multiple Coulomb excitation (MCE) studies with ^{208}Pb projectiles,
which have been performed at the UNILAC to investigate high spin
states in deformed nuclei. These studies represent only part of
the Coulomb excitation program presently carried out at GSI, however they examplify rather nicely some of the possibilities opened
up by the availability of this unique projectile. Moreover, they
allow me to discuss most of the experimental problems inherent in
MCE-experiments with very heavy projectiles and the ways we choose
to handle them.

The excitation of high spin states in Coulomb excitation studies
with very heavy ions is the result of a multistep process, which is
predominantly induced by the Coulomb interaction between the projectile and the target nucleus. Magnetic excitations as well as
relativistic contributions to the electric transitions are usually
negligibly small. Moreover, the electric excitation amplitudes due
to the Coulomb interaction decrease drastically with increasing
multipolarity. Thus e.g. the members of the ground state band (g-
band) as well as those of the various sidebands of a deformed
nucleus are mainly excited via several consecutive E2-transitions.
(fig. 1). E1-excitations, which are in principle possible between
states of opposite parity, are usually small as the B(E1)-values
between states not involving the giant dipole resonance are in most
cases strongly hindered compared to an E1 single-particle unit.

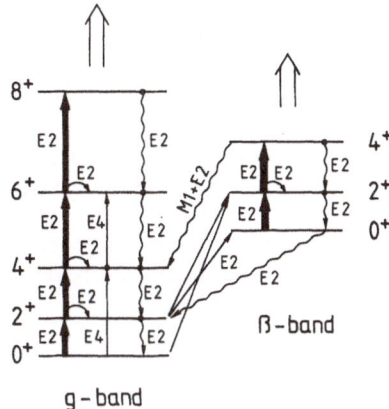

Fig. 1 *Typical excitation and de-excitation pathes observed in MCE-*
         studies on well deformed nuclei.

Thus the excitation probability of a particular level, which can be
deduced from the intensities of its de-excitation γ-rays, is determined in a rather complicated way by mainly the dynamic and static
E2-transition matrix elements connecting this state via several
intermediate states to the ground state of the nucleus. Nevertheless
it is possible - as I will show later - to deduce individual B(E2)-
values at least for the ground state band in a model independent way.

       Presently, the Multiple Coulomb excitation and the heavy ion
induced fusion reactions of the types (HI,xn) are the most suited
reaction modes to study the properties of heavy nuclei at high
angular momenta. The two production modes complement each other
rather well as one can exploit the different population mechanisms
of the two processes: In (HI,xn)-reactions one observes mainly
states in n-deficient nuclei which belong to the yrast-line or
which are energetically close to it; the feeding pattern of a
particular state is complicated and not known in detail, but the
well defined kinematic of the evaporation residues, which recoil
in a narrow cone around the beam direction, allows for the application of special γ-spectroscopic methods as e.g. the recoil distance
technique to determine the lifetimes of the yrast-states. In MCE-
experiments, on the other hand, mainly states are populated which
are connected to the ground state via a string of large B(E2)-
values and which do not necessarily belong to the yrast-sequence;
the feeding pattern can be controlled and the relevant B(E2)-
values can be deduced from the observed γ-ray yields. The seeming
disadvantage of MCE as being limited to stable or very longlived
nuclei is compensated by the fact that the n-rich stable nuclei
usually are not accessible by (HI,xn)-reactions. In the actinide
region, moreover, where heavy ion fusion reactions lead mainly to
fission, MCE is at the time being the only available tool to study
the high spin behaviour of these nuclei.

In the following I will first discuss some of the basic concepts used in the theoretical description of the Coulomb excitation process and give a few simple relations which are very useful in planning MCE-experiments. I will then describe the methods we developed at GSI to perform MCE experiments with very heavy projectiles. Finally, I will present some results obtained for the actinide nuclei to illustrate the special role of MCE experiments with ^{208}Pb-ions in the spectroscopy of high spin states and examplify the importance of having access to both the level energies and the E2-transition matrix elements when discussing the possible structure of these states.

2. THEORETICAL BACKGROUND

Only a short outline of the theory of Coulomb excitation with heavy ions will be given in the following; a comprehensive view of the subject can be found in the monograph of K. Alder and A. Winther "Electromagnetic Excitation"[3].

In a full quantum mechanical treatment of the Coulomb excitation process one has to describe the projectile-target system by a wave function which is the solution of the Schrödinger equation for the two nuclei in their mutual electromagnetic field, and which behaves asymptotically as a plane wave. Although the general solution of this problem through the use of a partial wave expansion is well known, the numerical problems encountered in the calculation of the scattering amplitude for MCE-experiments with very heavy ions is tremendous and could so far only be handled exactly for cases where less than 10 nuclear states are involved in the excitation process[4]. The main reasons for these difficulties are due to the large number of open channels and the long range of the Coulomb force, which requires that partial waves up to very large orbital angular momenta ($\ell > 1000$) have to be considered.

Almost all Coulomb excitation experiments are therefore analysed using the semi-classical approach, where the relative motion of the two nuclei is described by a classical trajectory; actually it is assumed that the nuclei move on hyperbolic orbits characteristic for elastic scattering of two classical point charges. In such a picture one has, for a given c.m. scattering angle Θ_1 of the projectile (see fig. 2), a well defined relation between the relative distance r_{12} of the two nuclear centres of mass and the time t, and the excitation is the result of the time-dependent electromagnetic field between the two nuclei as they move along the classical orbit. Thus the relative motion of the two centres of mass is separated from the intrinsic degrees of freedom of the nuclei and their intrinsic wave function is determined by the time-dependent Schrödinger equation containing the Hamiltonian of the free nuclei and the mutual electromagnetic interaction Hamiltonian.

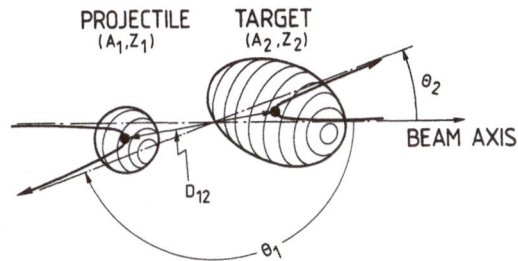

PROJECTILE   TARGET
(A_1, Z_1) (A_2, Z_2)

θ_2

BEAM AXIS

D_{12}

θ_1

REDUCED MASS	$\mu = \dfrac{A_1 A_2}{A_1 + A_2}$ [amu]
REDUCED BEAM-ENERGY	$\varepsilon = E_1^{lab} / A_1$ [MeV/amu]
SOMMERFELD PARAMETER	$\eta = 0.158 Z_1 Z_2 / \sqrt{\varepsilon}$
ADIABATICITY PARAMETER	$\xi_{of} = \eta \Delta E_f / (2\mu\varepsilon)$
$D_{12}(\theta_1 = 180°) \equiv$	$2a = 1.44 Z_1 Z_2 / (\mu\varepsilon)$ [fm]

*Fig. 2   Definition of some quantities used in the theoretical des-
cription of the MCE process. $\theta_1 (\theta_2)$ denotes the scattering
angle of the projectile (target) in the c.m. system; ΔE_f
is the excitation energy (in MeV) of the state $|f\rangle$.*

The solution of the Schrödinger equation is considerably simplified - although mainly from a practical rather than a theoretical
point of view - if one of the nuclei, let say the projectile, can
be assumed to be structureless (which is the case for e.g. ^{208}Pb).
The Schrödinger equation then reads

$$i\hbar \frac{\partial}{\partial t} |\Psi(t)\rangle = [H_o + V(t)] |\Psi(t)\rangle \qquad (1)$$

where $|\Psi(t)\rangle$ denotes the intrinsic wave function of the target
nucleus, H_o the Hamiltonian of the free nucleus and V(t) the electromagnetic interaction between the point charge and the charge-cur-
rent distribution of the target nucleus. Expanding $|\Psi(t)\rangle$ in terms
of the complete set of orthogonal eigenstates $|n\rangle$ of the free
nuclear Hamiltonian,

$$|\Psi(t)\rangle = \sum_n a_n(t) e^{-\frac{i}{\hbar} E_n t} |n\rangle$$

with $H_o |n\rangle = E_n |n\rangle$, a set of coupled differential equations in the
excitation amplitudes $a_n(t)$ is obtained from eq.(1)

$$i\hbar \frac{\partial}{\partial t} a_n(t) = \sum_m \langle n | V(t) | m \rangle \, e^{+\frac{i}{\hbar}(E_n - E_m)t} a_m(t) \qquad (2).$$

The matrix elements occuring in eq.(2) have the general structure

$$\langle n|V(t)|m\rangle = \sum_{\sigma\lambda,\mu} f_{\sigma\lambda\mu}(\vec{r}_{12}(t)) \cdot C_{\lambda\mu}(I_m I_n) \langle I_m \| \mathfrak{M}(\sigma\lambda)\| I_n\rangle^* \quad (3),$$

where I denotes the spin of the nuclear state, $\sigma\lambda$ the type and multipolarity of the transition and μ the z-component of λ. The nuclear information is thus contained solely in the reduced electro-magnetic matrix elements $\langle I_m \| \mathfrak{M}(\sigma\lambda)\| I_n\rangle$; the off-diagonal elements are connected with the reduced transition probabilities $B(\sigma\lambda)$ via

$$B(\sigma\lambda; I_n \rightarrow I_m) = \frac{1}{2I_n+1} |\langle I_m \| \mathfrak{M}(\sigma\lambda)\| I_n\rangle|^2 \quad (4),$$

while the diagonal elements are given (besides some historical factors) by the static multipole moments of the target nucleus in state $|n\rangle$.

For the initial condition $a_m(t = -\infty) = \delta_{I_m I_i} \delta_{M_m M_i}$, where I_i, M_i denote the spin and the magnetic quantum number of the ground state, and a given scattering angle Θ_1 of the projectile, the ex-citation amplitude of a particular nuclear state $|f\rangle = |\alpha_f, I_f, M_f\rangle$ after the collision is given by $a_{I_f M_f}(M_i; \Theta_1) \equiv a_f(t = +\infty)$ and ob-tained by solving eq.(2) including all relevant states and corres-ponding reduced matrix elements. This can be done numerically by using the famous Winther-de Boer program[5]. We actually use the version of Lell[6], which we adjusted to the GSI computer facility. This program allows to include electric and magnetic transitions with multipolarities $\lambda = 1$ to 6 and has virtually no restriction other than time limitations on the number of nuclear states in-cluded.

From the excitation amplitudes $a_{I_f M_f}(M_i; \Theta_1)$ all observables can be calculated. In particular, for an initially unorientated target nucleus the excitation probabilities $P_f(\Theta_1)$ of the state f is determined by

$$P_f(\Theta_1) = \frac{1}{2I_i+1} \sum_{M_i M_f} |a_{I_f M_f}(M_i; \Theta_1)|^2 \quad (5)$$

and the differential cross-section in the c.m. system is thus given by

$$\frac{d\sigma_f(\Theta_1)}{d\Omega_1} = \frac{d\sigma_R(\Theta_1)}{d\Omega_1} P_f(\Theta_1) \quad (6)$$

where $d\sigma_R(\Theta_1)/d\Omega_1 = 0.25 a^2 \sin^{-4}(\Theta_1/2)$ denotes the classical Rutherford cross-section. Moreover, the double-differential cross-section for the scattering of the projectile into $d\Omega_1$ leading to the excitation of the state f and the emission of a γ-ray into $d\Omega_\gamma$ from the subsequent decay of a state g to a state h is given by

$$\frac{d^2 \sigma_{f;gh}(\Theta_1;\Theta_\gamma \phi_\gamma)}{d\Omega_1 d\Omega_\gamma} = \frac{d\sigma_f(\Theta_1)}{d\Omega_1} \cdot \frac{dW_{f;gh}(\Theta_1;\Theta_\gamma \phi_\gamma)}{d\Omega_\gamma} \qquad (7a)$$

with

$$\frac{dW_{f;gh}(\Theta_1;\Theta_\gamma \phi_\gamma)}{d\Omega_\gamma} = \frac{1}{\sqrt{4\pi}} \sum_{K\varkappa} \hat{S}^f_{K\varkappa}(\Theta_1) H_K(f \to g) F_K(I_h,I_g) Y_{K\varkappa}(\Theta_\gamma,\phi_\gamma) (7b)$$

Here $\hat{S}^f_{K\varkappa}(\Theta_1)$ are the statistical tensors describing the population of the state f

$$\hat{S}^f_{K\varkappa}(\Theta_1) = \frac{1}{P_f(\Theta_1)} \frac{\sqrt{2I_f+1}}{2I_i+1} \sum_{M_f M_{f'}} (-1)^{I_f - M_{f'}} \begin{pmatrix} I_f & K & I_f \\ -M_{f'} & \varkappa & M_f \end{pmatrix} \cdot$$

$$\cdot \sum_{M_i} a^*_{I_f M_{f'}}(M_i;\Theta_1) a_{I_f M_f}(M_i;\Theta_1) \qquad (7c)$$

with $\hat{S}^f_{oo}(\Theta_1) = 1$, $H_K(f \to g)$ accounts for all unobserved transitions leading from the state f to the state g, the γ-decay of which is observed, $F_K(I_h,I_g)$ are the angular correlation coefficients describing the γ-decay of the state g to the state h, and $Y_{K\varkappa}$ are the spherical harmonics. In heavy ion induced Coulomb excitation experiments one can usually only measure the scattering angle Θ_1; the energy resolution is in most cases not sufficient to determine the final state f which has been reached in the excitation process. In this case the relevant double differential cross-section is obtained by summing eq.(7) over all states f with excitation energies $\Delta E_f \geq \Delta E_g$, i.e.

$$\frac{d^2 \sigma_{gh}(\Theta_1;\Theta_\gamma \phi_\gamma)}{d\Omega_1 d\Omega_\gamma} = \sum_{\substack{f \\ (\Delta E_f \geq \Delta E_g)}} \frac{d^2 \sigma_{f;gh}(\Theta_1;\Theta_\gamma \phi_\gamma)}{d\Omega_1 d\Omega_\gamma} \qquad (8)$$

Note that the polar angles $\Theta_\gamma, \phi_\gamma$ describing the direction of the γ-quantum in eqs.(7),(8) are given in a coordinate system which has the same orientation as the c.m. coordinate system used to evaluate the excitation amplitudes but which has its origin in the centre of mass of the γ-emitting nucleus (rest coordinate system).

In fig. 3a the calculated differential cross-section $d\sigma_I(\Theta_1)/d\Theta_1 d\phi_1 = |d\sigma_I(\Theta_1) / d\Omega_1| \sin\Theta_1$ for the excitation of some states of the g-band of ^{164}Dy by 4.7 MeV/amu ^{208}Pb projectiles is shown as a function of the c.m. scattering angle Θ_1 of the projectile, respectively as a function of the recoil angle ϑ_2 of the target in the laboratory system. The recoil angles ϑ_2 are simply given by the elastic scattering relation $\vartheta_2 = (\pi - \Theta_1)/2$ as the excitation energies

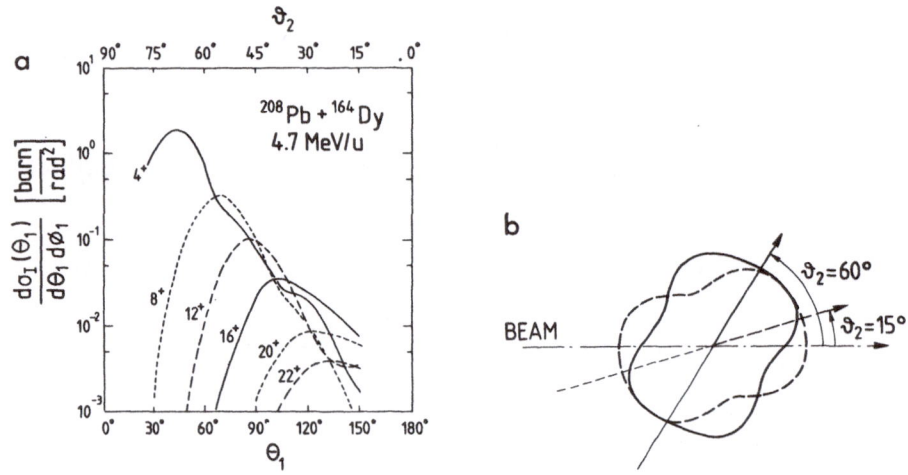

Fig. 3 (a) *Calculated differential cross-section for some states of the ground state band of ^{164}Dy excited by 4.7 MeV/amu ^{208}Pb. Rigid rotor matrix elements and an intrinsic quadrupole moment of $Q_O = 7.5$ eb were assumed (see also text).*
(b) *Schematic drawing of the angular correlation function $dW_{I;I-2}(\theta_1;\theta_\gamma,\phi_\gamma)/d\Omega_\gamma$ for I >> 1 and two lab. recoil angles ϑ_2 of the target nucleus ($\theta_1 = \pi - 2\vartheta_2$).*

ΔE_I are very small compared to the projectile energy. Note that the $I^\pi = 22^+$ state is still excited with a cross section of several mb/rad^2 at scattering angles around $\theta_1 \approx 135°$. Fig. 3b shows a schematic drawing of the angular correlation function $dW_{I;I-2}(\theta_1;\theta_\gamma\phi_\gamma)/d\Omega_\gamma$ for I>>1 and two scattering angles $\theta_1 = 150°$ ($\hat{=}\vartheta_2 = 15°$) and $\theta_1 = 60°$ ($\hat{=}\vartheta_2 = 60°$). In the scattering plane the correlation functions are roughly symmetric around the recoil axis of the target nuclei, which is a distinct axis because of the symmetry of the classical trajectory around $(\pi - \theta_1)/2$. Moreover, in the high spin limit the correlation functions depend mainly on the spin change ΔI and the multipolarity of the transition and only slightly on I.

The semi-classical approximation is justified only if the de-Broglie wavelength $\lambdabar$ connected with the relative motion of the two nuclei is much smaller than the size of the scattering system. This condition is satisfied if the Sommerfeld parameter $\eta = a/\lambdabar \gg 1$ (see also fig. 2). Moreover, the use of classical hyperbolic orbits is only justified if the two nuclei do not penetrate each other and if the change of the orbit due to the excitation of the nuclei is very small, i.e. $\Delta E_f/\mu \cdot \varepsilon = 2\xi_{of}/\eta \ll 1$ and $|I_f-I_i| \ll \ell/\hbar$, where $\ell \approx \sqrt{2\mu\varepsilon} \cdot a = \eta\hbar$ is the angular momentum connected with the relative motion. Although in actual calculations within the semi-classical approach the change of the orbit due to the nuclear excitation is taken into account at least approximately by symmetrizing the orbital parameters with respect to initial and final relative velocities, the last two inequalities are still important in MCE experiments. It follows that - besides the requirement of non-penetration - the necessary conditions for the applicability of the semi-classical approach can be written as

$$\eta \gg Max\left(1, 2\,\xi_{of}\,, |I_f-I_i|\right) \tag{9}$$

As for a state f to be excited with a reasonable probability the adiabaticity parameter, which measures the collision time relative to the characteristic transition time $\Delta E_f/\hbar$, has to be of the order of 1 or less ($\xi_{of} \lesssim 1$), eq.(9) reduces for MCE-studies of high spin states to $\eta \gg |I_f - I_i|$. This indicates that in these studies possible deviations of the semi-classical from the full quantum-mechanical calculations must be predominantly due to the insufficient handling of the angular momentum transfer.

In Coulomb excitation experiments with very heavy projectiles η is of the order of 100 or more as long as the basic requirements for pure Coulomb excitation - that is that the two nuclei stay well outside the range of the nuclear forces - is fulfilled. This condition requires that the projectile energy must satisfy the inequality (see also fig. 2)

$$D_{12}(\Theta_1=180°)=2a \geq r_0\left(A_1^{1/3}+A_2^{1/3}\right)+d_0 \tag{10}$$

where the right hand side is the commonly used parametrization of the "sum of the nuclear interaction radii". The largest energy consistent with eq.(10) is the so-called safe energy ε_s. People usually disagree on what to choose for the two parameters r_0 and d_0 in order for ε_s to be really "safe". This quarrel will very likely never be solved as it depends on the sensitivity of the experiment and on the kind of information one wants to deduce. As long as we are only interested in B(E2)-values we usually adopt $r_0 = 1.25$ fm and $d_0 = 4$ fm which lead to safe energies consistent with values deduced from excitation functions measured for heavy ions impinging on [238]U (ref.7).

For [208]Pb projectiles these parameters lead to safe energies of $\varepsilon_s = 4.7$ MeV/amu and $\varepsilon_s = 5.3$ MeV/amu for $A_2 = 160$ and $A_2 = 240$, respectively. The corresponding Sommerfeld parameters are $\eta \approx 400$ and $\eta \approx 500$, indicating that the semi-classical approach is indeed very well justified in MCE studies with such heavy projectiles. We will even go one step further and estimate the maximum angular momentum which can be transferred in MCE-experiments with heavy ions to a well deformed target nucleus (a rigid rotor) from a classical picture of the excitation process. This leads to a rather simple expression which is very useful for planning MCE experiments on deformed nuclei.

In a classical picture an axially deformed target nucleus will start to rotate due to the torque produced by the interaction between the intrinsic quadrupole moment Q_O of the target nucleus and the charge of the passing projectile. In the limit that the collision time is very short compared to the rotation time of the nucleus one obtains the following classical estimate $L_c(\theta_1)$ for the maximum angular momentum transferred to the target nucleus[8]

$$L_c(\theta_1) = L_c^m \cdot f(\theta_1) \tag{11a}$$

and

$$L_c^m \approx 26 \frac{Q_0 [ebarn]}{(2a [fm])^{3/2}} \sqrt{\frac{Z_1}{Z_2} \cdot \mu [amu]} \quad [\hbar] \tag{11b}$$

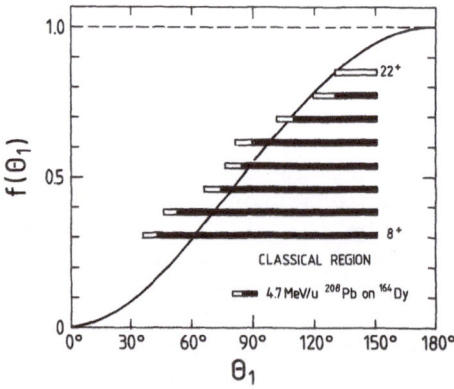

Fig. 4   The function $f(\theta_1)$ appearing in the classical estimate of the maximum angular momentum transfer (eq.(11)). Also shown are the scattering angles (open bar: width of the last $\Delta\theta_1$-window) for which the states of the ground state band $8^+ \leq I^\pi \leq 22^+$) of ^{164}Dy could actually be observed in the MCE-experiment performed with 4.7 MeV/amu ^{208}Pb projectiles ($L_c^m = 26\ \hbar$).

The function $f(\theta_1)$ is plotted in fig.4. For ^{208}Pb projectiles impinging on ^{164}Dy $(^{238}$U$)$ and projectile energies $\varepsilon = \varepsilon_s$, eq.(11b) results in $L_c^m \approx 26\hbar$ $(36\hbar)$ assuming $Q_0 = 7.5$ eb $(11.1$ eb$)$.

According to this estimate we expect to populate the ground state band of deformed even-even nuclei up to spins $I = L_c(\theta_1)/\hbar$ provided the nuclei really behave like rigid rotors. Although for a given scattering angle higher states are populated due to quantum mechanical effects (compare e.g. the results obtained from eq.(11) for ^{164}Dy to the semi-classical calculation shown in fig.3a) their excitation probabilities are steeply decreasing; it is our experience that eq.(11) agrees well with what can be observed in actual MCE experiments on "good rotors" (see fig.4)

3. EXPERIMENTAL METHODS

In MCE experiments with heavy ions the measurement of the energy and direction of the scattered projectiles (target nuclei) alone is not sufficient to distinguish between different excited states as the energy resolution one has to reach is outside the possibilities of present experimental techniques. Therefore one usually observes the de-excitation γ-rays either in a singles mode or in coincidence with the direction of the scattered projectile or the recoiling target nucleus.

The main experimental problem one has to cope with in these studies is the large Doppler broadening of the γ-ray lines, which is due to the magnitude and spread of the velocity of the γ-emitting nuclei; for e.g. 4.7 MeV/amu ^{208}Pb on ^{164}Dy the recoil velocities in the lab. system range between $v/c \approx 0$ % and $v/c \approx 12$ % with recoil angles between $\vartheta_2 = 0^{\circ}$ and $\vartheta_2 \approx 90^{\circ}$. For these large recoil velocities the relativistic transformation of the γ-rays from the rest system of the γ-emitting nucleus into the lab. system results in sizeable effects. In the following the discussion of this transformation will be restricted to first order terms in v/c, although in actual experiments second order terms have to be considered as well (Note that a second order treatment of the γ-transformation is still consistent with a non-relativistic **treatment** of the reaction kinematic).

In first order in v/c the projection angle $\hat{\vartheta}_\gamma$ (see fig.5), the γ-energy E_γ and the solid angle element $d\Omega'_\gamma$ in the lab. system are connected with the corresponding quantities $\hat{\Theta}_\gamma$, $E_{\gamma 0}$ and $d\Omega_\gamma$ in the rest coordinate system via

$$\cos \hat{\vartheta}_\gamma = \frac{\cos \hat{\Theta}_\gamma + v/c}{1 + v/c \cos \hat{\Theta}_\gamma} \approx \cos \hat{\Theta}_\gamma + \frac{v}{c} \sin^2 \hat{\Theta}_\gamma \qquad (12a)$$

$$E_\gamma = E_{\gamma 0} \frac{(1 - (v/c)^2)^{1/2}}{1 - v/c \cos \hat{\vartheta}_\gamma} \approx E_{\gamma 0} \left(1 + \frac{v}{c} \cos \hat{\vartheta}_\gamma\right) \qquad (12b)$$

$$d\Omega_\gamma' = d\Omega_\gamma \left(\frac{E_\gamma}{E_{\gamma 0}}\right)^2 \approx d\Omega_\gamma \left(1 + 2\frac{v}{c}\cos\hat{\vartheta}_\gamma\right) \qquad (12c)$$

For v/c = 12 % and $E_{\gamma 0}$ = 500 keV, for example, one estimates $\hat{\theta}_\gamma - \hat{\vartheta}_\gamma \approx 70°$ for $\hat{\theta}_\gamma = 90°$, $\Delta E_\gamma = E_\gamma - E_{\gamma 0}$ = 60 keV and $(\Omega_\gamma' - \Omega_\gamma)/\Omega_\gamma$ = 24 % for $\hat{\vartheta}_\gamma = 0°$. As the angular change is rather small compared to the typical structure of γ-angular correlation functions and as the change of the solid angle elements can be obtained from the measured γ-energy, the main experimental problem is due to the energy shift (eq.(12b)). Because of the kinematic spread of $\vec{v}$, a $E_{\gamma 0}$ = 500 keV transition - produced in a MCE-experiment with a thin target such that the excited nuclei recoil into vacuum - results in a 60 keV broad line when observed in a singles mode. But even if $\vec{v}$ is accurately known for each γ-quant detected, the finite size of the γ-detector (say $\Delta\hat{\vartheta}_\gamma \approx \pm 10°$) produces an energy spread of 20 keV (1.0 keV) for $\hat{\vartheta}_\gamma = 90°$ (0°) as compared to the energy resolution of a Ge-detector of $\sim$ 1.7 keV.

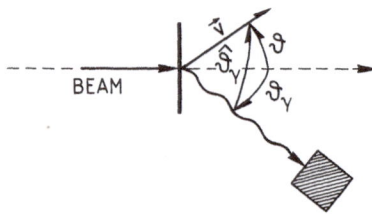

Fig. 5 Definition of the projection angle $\hat{\vartheta}_\gamma$.

The method usually employed in light ion induced Coulomb excitation measurements, where one slows down the recoiling nuclei in a target backing before their decay, is only of limited use in MCE-experiment with very heavy projectiles: For initial velocities of the order of 10 % of the speed of light the stopping times amount to 3 - 5 psec and are thus much longer than the lifetimes of most high spin states (e.g. $\tau \lesssim 1$ psec for states of the ground state band in ^{164}Dy with $I^\pi \geq 10^+$). Therefore these states still decay in flight preventing their identification if only weakly excited. Only in the actinide region one does succeed in special cases by using this method - in conjunction with a sum-spectrometer and by inverting the role of projectile and target - to observe very high spin states (see sub-section 3.3).

To avoid - or at least to minimize - the Doppler-spread problems encountered in MCE-experiments with heavy projectiles we therefore developed an experimental technique, which can be rather generally applied in these kind of experiments and which has been used in most of the MCE-measurements performed so far at GSI. The basic idea of this method - which will be described in detail in the following subsection 3.1 - is to use thin, unbacked targets and to determine for each γ-event the corresponding velocity projection v/c cos$\hat{\vartheta_\gamma}$. The method has the additional advantage that it allows to determine the γ-yields as a function of the c.m. scattering angle of the projectile over a large angular range. This is of particular importance if one aims at a model-independent analysis of the γ-yields in terms of individual B(E2)-values (see subsection 3.2).

3.1 MCE-Experiments with Thin Targets

In MCE-studies with thin ($\leq$ 1 mgr/cm^2), unbacked targets the excited target nuclei recoil into vacuum with velocities which are determined by the reaction kinematic only. As the excitation energy for states with reasonable cross-sections is much smaller than the projectile energy ($\Delta E_f \ll \mu \cdot \varepsilon$), the kinematic can be well approximated by elastic scattering relations. Thus the recoil velocity v_2 of the target nuclei in the lab. system (oriented with the z-axis pointing in the beam direction) is given by

$$v_2 = 2\, v_{cm}\, \cos\vartheta_{\frac{9}{2}}^{\frac{9}{2}} \tag{13a}$$

where v_{cm} is the velocity of the centre of mass,

$$\frac{v_{cm}}{c} = 0.04635 \frac{A_1}{A_1 + A_2} \sqrt{\varepsilon\left[\frac{MeV}{amu}\right]} \tag{13b}.$$

Together with eq.(12b) and the trigonometric relation (see also fig.5)

$$\cos\hat{\vartheta_\gamma} = \cos\vartheta_{\frac{9}{2}} \cos\vartheta_\gamma + \sin\vartheta_{\frac{9}{2}} \sin\vartheta_\gamma \cos(\varphi_2 - \varphi_\gamma)$$

where φ_2 and φ_γ describe the azimuthal angle of the recoiling target and of the emitted γ-ray in the lab. system, respectively, one obtains

$$E_{\gamma_0} = E_\gamma \left[1 - \frac{v_{cm}}{c} f(\vartheta_2 \varphi_2; \vartheta_\gamma \varphi_\gamma) \right] \tag{14a}$$

Note that $f(\vartheta_2\varphi_2; \vartheta_\gamma\varphi_\gamma)$ is an universal function in the sense that it is independent of the specific projectile-target combination; the system-dependence of the Doppler shift is completely contained in v_{cm}/c appearing as a constant in eq.(14). The f-function has the property $f(\vartheta_2\varphi_2; \pi-\vartheta_\gamma, \pi+\varphi_\gamma) = -f(\vartheta_2\varphi_2; \vartheta_\gamma\varphi_\gamma)$. Moreover, for a given position of the γ-detector ($\vartheta_{\gamma_0}, \varphi_{\gamma_0}$) and a limited range of ϑ_2, φ_2, namely $|\vartheta_2 - \vartheta_{\gamma_0}| \leq 30°$, $|\varphi_2 - \varphi_{\gamma_0}| \leq 30°$ or $|\vartheta_2 + \vartheta_{\gamma_0} - \pi| \leq 30°$, $|\varphi_2 - \varphi_{\gamma_0} - \pi|$

$\leq$ 30°, which is required in order to keep the finite size effect of the γ-detector reasonably small, the φ_2-dependence of f can be neg-lected i.e.

$$E_{\gamma_0} = E_\gamma \left[1 - \frac{v_{cm}}{c} f(\vartheta_2 \overline{\varphi_2} ; \vartheta_{\gamma_0} \varphi_{\gamma_0}) \right] \quad \text{(14b)}$$

As we succeeded to construct a detector which provides a signal directly proportional to f, we are able to calculate on-line for each γ-event the corresponding transition energy by a simple arith-metic procedure.

A schematic view of the experimental set-up is shown in fig.6 De-excitation γ-rays are recorded by two Ge-detectors in coincidence with the recoiling target nuclei, which are observed in a large-area position-sensitive avalanche detector (L) positioned at a distance of 15 cm from the target. The detector covers an angular range of 15° $\leq \vartheta_2 \leq$ 60°, and -25° $\leq \vartheta_2 \leq$ +25°, corresponding to pro-jectile c.m. scattering angles of 150° $\geq \theta_1 \geq$ 60°; the recoil detec-tor thus covers the complete angular region relevant for the exci-tation of high spin states (cf. fig. 3a). A schematic drawing of the avalanche detector is shown in fig.7. While the anode foil consists of a thin (2 μm) aluminized plastic foil, the cathode of the detec-tor is made out of a 0.13 mm thick printed circuit board, which is subdivided in ≈ 1 mm broad slices corresponding to curves of equal polar angles ϑ with respect to the beam. These slices are connected to from a kind of meander; together with a similar pattern printed on the back side of the circuit board, however connected to form a meander of opposite phase, a continuous delay line[9] is obtained, the delay of which is determined by the envelope. Thus by a proper choice of this boarder line the delay of the cathode signal, mea-sured relative to the prompt anode pulse, can be made to be direct-ly proportional to $f(\vartheta_2, \overline{\varphi_2} ; \vartheta_{\gamma_0}, \varphi_{\gamma_0})$. Note, moreover, that for the geometry chosen f is a monotonic function of ϑ_2 which allows to determine the γ-yields as a function of the c.m. scattering angle θ_1. A precise angle calibration of the delay line is achieved by (i) investigating the delay line with an electronic pulser and (ii) by bombarding medium mass targets with [208]Pb and analyzing the maximum scattering angles of the [208]Pb projectiles.

In order to be able to decide whether the recoiling target nucleus (and not the scattered projectile) has been observed in the recoil detector, the corresponding projectile is detected in a se-cond position-sensitive detector (R). By measuring the time-of-flight difference and/or the kinematic correlation between the two coinci-dent particles, both types of events can be separated. The additional detector (R) is the only part of the set-up which depends to some extent on the projectile-target combination actually studied: For [208]Pb ions impinging on light rare earth nuclei such as [164]Dy one

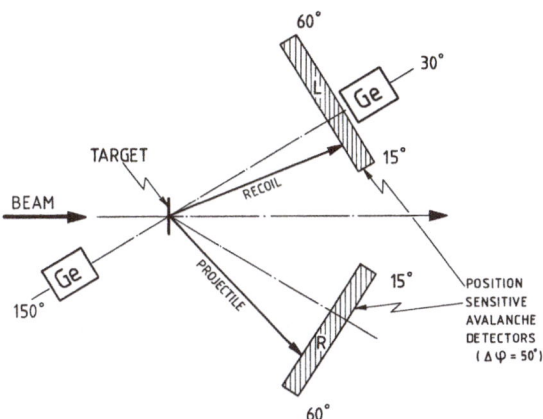

Fig. 6   *Schematic view of the experimental set-up used in MCE-experiments with thin targets and* [208]*Pb-beams. While the positions of the "recoil" detector (L) and of the Ge-detectors remain fixed, that of the "projectile" detector (R) has to be adjusted to the projectile-target mass ratio. Shown is the set-up used in the MCE-studies on light rare earth nuclei, where for detector (R) a detector identical to (L) can be used.*

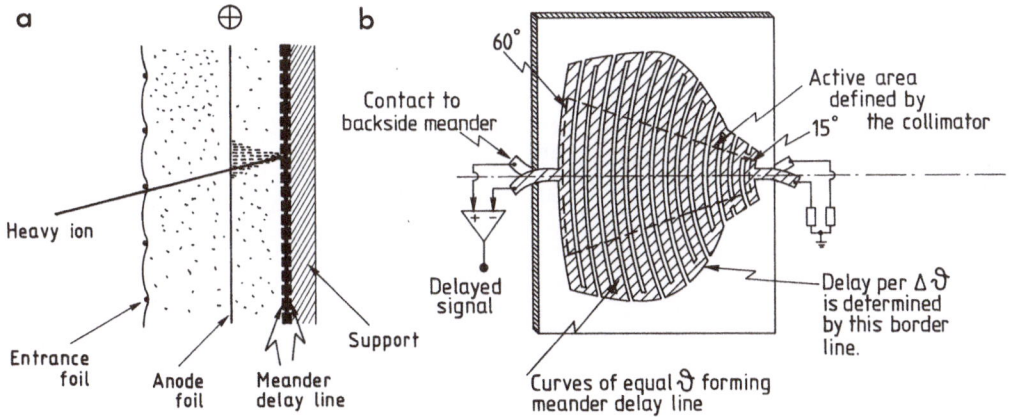

Fig. 7   *Schematic drawing of the position-sensitive avalanche detector used to detect the recoiling target nuclei.*
(a) Cross-section. The detector is filled with 10 Torr of Isobutan. The anode-cathode distance is 5 mm.
(b) Front view of the cathode. Note that the cathode between 15⁰ *and* 60⁰ *is actually devided into 120 slices which are approximately 1 mm broad (For more details see text).*

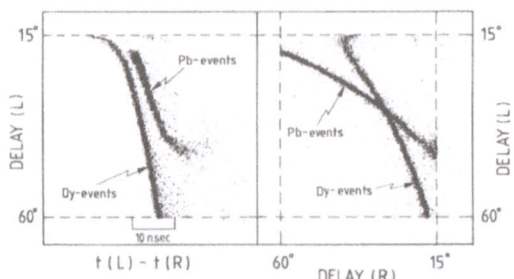

Fig. 8   *Two-dimensional plot of the time-of-flight difference vs.
the position (delay) in detector (L) [left panel] and of
the position (delay) in detector R vs. the position (delay)
in detector (L). [right panel] for 4.7 MeV/amu ^{208}Pb projectiles impinging on a 800 μgr/cm^2 ^{164}Dy-target. (Dy-events:
detection of Dy in the recoil detector (L). Pb-events: detection of Pb in the recoil detector (L)).*

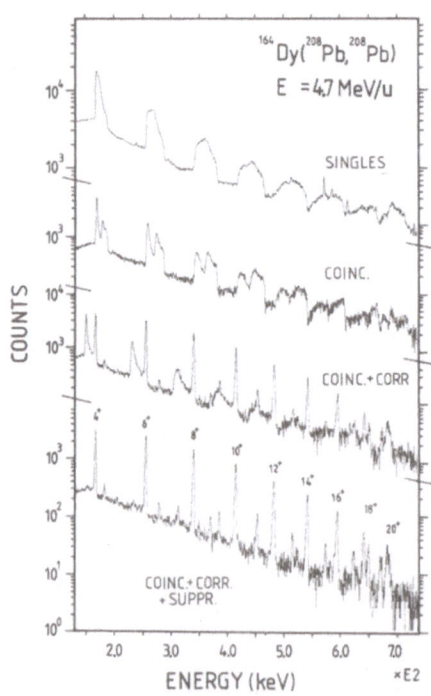

Fig. 9   *Various stages in the data reduction for the γ-rays observed
in the forward Ge-detector ($\vartheta_{\gamma_0} = 30^0$). The spectrum at the
bottom displays the Doppler-corrected γ-spectrum for Dy-
events with $17^0 \leq \vartheta_2 \leq 58^0$. The marked lines are due to the
ground state band transitions I → I-2 in ^{164}Dy.*

can actually use a detector identical to the recoil detector without restricting the efficiency of the set-up by the additional coincidence requirement (see also fig.6). In the MCE experiments of the actinides with ^{208}Pb ions the corresponding detector was positioned between $50^{\circ} \leq \vartheta_1 \leq 90^{\circ}$; its cathode was subdivided to correspond again to curves of equal polar angle ϑ with respect to the beam and the delay was adjusted to be proportional to the lab. scattering angle ϑ.

As shown in fig.8 for the case of 4.7 MeV/amu ^{208}Pb ions impinging on an 800 µgr/cm^2 thick ^{164}Dy-target, the two types of events can be clearly separated. The time-of-flight difference and position resolution which was obtained in this set-up with ^{208}Pb projectiles is sufficient to achieve a kinematic separation for targets with $\left|208 - A_2\right| \gtrsim 20$. Moreover, γ-events due to impurities and supporting foils (which might be necessary in some cases to stabilize the targets) are strongly suppressed by the kinematic coincidence. In fig.9 some intermediate steps in the data reduction are shown; although the Doppler-corrected and purified γ-spectrum displayed at the bottom part of the figure is obtained in an "on-line" mode the data are also stored event by event on magnetic tape to allow for a more sophisticated "off-line" analysis. The energy resolution which can be obtained in these spectra is of the order of 3 - 4 keV for $E_{\gamma 0}$ = 500 keV and is mainly due to finite size effect of the γ-detector. To reduce possible errors in the transition energies arising from uncertainties in the determination of the Doppler correction, we in

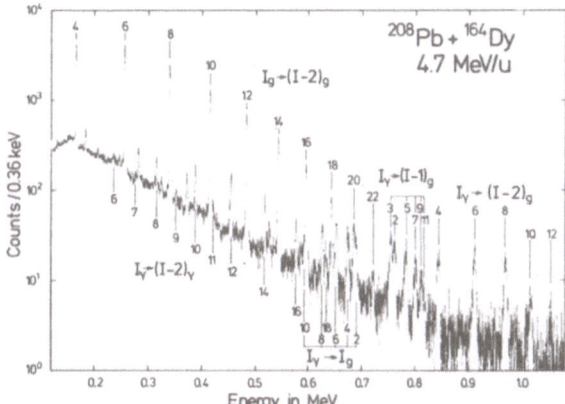

Fig. 10   *Total Doppler-corrected γ-spectrum following MCE of ^{164}Dy with 4.7 MeV/amu ^{208}Pb-ions ($64^{\circ} \leq \theta_1 \leq 146^{\circ}$). The unmarked γ-lines are due to the excitation of the target-impurity ^{162}Dy.*

general calibrate the corrected spectra obtained for the various
ϑ-slices via the usually known transition energies of the low spin
states. Moreover, by averaging the energies observed in the two γ-
detectors 180° apart several correction factors cancel each other.
In this way ultimate accuracies for $E_{\gamma 0}$ of up to 0.3 keV can be
achieved.

The total Doppler-corrected γ-ray spectrum for ^{164}Dy is shown
in fig.10. All observed γ-lines can be attributed to the decay of
states belonging to the g- and γ-band of ^{164}Dy and to the g-band of
^{162}Dy contained as a ≃ 3 % impurity in the targets used; the g- and
γ-band of ^{164}Dy can be followed up to $I^{\pi} = 22^+$ and 18^+, respectively.
The assignments are primarily based on energy systematics and the
Θ_1-dependence of the γ-yields. Neither members of the β-band (not
known in ^{164}Dy) nor of the lowest lying, known octupole band are
observed. Prior to our measurements the g- and γ-band of ^{164}Dy was
known[10] from light ion induced Coulomb excitation and inelastic
scattering experiments up to 14^+ and 6^+, respectively.

3.2 Determination of B(E2)-values from γ-ray yields

As will be shown below, a model independent determination of
B(E2)-values between the various states is only possible if the γ-
ray yields are measured over a wide range of c.m. scattering angles
Θ_1. In our particle-γ-set-up described above the γ-intensity $Y_{I_1 \to I_2}(\Theta_1)$,
which contains contributions from the direct excitation of the state
with spin I_1 as well as from higher lying levels feeding it, can
be determined from the triple-coincidence data for $150^\circ \lesssim \Theta_1 \lesssim 60^\circ$.
For the B(E2)-value determination of the low spin members of the
ground state band it is necessary, however, to even enlarge the angular range to lower Θ-values. This is achieved by requiring only a
double coincidence between the γ-ray and the recoil detector (L) and
by analysing also those events which correspond to the detection of
the scattered Pb-projectiles in (L). As the transition energies are
known from the triple-coincidence measurements, the observed Doppler-
shift can now be exploited to distinguish these events from those
corresponding to the detection of the recoiling nuclei in (L). Thus
it is possible to determine γ-yields at least for transitions between the low spin members of the ground state band also for projectile
c.m. angles as small as 30°; the actual lower limit depends, of course,
on A_2.

In fig. 11 the γ-yields for the ground state band transitions
in ^{164}Dy are shown including both event types. As for the two event
types the angular correlation function is viewed from ≈ 0° and ≈ 90°
with respect to the symmetry axis (see fig. 3b), that is in its
maximum and minimum, respectively, the good agreement between the
normalized γ-yields in the overlap region indicates that effects due
to the possible attenuation of the angular correlation caused by
the hyperfine interaction between the nucleus and the highly ionized
Dy-nuclei are negligibly small.

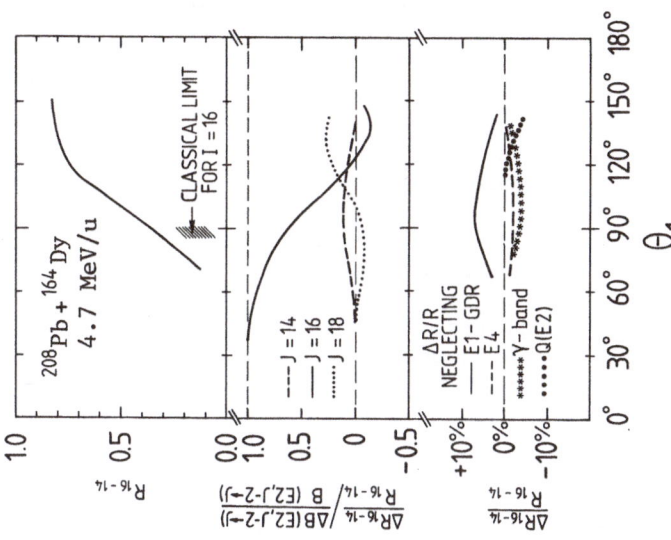

Fig. 12

Sensitivity of the intensity ratio $R_{I \to I-2}(\theta_1)$ defined by eq. (15) on various electromagnetic matrix elements for the $16^+ \to 14^+$ ground state band transition in ^{164}Dy.

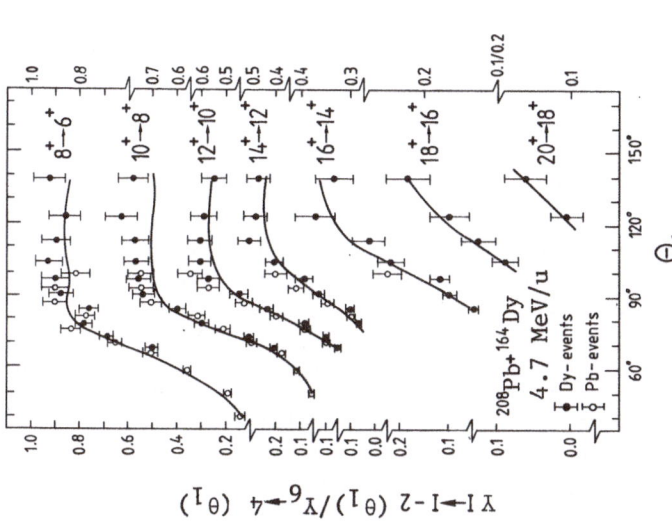

Fig. 11

Normalized γ-ray yields for the ground state band transitions in ^{164}Dy as a function of the c.m. scattering angle θ_1 of the ^{208}Pb-projectile. The solid lines represent the best fit of the data obtained by adjusting the corresponding $B(E2; I-2 \to I)$-values as discussed in the text.

To visualize how we determine B(E2)-values for the g-band transitions
in even-even nuclei from the corresponding yield data let's first
consider a description of the multiple Coulomb excitation process
in n-th order perturbation theory, where the g-band can be excited
by n E2-excitations up to spin I = 2n. In n-th order the yield ratio

$$R_{I \to I-2}(\Theta_1) = \frac{Y_{I \to I-2}(\Theta_1)}{Y_{I-2 \to I-4}(\Theta_1)} \tag{15}$$

for I = 2n and scattering angles below a certain limiting angle Θ_n
is directly proportional to the corresponding B(E2)-value, i.e.
$R_{I \to I-2}(\Theta_1) \propto B(E2; I-2 \to I)$. This proportionality, which is the
generalization of the well known first order perturbation expansion
$\sigma_{2+} / \sigma_{0+} \propto B(E2, 0^+ \to 2^+)$, has been used in the literature to determine B(E2)-values from multiple Coulomb-excitation experiments;
its applicability to actual experiments, however, is highly questionable as for scattering angles where the level with spin I = 2n is
observed with a reasonable yield, n-th order perturbation theory is
no longer valid. Nevertheless, the perturbation theory suggests,
that the B(E2)-value determination should be based on the yield
ratios $R_{I \to I-2}(\Theta_1)$ as defined by eq.(15) rather than on the γ-ray
yields $Y_{I \to I-2}(\Theta_1)$ themselves.

   To explore the correlation between the yield ratio within the
ground state band and the corresponding B(E2)-value, we use the semi-
classical Winther-de Boer code and proceed in the following way:
Starting from rotational model B(E2)-values we vary all B(E2)-values
in the ground state band one after the other and study the theoretical
R-values for different Θ_1. We define a sensitivity matrix $C_{IJ}(\Theta_1)$,
which describes the correlation of $R_{I \to I-2}(\Theta_1)$ with a certain B(E2;
J-2 → J)-value within a linear expansion,

$$C_{IJ}(\Theta_1) = \frac{\Delta R_{I \to I-2}(\Theta_1)}{R_{I \to I-2}(\Theta_1)} \Bigg/ \frac{\Delta B(E2; J-2 \to J)}{B(E2; J-2 \to J)} \tag{16}$$

In fig. 12 the scattering angle dependence of some elements of the
sensitivity matrix are shown for the case of 4.7 MeV/amu ^{208}Pb ions
on ^{164}Dy. It is obvious that the result of the n-th order perturbation theory, $C_{IJ}(\Theta_1) = \delta_{IJ}$, is only valid in an angular range,
where the yield of the corresponding γ-line is so small that it cannot be determined experimentally(see fig.11 and the top part of fig.
12). On the other hand there is almost no correlation between
$R_{I \to I-2}(\Theta_1)$ and the corresponding B(E2; I-2 → I)-value in the region
of large Θ_1, where the $R_{I \to I-2}$-values have reached their maximum;
here $R_{I \to I-2}$ depends on almost all B(E2; J-2 → J)-values including
those with J > I. Therefore, if one aims at a determination of individual B(E2)-values, the intermediate angular range is most im-

portant where the sensitivity $C_{II}(\theta_1)$ is about 0.5. For a given
state I of the ground state band the corresponding scattering angles
are those around and closely above the classical limit θ_I, i.e. for
angles with $f(\theta_I) \approx I\hbar/L_C^m$ (see figs.4, 12); they vary from $\theta_{2+} \approx$
30° to $\theta_I \to 180^\circ$ for $I \to L_C^m/\hbar$.

In the actual determination of the B(E2)-values from the experimental yield ratios we use a two-fold iterative procedure: (i)
B(E2; I-2 → I)-values together with their statistical errors are
deduced from a least-squares fit of the theoretical to the experimental $R_{I \to I-2}(\theta_1)$-values using eq.(16) but retaining only the diagonal elements of $C_{IJ}(\theta_1)$. Then new theoretical R-values are calculated using eq.(16) together with the new set of B(E2)-values and
considering the <u>complete</u> sensitivity matrix. The procedure is then
repeated until a stationary solution for the B(E2)-values is obtained.
(ii) With this set of B(E2)-values new theoretical R-values are calculated performing a completely new semi-classical calculation. The
total procedure is repeated until convergence is reached. Note that
the procedure ensures that the $R_{I \to I-2}(\theta_1)$-values which display the
largest sensitivity to the corresponding B(E2; I-2 → I)-values are
most heavily weighted in the fit, while those R-values influenced
by many B(E2)-values have only a very small weight. Correspondingly,
the correlations between the experimental B(E2)-values are weak and
usually lead to only slightly larger errors in the extracted B(E2)-
values when taken into account in the final step of the analysis.
The weak correlation simultaneously assures that the influence of
the B(E2)-values between the very low spin states, which usually
have to be taken from the literature, onto the B(E2)-values extracted for the high spin states is small.

The ground state B(E2)-value in ^{164}Dy resulting from the fit
of the data shown in fig.11 are displayed in fig.13a. In the semi-
classical calculations of the γ-yields the influence of the static
quadrupole moments and of the E4-matrix elements (both estimated
using the rigid rotor model) as well as of the giant dipole resonance and of the γ-band (see below) was taken into account. The extracted B(E2)-values are in agreement with the rigid rotor predictions for $Q_{og} = 7.5$ ebwith a statistically significant, theoretically
not understood exception at low spins.

Possible systematic uncertainties of the extracted B(E2)-values
stemming e.g. from the efficiency calibration of the Ge detector,
the angle calibration of the particle detector, the target thickness,
the beam energy etc. are greatly reduced since only yield ratios are
considered. Taken together, they typically amount to a few percent
error in the B(E2)-values. Another source of systematic error is
due to an incomplete consideration of other electromagnetic matrix
elements influencing the excitation of the g-band levels. As can be
seen in the bottom part of fig.12, their influence on the yield
ratios is small in the relevant θ_1-region. Since at least reasonable

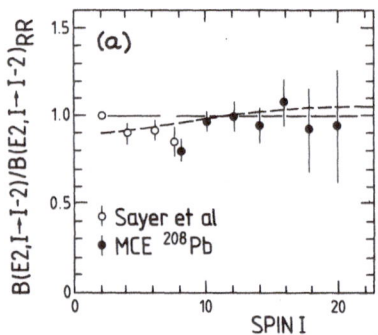 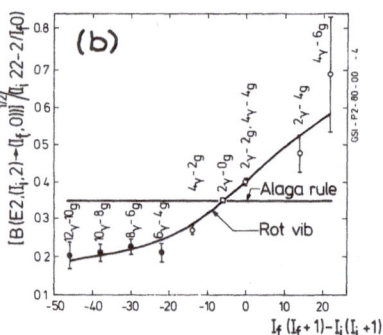

Fig. 13 (a) Inband B(E2)-values for the ground state band of ^{164}Dy, normalized to the rigid rotor predictions (the open points are from ref. 11). The dashed line presents the prediction of the modified rotation-vibration model (see text). (b) Comparison of the interband E2-matrix elements between the γ- and ground state band in ^{164}Dy with the prediction of the modified rotation-vibration model (see text). The open points are taken from ref. 12.

model estimates for these matrix elements are included in our semi-classical calculations, even large uncertainties in these estimates do only weakly influence the extracted g-band B(E2)-values; for ^{164}Dy these errors were estimated to be less than 5 % in the B(E2)-values. Finally we have to estimate the effect of the semi-classical approximation onto the B(E2)-values deduced. Scaling the calculations of Tolsma[13] to e.g. the ^{164}Dy-measurement we find that in the angular region of interest the quantal R's differ from the semiclassical R's by only a few percent. We can thus conclude that the systematic uncertainties amount e.g. in the case of ^{164}Dy to a combined error of 5 % - 10 % in the B(E2)-values.

The procedure described above can be used to determine in a model independent way individual B(E2)-values between states belonging to the main excitation path. The determination of B(E2)-values to or within sidebands from the observed γ-yield can in general not be performed without explicit model assumptions concerning the relative signs of the reduced matrix elements and, quite often, of even the size of some of them.

In ^{164}Dy, for example, we described the γ-intensities observed for transitions in the γ-band and between the γ-band and the g-band in a modified rotation-vibration (RV) model. In this analysis the intrinsic quadrupole moment in the γ-band as well as a factor reducing the interaction Hamiltonian of the original RV model[14] were adjusted to get agreement between calculated and observed γ-yields for <u>all</u> transitions. The best fit was obtained for $Q_{0\gamma}$ = 8.2 eb and a reduction factor of 2.7, making use of the parametrization of

ref. 15. The magnitude of the band mixing is mainly determined by
the E2-branching ratios of the $I_\gamma \to (I-2)_g$-relative to the $I_\gamma \to (I-2)_\gamma$
transitions for I_γ = 6, 8, 10, and 12. The deviation of the deduced
interband E2-matrix elements from the Alaga-rule (which neglects
the mixing between the K = 0 and the K = 2 band) is shown in fig.13b.
A first order perturbation treatment using an effective ΔK = 2 interaction [16] predicts a straight line through the points at low spin
and does not reproduce the high spin data. The solid line in fig.13b
shows the result of the calculation within the modified RV model.
For the g-band B(E2)-values it predicts only small deviations from
the rigid rotor values; satisfactory agreement between these predictions and the measured values can be obtained by a slight adjustment of the intrinsic quadrupole moment Q_{og} (see fig.13a). Thus our
modified RV calculation yields a consistent description of the collective properties of ^{164}Dy for spins up to 20^+. One should bear in
mind, however, that agreement could only be obtained by reducing
the strength of the RV-interaction Hamiltonian by almost a factor
of 3. Obviously the usual ansatz[14] is overestimating the coupling
between the ground state and vibrational bands, or equivalently, is
overestimating[17] the stretching of the nucleus due to the centrifugal
force acting at high angular momenta, i.e. large rotational frequencies.

3.3 MCE-Experiments with Thick Targets

    Although in MCE experiments performed with heavy ions and thick
(or backed) targets the identification of the highest states populated is hindered by the large recoil velocities of the excited
nuclei and the correspondingly long slowing down times, the γ-lineshapes observed in these measurements carry information about the
time-development of the decaying state. The γ-lineshape may thus
be used to extract the lifetime τ of the state the decay of which
is observed (Doppler-Shift-Attenuation method) if one succeeds to
suppress the feeding of the state through the γ-decay of higher
lying states; this feeding produces an additional time delay which
might completely mask the dependence of the γ-lineshape on the lifetime τ.

    As in experiments with very heavy projectiles most of the scattered
projectiles have lab. angles < 90°, one cannot control the feeding
in these experiments by coincident measurement of the scattered
particles. We therefore developed a technique which employs the
measurement of the total γ-energy released in the de-excitation
process. The experimental set-up (see fig.14) consists of a large
annular NaI-detector radially segmented into 6 separated parts.
The target is positioned in the centre of the axial bore. Individual
γ-transitions are recorded by a Ge(Li) detector at 0°, while the
rest of the γ-cascade is detected in the NaI-detector. For an ideal
set-up the sum of the energy-pulses observed in the Ge- and the NaI-
detector is equal to the excitation energy ΔE_f of the state reached

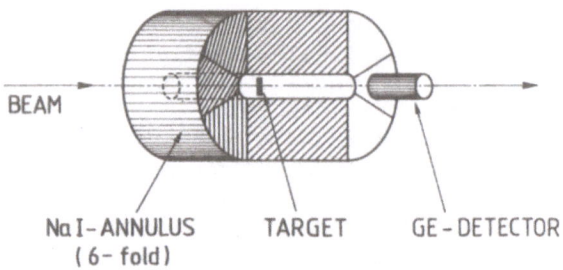

BEAM

NaI–ANNULUS   TARGET   GE–DETECTOR
(6– fold)

Fig. 14   *Schematic view of the sum-spectrometer set-up used in MEC-*
          experiments with thick targets. The size of the NaI-annulus
          was 25 cm Ø x 20 cm for the experiment shown in fig. 15 and
          40 cm Ø x 30 cm for the measurement shown in fig. 16.

238U

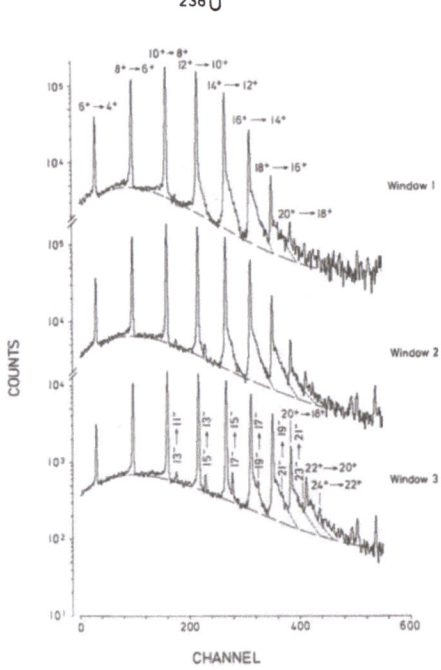

Fig. 15   *γ-spectra observed in the sum-spectrometer set-up at 0°*
          with respect to the beam by bombarding a thick 238U *target*
          with 4.7 MeV/amu 208Pb *ions. The spectra were gated with*
          three different sum-energy windows (Window 1, 2 and 3 cor-
          respond to low, medium and high sum-energies, respectively).
          The g-band (more precisely the α-band, see sect. 4) of 238U
          can be followed up to spin 24⁺, the K = 0⁻ octupole band
          up to spin 23⁻.

in the excitation process. Thus by gating on the total sum-energy
it is possible to select the starting point of the γ-cascade corresponding to the γ-event observed in the Ge-detector.

Due to the fact, however, that the NaI-annulus does not cover
the full solid angle and does not have 100 % efficiency for the
full energy peak, the resolution obtained in ΔE_f is limited. Nevertheless, the set-up is still rather effective in reducing the cas-
cade feeding. This is demonstrated in fig.15 displaying three Ge-
spectra observed in the bombardment of a thick ^{238}U-target with
4.7 MeV/amu ^{208}Pb ions, which where gated with different sum-energy
windows. Consider, for example, the $14^+ \rightarrow 12^+$ g-band transition ($\tau(14^+) \approx$
4 psec); its lineshape is most pronounced if the sum-energy selected
is low (corresponding to excitation energies close to ΔE_{14^+}), while
almost no lineshape is observed in the γ-spectrum gated on a high
sum-energy, indicating that under this condition the direct population of the 14^+ state is small and the time involved in the feeding
process is so long that the U-nuclei are slowed down before the
$14^+ \rightarrow 12^+$ decay takes place. The lifetimes deduced by analysing the
γ-lineshapes shown in fig.15 will be discussed in section 4.

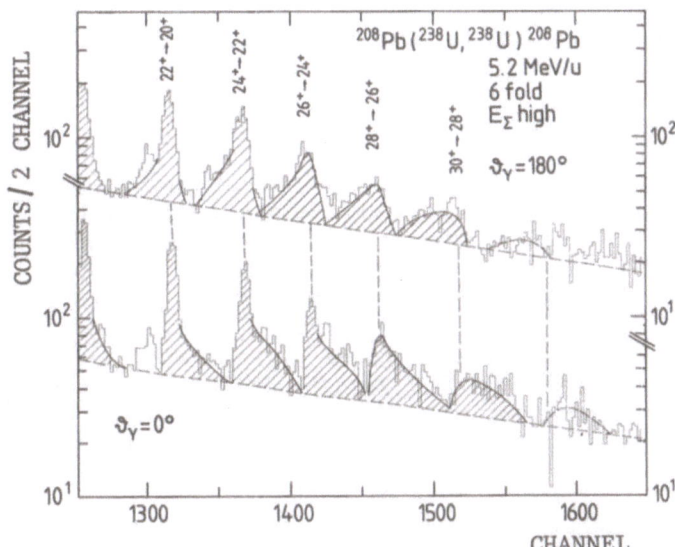

Fig. 16 Projectile excitation of ^{238}U observed in the sum-spectro-
         meter set-up by bombarding a thick ^{208}Pb-target with 5.2
         MeV/amu ^{238}U-ions. The γ-spectra were recorded at 0° and
         180° (using an annual Ge-detector) with respect to the beam
         and were gated on high sum-energies and 6fold events. The
         g-band (c-band) of ^{238}U can be followed at least up to spin 30.

In the actinide region, where even the $I^{\pi} = 30^+$ state of the g-band is expected to live still for 0.5 psec, the sum-energy technique can also be applied to observe and identify very high spin states if projectile excitation is studied: The slowing down times are considerably shorter by inverting the role of projectile and target. For e.g. 5.2 MeV/amu ^{238}U on ^{208}Pb the recoil velocity of the excited ^{238}U-projectiles in head-on collisions ($\theta_1 = 180^o$), being the most important ones for the excitation of the highest spin states, amounts to v/c ($\theta_1 = 180^o$) = 0.7 % as compared to an U-recoil velocity of v/c ($\theta_1 = 180^o$) = 11 % obtained when bombarding an ^{238}U-target with 5.2 MeV/amu ^{208}Pb ions. The stopping time for ^{238}U of v/c = 0.7 % recoiling in Pb is about 0.5 psec as compared to about 2 psec for ^{238}U of v/c = 11 % recoiling in U. Moreover, by triggering on very high sum-energies and high multiplicities (e.g. by requiring all 6 parts of the annulus to fire) one can enhance events belonging to head-on collisions and simultaneously maximize the feeding time.

Obviously the method just described can only be applied in cases where the nuclei of interest are available as beams with decent intensities and the lifetimes of the excited states are still of the order of a psec. Fig.16 shows the result of such a measurement for 5.2 MeV/amu ^{238}U impinging on a thick ^{208}Pb-target using in addition to the Ge-detector at 0° a Ge-annual detector at 180°; the g-band of ^{238}U (more precisely: the c-band, see below) can be followed up to at least spin 30^+, which is also the highest spin state we observed in a thin target experiment (cf. fig. 17).

4. HIGH SPIN STATE STUDIES IN ACTINIDE NUCLEI

Having discussed in some detail the general procedure adopted by us at GSI to measure and analyse MCE-experiments with very heavy projectiles I shall finally discuss some MCE-results obtained for actinide nuclei. As already pointed out, MCE with very heavy ions like ^{208}Pb is at the time being the only suitable way to study high spin states in deformed actinides. Thus very little was known so far about the bahaviour of these nuclei at high angular momenta; while the g-band of ^{232}Th and ^{238}U was studied at the HILAC at Berkeley with Xe-projectiles up to spin $I^{\pi} = 20^+$ (ref.18) and $I^{\pi} = 22^+$ (ref.19) respectively, in the other even-even nuclei the g-bands are known at most up to spin 10 or 12 (ref.10).

A priori one does not expect the actinides to behave in a qualitatively different way than the equally well deformed rare earth nuclei. In particular, the structure of the high spin states of the actinide nuclei should be affected in a similar way as in the rare earth nuclei by the centrifugal and Coriolis forces acting at high rotational frequencies. It is generally believed that the backbending behaviour observed in the yrast sequence of many rare earth nuclei around spin $I^{\pi} \approx 14^+$ to 16^+ is due to an intersection of the g-band with a quasi-particle band (s-band) involving two rotation-aligned

$i_{13/2}$-neutrons[20]. In the actinide nuclei the highest j-orbits close to the Fermi surface, which are expected to give rise to the lowest lying s-bands, are the $j_{15/2}$ neutron and $i_{13/2}$ proton orbits. In contrast to the rare earth nuclei, however, where the lowest lying proton s-band, which is based at the $h_{11/2}$ orbit, crosses at a much higher frequency than the $i_{13/2}$ neutron s-band, in the actinides the lowest neutron and proton s-bands are predicted[21] to cross the g-band at almost the same frequencies ($\hbar\omega_c^n \approx \hbar\omega_c^p \approx 0.22$ MeV). But besides the exciting question if crossings of the g-band with rotation-aligned two quasi-particle bands do occur in the actinides as well, the collective structure of the g-bands of these nuclei is of equal interest as the validity of collective model approaches, applied so far mainly to low spin states, can be tested in the high-spin regime.

So far we studied ^{232}Th, ^{234,236,238}U, and ^{248}Cm by bombarding thin (≤ 1 mgr/cm^2), enriched targets with 5.3 MeV/amu ^{208}Pb projectiles using the particle-γ-coincidence arrangement described above. In addition, ^{232}Th and ^{238}U were investigated using the sum-spectrometer technique described in subsection 3.3. In the nuclei

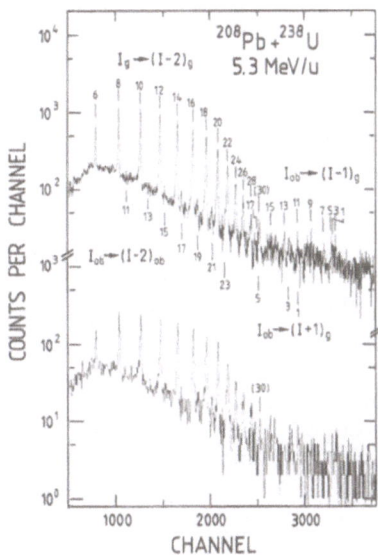

Fig. 17   Doppler-corrected γ-ray spectrum for ^{238}U observed in the bombardment of a 0.8 mgr/cm^2 ^{238}U-target with 5.3 MeV/amu ^{208}Pb ions for c.m. scattering angles $97° \leq \theta_1 \leq 146°$ (top) together with a 1-fold spectrum obtained by requiring in addition to the γ-recoil-projectile triple-coincidence a further γ-coincidence with a 4" x 5.5" NaI-detector (bottom). The g-band (c-band) and the K = 0⁻ octupole band (ob) can be followed up to spin 30⁺ and 23⁻, respectively.

studied we observe a collective (c-) band, which is at least at low spins identical to the g-band, up to $I^\pi = 28^+$ in ^{232}Th and ^{234}U, and $I^\pi = 30^+$ in ^{236,238}U and ^{248}Cm. The side band most strongly excited in these nuclei beside the c-band is the lowest lying octupole band, which could be followed e.g. in ^{238}U up to the 23^- state. The assignments of previously not observed γ-transitions are based on energy systematics and in particular on the θ_1-dependence of the γ-ray yields. In fig.17 the Doppler-corrected γ-ray spectrum observed for ^{238}U is shown (top) together with a 1-fold spectrum obtained by requiring in addition to the γ-recoil-projectile triple-coincidence a further γ-coincidence with a 4"x 5.5" NaI detector (bottom). The 1-fold spectrum , in which the high spin states of the c-band are enhanced due to their high multiplicity, supports the interpretation that the broad line occurring around channel 2500 in the top spectrum of fig.17 contains the 30^+-member of the c-band. This state is even more clearly observed in the sum-spectrometer measurement displayed in fig. 16. As an example for the power of the particle-γ -coincidence arrangement, the Doppler corrected γ-spectrum obtained for ^{248}Cm is displayed in fig.18. The target consisted of a 250 μgr/cm^2 isotopically enriched ^{248}Cm, which was electroplated onto a 1 mgr/cm^2 Ti foil pointing upstream, and the ^{208}Pb-energy was adjusted such that the beam energy after the foil was 5.3 MeV/amu. Note that the intense γ-background due to the excitation of the Ti-isotopes and the Ti-induced fission events could be completely suppressed by the kinematic coincidence requirement.

The alignment plot for the c-band transitions of the actinides studied is shown in fig. 19; displayed is the average spin $I^* = I-1$ for the transition $I \rightarrow I-2$ vs. the rotational frequency $\hbar\omega = (\Delta E_I - \Delta E_{I-2})/2$. At high rotational frequencies all nuclei show deviations from the two parameter Variable-Moment-of-Inertia (VMI) fit[22], which was adjusted in the low spin region. The deviations, which are most pronounced for ^{238}U and ^{248}Cm, resemble the upbending behaviour observed in rare earth nuclei and may be interpreted as a crossing of the g-band with a strongly interacting excited band. It should be noted, that the c-band at high spins does not necessarily correspond to the yrast-sequence of the nucleus as a crossing of the g(c-) band with a weakly interacting excited band would be difficult to detect in MCE experiments; in this case the main excitation path would follow the g(c-)band rather than the crossing band.

The in-band B(E2)-values play a key role in the discussion of the possible reasons for the observed frequency dependence of I^*. So far the B(E2)-values for transitions within the c-band were determined from the yield ratios $R_{I \rightarrow I-2}(\theta_1)$ for ^{232}Th, ^{234,236,238}U. Moreover, for ^{238}U the lifetimes of the states with $12 \leq I^\pi \leq 22^+$, being inversely proportional to the B(E2, $I \rightarrow I-2$)-values, were deduced from the γ-lineshapes observed in the sum-spectrometer measurements with thick targets (subsect. 3.3); the B(E2)-values extracted via the two different methods agree very well within their errors

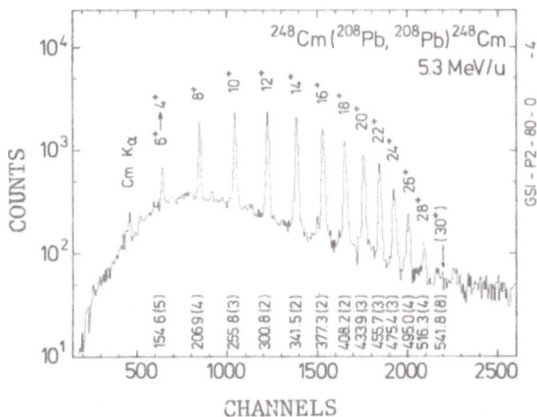

Fig. 18 *Doppler-corrected γ-ray spectrum observed in the bombardment of ^{248}Cm with 5.3 MeV/amu ^{208}Pb projectiles ($97^{o} \leq \theta_{1} \leq 146^{o}$). The transition energies within the g-band (c-band) are given in keV.*

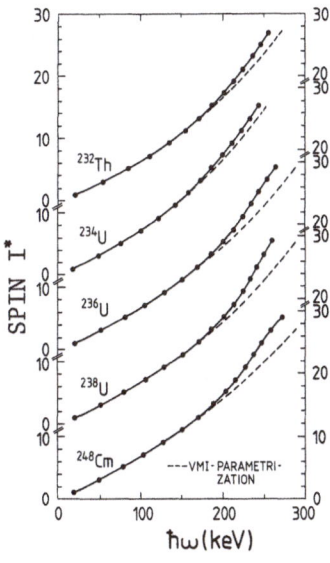

Fig. 19 *Plot of the average spin $I^{*} = I-1$ for the transitions $I \rightarrow I-2$ vs the rotational frequency $\hbar\omega = (\Delta E_{I} - \Delta E_{I-2})/2$ for the c-bands of the five even-even actinides studied. The dashed lines represent a two parameter VMI-fit to the data at low spins, where the c-band is identical to the g-band.*

of ±10 %. In all four nuclei analysed the B(E2)-values were found
to be in remarkably good agreement with the rigid rotor predictions
up to the highest spins observed. Instead of displaying the individual B(E2)-values, fig.20 shows the (normalized) γ-yields observed
within the c-band of ^{238}U as function of spin for the highest Θ_1-
windows. The solid curves are the intensities calculated with the
semi-classical approximation assuming rigid rotor values for the in-
band B(E2)-values and including E2- and E4-excitations as well as
the excitation of the β-, γ- and octupole band. Also shown in fig.20
are the intensities calculated in the same way but using B(E2)-values
predicted by the extended rotation-vibration (RV)- model[23]. In the RV-
model the deviation of I^* from the simple rigid rotor prediction
$I^* \alpha \omega$ is interpreted as being due to the RV-coupling, i.e. due
to the centrifugal stretching of the nucleus under the action of
the centrifugal force. This model leads to B(E2)-values, which are
for the highest spins observed up to 40 % larger than the rigid
rotor values; they are in obvious disagreement with our experimental
findings. This demonstrates again that the RV-coupling in real nuclei
is smaller than usually assumed in these calculations and may not be
adjusted to account for the observed increase of the moment of inertia, i.e. the observed frequency dependence of I^*.

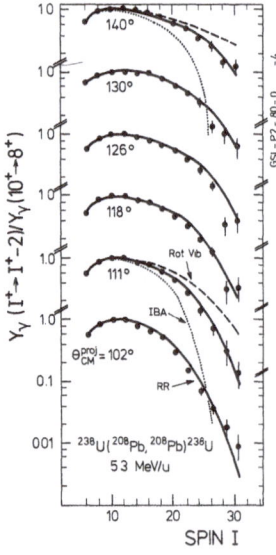

Fig. 20 *Comparison of the (normalized) γ-yields within the c-band
of ^{238}U with calculated yields using inband E2-matrix elements obtained from the rigid rotor (RR), the extended
rotation-vibration[23] (Rot.Vib.) and the interacting boson
(IBA, SU (3)-limit)[24] model. For more details see text.*

The rigid rotor behaviour found for the B(E2)-values excludes all models which connect the change of the moment of inertia with increasing rotational frequency to a deformation change; it excludes in particular the possibility that the upbending is caused by the interaction of the g-band with a band built on a much more strongly deformed configuration. Our data, i.e. the frequency dependence of I^* and the B(E2)-values, are consistent, however, with the assumption that the smooth increase of I^* described by the VMI-curve is mainly due to an increasing contribution of unpaired partially aligned configurations to the g-band and that the upbending is caused by a crossing of the g-band with a strongly interacting s-band. One way of getting more insight into the structure of the crossing s-band and to verify whether the predicted neutron or proton s-band is responsible for the observed upbending in the actinides is to study the alignment effect in odd-A actinides like ^{235}U and ^{237}Np having the odd particle in the neutron $j_{15/2}$ and in the proton $i_{13/2}$ shell, respectively. As the unpaired particle delays the alignment of a pair within its own subshell, these nuclei might provide a separate view of the proton and neutron system. Following the procedure of ref.21, the amount of aligned angular momentum $i(\omega)$ carried by the individual nucleons is given e.g. for the even U-isotopes by the difference between the measured $I^*(\omega)$ and the VMI-curve shown in fig.19, as the latter is believed to represent the $I^*(\omega)$-dependence of the core. The resulting $i(\omega)$ are plotted together with those for ^{235}U and ^{237}Np (ref.25) in fig. 21. Note that beside the alignment of the odd nucleon an additional alignment is observed for ^{235}U (but not for ^{237}Np), which starts at similar rotational frequencies as in the even U-isotopes and which indicates that the upbending observed might be due to the s-band based on an aligned $i_{13/2}$ proton pair, crossing the g-band at $\hbar\omega_c^p \approx 0.25$ MeV. MCE measurements on ^{235}U and ^{237}Np with ^{208}Pb-ions are presently under way at GSI to follow the g-bands of these nuclei to higher rotational frequencies.

Finally I would like to compare the present results for the actinides with the interacting boson model (IBA)[24], which is an elegant and successful model to describe the collective properties of low lying states all over the periodic table. However, the IBA predicts a spin cut-off of the g-bands at I = 2n, where n denotes the number of active bosons which is given by half the number of nucleons (holes) away from the nearest closed shell. This spin cut-off, which is essential to obtain the SU(6) group structure on which the IBA is based, is in obvious contrast to the geometrical model. Assuming ^{208}Pb to represent the core, the IBA predicts a cut-off of the g-band at spins 24^+, 26^+, 28^+, 30^+, for ^{232}Th, ^{234,235,236}U, respectively, and strongly reduced B(E2)-values already at moderate spins. These predictions are in contrast to our data (see also fig. 20); it is felt that even by including rotation-aligned bands in the model space the effect of the g-band cut-off on the B(E2)-values within the c-band cannot be removed and agreement be obtained with

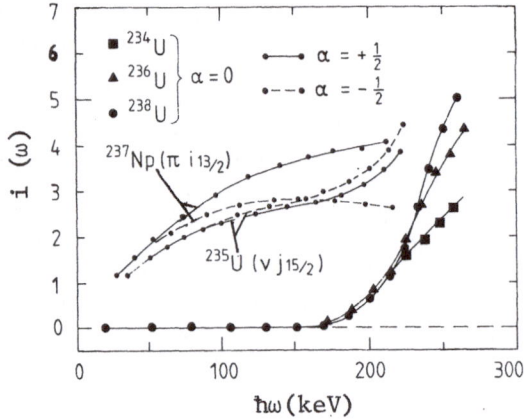

Fig. 21   Plot of the aligned angular momenta $i(\omega)$ for $^{234,236,238}U$
and ^{235}U, ^{237}Np, which were deduced as discussed in the
text (α denotes the signature[16] of the level).

the observed rigid rotor behaviour over the _entire_ spin region. As
the cut-off is mainly due to the restriction of the model space to
s- and d-bosons, our data indicates that this model space is not
sufficient to describe high spin states.

ACKNOWLEDGEMENT

     The Coulomb excitation program currently carried out at GSI
lives from the very close and fruitful collaboration of members of
the GSI Nuclear Spectroscopy Group (H. Emling, P. Fuchs, E. Grosse,
R. Kulessa[*], D. Schwalm, R.S. Simon, H.J. Wollersheim, R. de Vito[**])
with several outside user groups. The results discussed in sect. 4
were obtained within collaborations between GSI and the University
of Frankfurt (Th.W. Elze, J. Idzko, K. Stelzer, H. Ower), the University of Mainz (N. Kaffrell, N. Trautmann), the Vanderbilt University at Nashville (J.H. Hamilton, A.V. Ramayya, R.B. Piercey) and
the University of Munich (D. Evers). The author is indebted to all
his colleagues for the enjoyable collaboration and many stimulating
discussions.

[*] on leave from Dept. of Nucl. Physics, Jagiellonian Univ., Cracow,
     Poland

[**] on leave from Michigan State University, East Lansing, USA

REFERENCES

1. T. Huus and Č. Zupančič, Dan. Mat. Fys. Medd. 28 (1953) No. 1,
   reprinted in "Coulomb Excitation" ed. by K. Alder and A. Winther
   (Academic Press, New York, 1966) p. 33 (This reprint collection
   includes many important papers on Coulomb excitation up to 1965.)
2. G. Breit, R.L. Gluckstern, and J.E. Russel, Phys. Rev. 103 (1956)
   727; J. de Boer and J. Eichler, in Advances in Nuclear Physics,
   Vol. 1, ed. M. Baranger and E. Vogt (Plenum Press, New York, 1968)
   p. 1; O. Häusser, in Nuclear Spectroscopy and Reactions, part C,
   ed. J. Cerny (Academic Press, New York, 1974) p. 55.
3. K. Alder and A. Winther, Electromagnetic Excitation (North-Holland
   Publ. Comp., Amsterdam, 1975);
   see also K. Alder, A. Bohr, T. Huus, B.R. Mottelson, and A. Winther, Rev. Mod. Phys. 28 (1956) 432.
4. See e.g. F. Roesel, J.X. Saladin, and K. Alder, Comp. Phys. Com-
   mun. 8 (1974) 35.
5. A. Winther and J. de Boer, in Coulomb excitation, ed. K. Alder
   and A. Winther (Academic Press, New York, 1966) p. 303.
5. A. Lell, diploma thesis, University of Munich (1978).
7. E. Grosse, MPI Heidelberg, report V26 (1975).
8. K. Alder and A. Winther, Mat. Fys. Medd. Dan. Vid. Selsk. 32
   (1980) No. 8 (see also ref. 3);
   N. Rowley and P. Colombani, Phys. Rec. C 11 (1975) 648.
9. R. Bosshard, R.L. Chase, J. Fischer, S. Iwata, and V. Radeka,
   IEEE Transaction on Nucl. Science NS-22 (1975) 2053.
10. Tables of Isotopes, ed. by C.M. Lederer and V.S. Shirley (J.
    Wiley and Sons, New York, 1978).
11. R.O. Sayer et al., Phys. Rev. C17 (1978) 1026 (and references
    quoted therein)
12. R.N. Oehlberg et al., Nucl. Phys. A219 (1974) 543.
13. L.D. Tolsma, Phys. Rev. C20 (1979) 592.
14. A. Faessler, W. Greiner and R.K. Sheline, Nucl. Phys. 70(1965)33.
15. H.J. Wollersheim and Th.W. Elze, Nucl. Phys. A278 (1977) 87.
16. A. Bohr and B.R. Mottelson, Nucl. Structure (W.A. Benjamin Inc.,
    Reading, 1975) Vol. 2 (1975) p. 158
17. A similar reduction has been discussed by K. Neergard, priv. com.
18. M.W. Guidry et al., Nucl.Phys. A266 (1976) 228.
19. E. Grosse et al., Phys. Rev. Lett. 35 (1975) 565.
20. F.S. Stephens and R.S. Simon, Nucl. Phys. A183 (1972) 257;
    R. Bengtsson and S. Frauendorf, Nucl. Phys. A314 (1979) 27; and
    Nucl. Phys. A327 (1979) 139;
    B. Banerjee, H.J. Mang and P. Ring, Nucl. Phys. A 215 (1973) 366
    A. Faessler, M. Ploszajczak,and K.W. Schmid, Progress in Particle
    and Nuclear Physics, ed. by H. Wilkinson (Pergamon Press, Oxford,
    1980), Vol. 5; M. Diebel, A.N. Mantri, and U. Mosel, Nucl. Phys.
    A345 (1980) 72.

21. R. Bengtsson, Int. Conf. on Nuclear Behaviour at High Angular
    Momentum, Strasbourg (1980), to be published in J. de Physique
    (see also R. Bengtsson and S. Frauendorf, ref. 20).
22. G. Scharff-Goldhaber, C.B. Dover, and A.L. Goodman, Ann. Rev.
    Nucl. Science 26 (1976) 239.
23. M. Seiwert and P. Hess, Univ. Frankfurt, private communication.
24. A. Arima and F. Iachello, Ann. of Physics 111 (1978) 201.
25. R. S. Simon et al., to be published in Z. f. Physik.

INELASTIC ELECTRON SCATTERING AT LOW MOMENTUM

TRANSFER AND NUCLEAR STRUCTURE*

Achim Richter

Institut für Kernphysik der Technischen
Hochschule Darmstadt
6100 Darmstadt, Germany

I.   INTRODUCTION AND TERMINOLOGY

In these two lectures, I shall describe how inelastic electron
scattering can be used to study rather elementary problems of nuclear structure which in part may have some far-reaching consequences on fundamental questions of mesonic degrees of freedom in nuclear physics. Because of the limited time available and for pedagogical reasons (I have been told explicitly by the organizers of
this school to lecture for young graduate students in nuclear physics) I confine myself to (e,e') on nuclei at low momentum transfer. This means that we will discuss electric and magnetic transitions of low multipolarity (λ = 1 and 2) in light and heavy nuclei.
I will illustrate this subject by a few selected examples of some
very recent work at the Darmstadt electron linear accelerator
(DALINAC). The formalism and also often even the physics problems
are basically the same in electron scattering at higher energies
where nuclear transitions of high multipolarity are studied. Therefore we loose little by staying in the realm of (e,e') at low momentum transfer but gain by its conceptional simplicity and direct
connection to problems of nuclear structure. In all nuclear structure aspects we will rely on the shell model to be discussed by
Joe McGrory[1] at this school.

Although I will illustrate in these lectures mainly the terri-
tory of low energy electron scattering and nuclear structure I will
try to put the material to be presented in a much broader perspective of nuclear physics. Several of our experiments have been

---

*Supported by Deutsche Forschungsgemeinschaft

motivated by results of electromagnetic and mesonic processes at
much higher energies, by the behaviour of magnetic moments, by weak
interaction processes, by inelastic scattering and charge exchange
of hadronic probes and by the possible existence of pionic nuclear
phase transitions. As will become apparent particularly in the second half of these lectures we are able in low energy electron
scattering below the meson threshold to gain important information
on the behaviour of soft modes in pion condensation. Therefore, I
also should like to focus strongly your attention to the lectures
of Wolfram Weise[2] in this school.

In the   course of these lectures I shall draw heavily from two
very similar courses of lectures given[3] in 1979 and earlier this
year[4] updated by our present knowledge on the subject and by new
results both on the experimental and theoretical side.

After those few introductory remarks we shall now begin by defining some of the quantities entering the physics of inelastic
electron scattering and of nuclear structure. In the experiments
to be discussed later on we will always use this knowledge. Electrons are one of the best probes to study nuclear, and in particular collective nuclear excitations. As is shown schematically in
fig. 1, the doubly differential cross section $d^2\sigma/d\Omega dE_x$ can be
studied as well as a function of the energy as of the momentum transfer to the nucleus. The range of momentum transfer attain-

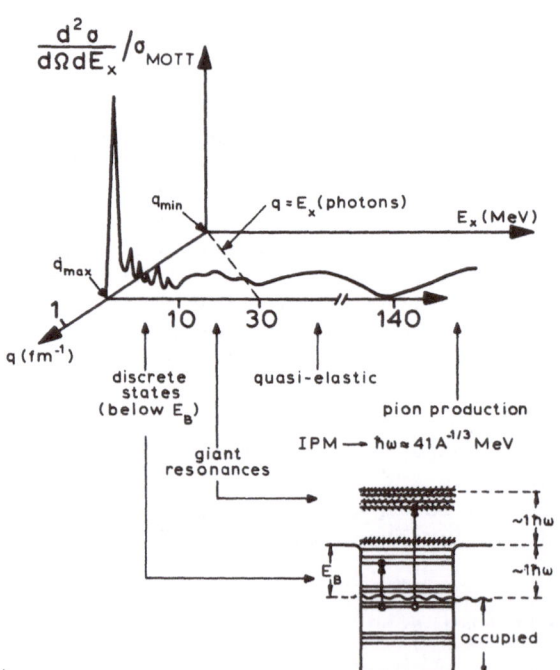

Fig. 1 The different
        regimes of nu-
        clear states
        excited in elec-
        tron scattering
        and the schema-
        tic explanation
        of discrete
        states and un-
        bound giant re-
        sonances in
        terms of the
        independent
        particle shell
        model (IPM).

able at the DALINAC is $q \lesssim 0.7$ fm^{-1} and nuclear states are excited up to $E_x \approx 30$ MeV in the experiments to be discussed later on. Following ref.[5] we may understand qualitatively what is shown in fig. 1. The discrete states below the binding energy E_B of a nucleon can be associated with simple particle-hole excitations between the different bound orbits in the independent particle shell model (IPM). The so-called unbound giant resonances correspond to the excitation of particles from bound orbits into the continuum. They therefore acquire broader widths. In these lectures we will primarily discuss the excitation of discrete nuclear states but occasionally embark on states in the unbound regime.

Those nuclear excitations depicted in fig. 1 can be characterized by the following properties:

(i)   There is some <u>clustering</u> of p-h strength at excitation energies (associated with alternating parity)

$E_x = 0\hbar\omega, \ 1\hbar\omega, \ 2\hbar\omega, \ 3\hbar\omega, \ \ldots\ldots$

$\Delta\pi = \ + \ , \ \ - \ , \ \ + \ , \ \ - \ , \ \ \ldots\ldots$

where in the IPM $\hbar\omega \approx 41A^{-1/3}$ MeV.

(ii)   In the presence of the nuclear residual interaction specific intraband effects occur, i.e. the excitation strength is concentrated at a certain excitation energy E_x by constructive superposition of p-h states. We may therefore speak of <u>collective excitations</u>.

(iii)   As illustrated schematically in fig. 2 the coupling of 1p-1h states to the underlying many particle – many hole states leads to a fragmentation of the excitation strength (or equivalently of the measured cross section) as a function of excitation energy E_x, i.e. to <u>fine structure</u> in the observables (even for states in the continuum). Therefore high energy resolution is a prerequisite in the experiments.

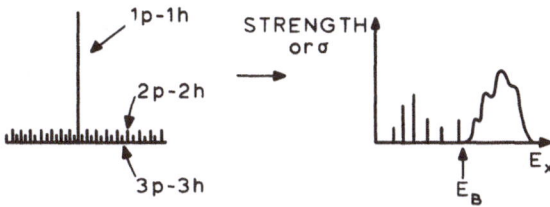

Fig. 2   Illustration of the coupling of 1p-1h states
         to more complicated configurations and its effect
         on the strength- (or cross section) distribution
         as a function of excitation energy.

The criteria (i) - (iii) may also be taken to characterize a
giant resonance (see also the lectures of Adrian van der Woude at
this school[6]):

a) It is a general feature of nuclei and its excitation energy E_x
   and width Γ (or strength) are smooth functions of mass number A.

b) The strength is localized narrowly around E_x.

c) It exhausts a large fraction of an energy weighted sum rule
   (EWSR). This again might be taken as a measure of collectivity.

Those features are well realized in case of the famous electric dipole giant resonance which is centered at $E_x \approx 78A^{-1/3}$MeV in heavy
nuclei. They are not always realized, however, for transitions
other than E1 and as we will see below in the case of ^{208}Pb the
hitherto called E2 giant resonance is lacking   collectivity. The
systematic study of magnetic transitions (M1 and M2) also shows
this lack of collectivity.

Before ending this brief first section let me remind you again,
that the advantages of electron scattering as a tool to study nuclear structure are well known (see e.g. ref.[7]), but that further
recent advance in experimental conditions of the accelerators, magnetic spectrometers and detection systems has led to an energy resolution of several parts in 10^{-4}. Electron scattering experiments
are now as sensitive to the details of nuclear level structure as
the corresponding experiments using hadronic probes, the (e,e')
experiments being more selective with respect to a certain multipolarity. This point will be illustrated through several experimental examples below and we will also remark upon the experimental
technique to obtain (e,e') spectra of high resolution.

II. ELECTROEXCITATION AND NUCLEAR STRUCTURE

The process of electroexcitation of nuclear levels is based
on the mechanism depicted schematically in fig. 3. An incoming
electron interacts with the charge-, current- and magnetization
density of the nucleus through the exchange of a virtual photon.
This virtual photon carries the momentum transfer. In elastic electron scattering those densities are pertinent to ground state properties of the nucleus while they are called transition densities
in the case of inelastic scattering. A determination of the latter

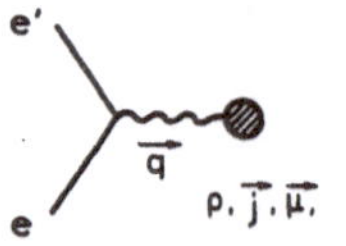

$$\sigma/\sigma_{MOTT} = |F_\lambda(q)|^2 \sim q^{2\lambda} B(\lambda, q)$$

Fig. 3   Interaction of the electron with the charge density ρ,
the current density $\vec{j}$ and magnetization density $\vec{\mu}$ of
the nucleus.

is the objective of (e,e'). Because of the fact that the momentum
transfer q obtainable in an actual experiment is distinctly finite
the transition densities can only be determined within some uncertainty.

Here we list and discuss only the relevant expressions which
are needed in order to understand the low momentum transfer experiments to be introduced below. The formulae given are simple ge-
neralizations of the expressions derived e.g. in ref.[7]. How are
the nuclear quantities which are used as a test for nuclear models
extracted from measured cross sections? In order to answer this
question we argue for the sake of clarity in first order plane
wave Born approximation (PWBA) although the data later on are all
analyzed in distorted wave Born approximation (DWBA). This is done
in a different notation than in ref.[7] but - partly for historical
reasons - in one which is used in ref.[8].

In (e,e') experiments at low q formfactors defined through

$$(d\sigma/d\Omega)/(d\sigma/d\Omega)_{MOTT} = |F_\lambda(q)|^2 \sim q^{2\lambda}B(\lambda,q) \tag{1}$$

are directly related to reduced transition probabilities B of a
transition of multipolarity λ. We get in particular for electric
(e) and magnetic (m) transitions the following expressions

$$\left(\frac{d\sigma}{d\Omega}\right)_{e\lambda} \sim q^{2\lambda}[B(C\lambda,q)V_L(\Theta) + B(E\lambda,q)V_T(\Theta)] \tag{2}$$

$$\left(\frac{d\sigma}{d\Omega}\right)_{m\lambda} \sim q^{2\lambda}[B(M\lambda,q)V_T(\Theta)]. \tag{3}$$

The nuclear physics is contained in the reduced transition
probabilities B while the angular behaviour of the longitudinal
and transverse parts (denoted by L and T) of the cross section is
described by the quantities V_L and V_T, respectively. The longitudinal part results from the interaction of the electron via the interaction of a virtual photon with the charge density of the nucleus, while the transverse part is due to the interaction with the
current and spin-magnetization density (see fig. 3). As noted above
those three densities are called transition densities. Three immediate consequences which are important with respect to the discussion of the experiments follow in connection with the relations
(2) and (3):

(i)   For $\Theta \rightarrow 180°$, $V_L \rightarrow 0$ and $V_T \rightarrow 0.25$; hence magnetic transi-
      tions are studied preferably under backward angles.

(ii) A measurement of $d\sigma/d\Omega$ at q = const. and under various scat-
      tering angles Θ allows in the manner of the Rosenbluth plot
      the separation of the longitudinal and transverse contribu-
      tion to the cross section, i.e. eq. (2) can be written in
      the form of a linear equation

$$(d\sigma/d\Omega)/V_T(\Theta) \sim B(C\lambda)[V_L(\Theta)/V_T(\Theta)] + B(E\lambda) = ax+b. \qquad (4)$$

(iii) For $q \to k$ where $k = E_x/\hbar c$ defines the photonpoint (E_x being the excitation energy of the level studied), $B(\lambda,q) \to B(\lambda,k)$ and we have in particular (if we neglect contributions from the magnetization density) the useful long-wavelength relation (Siegert-Theorem)

$$B(E\lambda,q) \to (k/q)^2 \, B(C\lambda,q). \qquad (5)$$

It is therefore at the photonpoint where results from inelastic electron scattering can be linked to results from other methods of nuclear spectroscopy like γ decay, photo- and Coulomb excitation. There is one further quantity which we need and which is directly related to the transition strength, and that is the so called ground state radiative width

$$\Gamma_\gamma^o \sim E_x^{2\lambda+1} (B\lambda,k). \qquad (6)$$

As is seen from eqs. (2) and (3) the measured differential cross sections on the l.h.s. yield directly the reduced transition probabilities $B(\lambda,q)$. Customarily the square root of the latter is plotted as a function of q^2 as shown schematically in fig. 4. In such a representation, $B(\lambda,q)^{1/2}$ decreases with q^2 for normal strong transitions, as is indicated for an M2 transition. If this M2 transition would have been erroneously identified as an M1 transition it would not only show an irregular increase in the $B(\lambda,q)^{1/2}$ plot (see fig. 4) but also have in general a totally unreasonable transition strength. Therefore, a highly model independent analysis (ex-

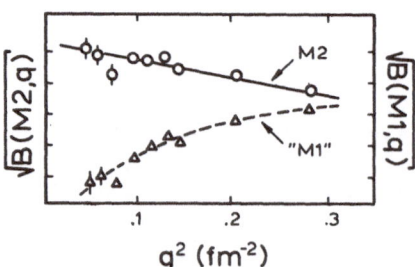

q^2 (fm^{-2})

Fig. 4   Schematic illustration of the regular behaviour of the transition strength of an M2 transition. If this M2 transition would have been wrongly identified as an "M1" transition it would display the irregular behaviour shown by the dashed line.

cept for small DWBA corrections) of measured differential
cross sections at low electron energies results in model independent determinations of transition strengths and an un-
ambiguous assignment of the multipolarity of the transition
(if the electrons are scattered off a spin-zero ground state,
otherwise we are dealing with mixed transitions).

(iv) In case of small momentum transfer in an electron scattering
experiment or, more precisely, if qR << 1 (where R measures
the radial extension of the charge-, current- and magnetization density) it is useful to recall the expansion[8]

$$[B(\lambda,q)/B\lambda,0)]^{1/2} = 1 - ...q^2 R_{tr}^2 + ...q^4 R_{tr}^4 - .... \qquad (7)$$

The quantity R_{tr} is called transition radius and is defined
in terms of the transition densities ρ (this symbol denotes
now all three transition densities introduced above) as the
ratio of two matrix elements

$$R_{tr}^{\nu}(\lambda) \sim [\int \rho_{\lambda\lambda}(r)r^{\lambda+\nu}r^2 d^3 r / \int \rho_{\lambda\lambda}(r)r^{\lambda}r^2 d^3 r] = <r^{\lambda+\nu}>/<r^{\lambda}>. \qquad (8)$$

If the left hand side of eq. (7) is plotted as a function of
q^2 the behaviour shown in fig. 5 is characteristically observed. The shape for transitions of different multipolarities is different, a signature which again is useful in identifying certain multipolarities and strengths.

Fig. 5

Characteristic
shapes of the
quantity on the
l.h.s. of eq. (7)
as a function of
different multi-
polarities. The
shapes yield di-
rectly the so call-
ed transition ra-
dius.

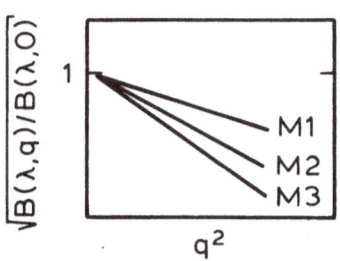

(v)   We have found that a model independent analysis in PWBA (with
proper DWBA corrections applied) works at our incident electron energies $E_o \leq 70$ MeV very satisfactorily up to the Ni-
isotopes (Z = 28). In heavy nuclei transition strengths can
only be evaluated with the help of a nuclear model. We have
to calculate transition densities $\rho_{\lambda\lambda}(r)$, insert them into
a DWBA code, compute the formfactors $|F_{\lambda}(q)|^2$, convert them

into formfactor like quantities $|F_\lambda(\Theta, E_o, E_x)|^2$ and compare those with the measured differential cross sections given in the form of $(d\sigma/d\Omega)/(d\sigma/d\Omega)_{MOTT}$. In the low momentum transfer region which we are concerned with in these lectures, the shape of $|F_\lambda|^2$ for different transition densities is only a strong function of the multipolarity λ and not of the particular wave function used in the calculation of the transition densities. Hence the calculated and the theoretical cross sections and the corresponding transition strength scale to a good approximation. A more detailed discussion of this scaling may be found in a forthcoming article [9].

In summary the <u>measured quantities</u> $d\sigma/d\Omega$, $(d\sigma/d\Omega)/(d\sigma/d\Omega)_{MOTT}$ $= |F\lambda(q)|^2$ in inelastic electron scattering and the <u>derived quantities</u> $B(E\lambda, M\lambda, q, k)$, Γ_γ^o, J^π and R_{tr} are then the test benches for the <u>nuclear models</u>.

III. EXAMPLES OF ELECTRIC TRANSITIONS

We start with an example of weak electric dipole transitions in the light nucleus ^{11}B merely to illustrate some of the features introduced in the first two sections of these lectures. This particular example may also show the powerful physics which occasionally rests in weak transitions. We shall then proceed to a discussion of a second example: the very new measurements of electric dipole and quadrupole strength distributions in the heavy nucleus ^{208}Pb. As I alluded to in the previous section, we are therefore dealing with an almost model independent determination of transition strength in a light and a clearly model dependent determination in a heavy nucleus.

Example 1: E1 excitation of non-normal parity states in ^{11}B

As fig. 6 shows, the positive parity states in ^{11}B at $E_x \approx 6.8 - 9.3$ MeV can be reached from the $3/2^-$ ground state by an E1/M2 transition, with the M2 part being much weaker [10]. Those positive parity states are most likely of the p-h structure [10]

$$\alpha\{(s)^4(p)^6(sd)^1\} + \beta\{(s)^3(p)^8\}$$

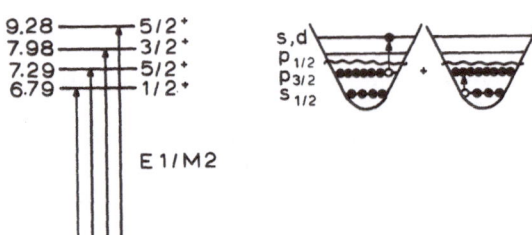

Fig. 6. Experimentally known non-normal parity states in ^{11}B and their p-h structure in a $1\hbar\omega$ configuration space.

within a 1ℏω configuration space, and α >> ß. Therefore Op → 1s, Od and Os → Op contributions to the E1 matrix element might interfere. This interference may be used for a <u>quantitative</u> test of the rôle of the Os hole in the wave function of those positive parity states. <u>Qualitative</u> evidence[11] for the presence of such an s-hole component is known from the Saclay experiment $^{12}C(e,e'p)^{11}B$ at E_O = 500 MeV, in which the $1/2^+$ state in ^{11}B at E_x = 6.79 MeV is strongly excited. It is interesting to note that the sequence of the non-normal parity states (fig. 6) exhibits collective features[12] and can be classified[13] in an SU(3) representation to belong predominantly to (λ,μ) = (4,2). An approximate SU(3) selection rule inhibits their excitation from the negative parity g.s. of ^{11}B. Sample spectra taken with a BN target are displayed in fig. 7. The states of interest (marked by arrows) are weakly excited and - as is evident from the line corresponding to the $1/2^+$ state which is only separated by 50 keV from a natural parity state in ^{11}B - good energy resolution is needed.

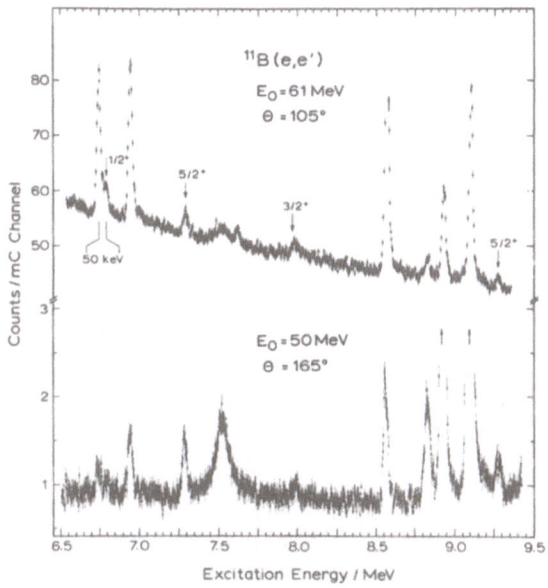

Fig. 7 Two high resolution spectra from ^{11}B. The target consisted c^c boron nitride. Several of the strong transitions correspond therefore to the electroexcitation of states in ^{14}N.

In order to extract the pure longitudinal contribution of the reduced transition probability as described in connection of eq. (4) various (e,e') measurements at q = const. have been performed. Two examples of angular distributions are displayed in fig. 8 where the measured cross section is multiplied by $k_O^2/V_T(\Theta)$ and plotted as a function of $V_L(\Theta)/V_T(\Theta)$. The quantity k_O denotes the momentum of the bombarding electron and $V_T(\Theta)$ and $V_L(\Theta)$ describe the transverse and longitudinal dependence of the cross section. In this particular representation, the quantity $(d\sigma/d\Omega)k_O^2/V_T(\Theta)$ is a linear function of $V_L(\Theta)/V_T(\Theta)$ and its slope and its intercept with

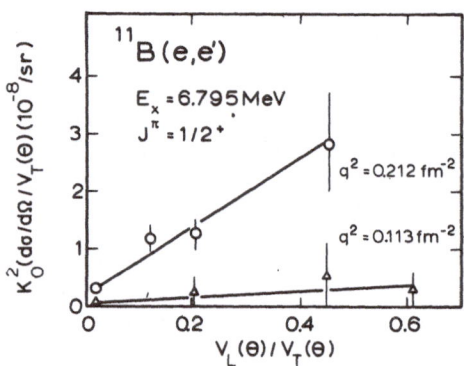

Fig. 8   Angular dependence
of the cross sec-
tion for the exci-
tation of the state
at E_x = 6.795 MeV
measured at con-
stant momentum
transfer q=0.46fm^{-1}
(circles) and
q=0.34fm^{-1} (tri-
angles). The shape
of the angular di-
stribution is di-
rectly proportion-
al to B(C1,q),
(see eq. (4) in
the main text).

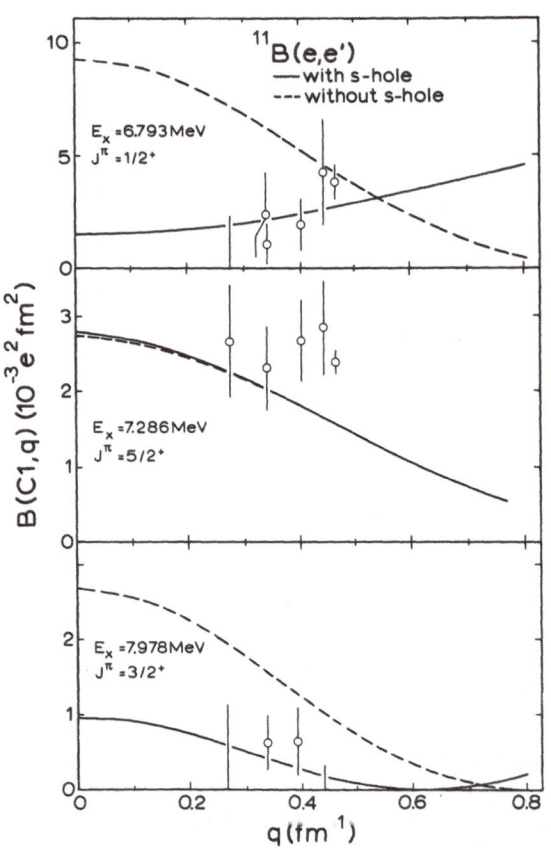

Fig. 9   Comparison of the
measured electric
dipole transition
strength from the
g.s. to states in
^{11}B of non-normal
parity with theo-
retical predic-
tions from ref.[10].

the ordinate yield directly the longitudinal and transverse contribution, respectively, to the cross section.

The so determined electric dipole transition strength to states of non-normal parity in ^{11}B is compared with theoretical predictions[10] in fig. 9. Clearly, the experimental data (noticeably the $1/2^+$ and $3/2^+$ states) indicate the necessity for the inclusion of a Os-hole (with $\beta^2 \approx 0.10$) in the positive parity wave functions. They also demonstrate the expected cancellation effects between Os $\rightarrow$ Op and Op $\rightarrow$ 1s, Od matrix elements which cause a rather dramatic effect in the momentum transfer dependent reduced transition strength. Hence we have here a prime example where an effect seen qualitatively in the Saclay experiment at $E_o = 500$ MeV could be studied quantitatively in an experiment at a small electron accelerator of electron energies an order of magnitude lower.

Example 2: E1 and E2 strength distribution in ^{208}Pb below the giant resonance

We have been able to work in PWBA (with minor DWBA corrections) in the example of E1 transitions in a light nucleus. Now we will focus our attention in this section onto the heavy nucleus ^{208}Pb and in particular onto a discussion of E1 and E2 strength distributions in the excitation energy region below the E1 giant resonance. The underlying experiment is an extension of our previously performed high resolution ^{208}Pb(e,e') experiment[14], in which we

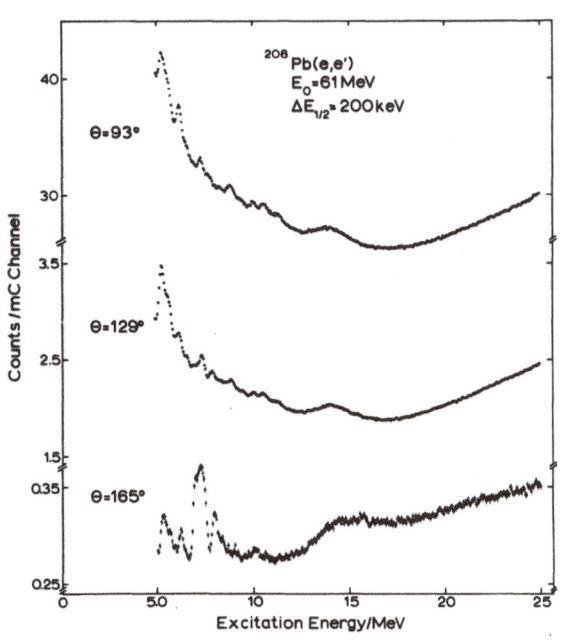

Fig. 10 Medium energy resolution ($\Delta E \approx 200$ keV FWHM) inelastic electron scattering spectra on ^{208}Pb. The radiation tail has not been subtracted. Below the E1 and E0 giant resonances hidden in the broad bump at $E_x \approx 14$ MeV, four peaks at $E_x \approx 8.9$, 10.2, 10.6 and 11.2 MeV are especially evident in the spectra at $\Theta = 93^o$ and 129°.

searched for magnetic transitions between $E_x \approx$ 6-8 MeV. As fig. 10
shows, four broad peaks at $E_x \approx$ 8.9, 10.2, 10.6 and 11.2 MeV are
seen in electron scattering experiments with medium energy resolution ($\Delta E \approx$ 200 keV FWHM) in the region of interest. Those peaks
were previously identified with the isoscalar E2 giant resonance[15,16]. Associated with this identification are at least two problems, however, which still wait for a solution. In the hadron
scattering[17-19] experiments ^{208}Pb(α,α') and ^{208}Pb(p,p') with medium energy resolution only <u>one</u> broad peak ($\Gamma \approx$ 3 MeV) is seen at
$E_x \approx$ 11 MeV carrying approximately 100% EWSR strength. A high resolution inelastic proton scattering[20] experiment, however, revealed many fine structure peaks in the same region of excitation
and associated them with states of various multipolarities. Furthermore, photon scattering and photoneutron experiments[21] provide
strong evidence for the existence of E1 strength in just this excitation energy region below the giant resonance. We therefore set
out to study this energy region in a high resolution (e,e') experiment.

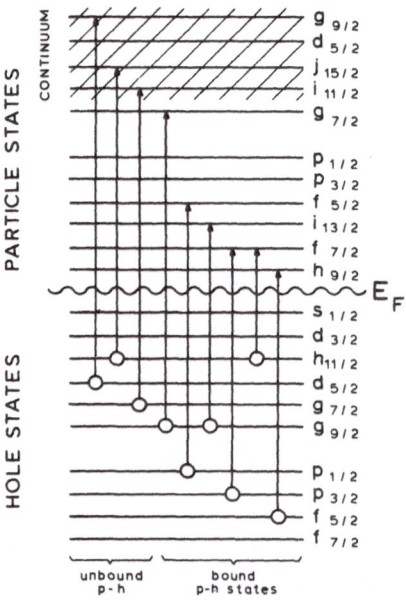

Fig. 11.   Dominant 1p-1h shell model excitations (only for protons)
responsible for the E2 giant resonance in ^{208}Pb.

Before we discuss this experiment, let us ask - totally in the spirit of our qualitative arguments brought forward in connection with fig. 1 - what kind of an E2 strength distribution in ^{208}Pb we might expect theoretically. For that, let us look at fig. 11. There the dominant 1p-1h shell model excitations (only for protons) are shown which altogether are responsible for the E2 giant resonance. An actual RPA shell model calculation[22] using the MSI force[23] puts this E2 giant resonance (E2 GR) at an excitation energy $E_x \approx 8.9$ MeV. We notice 2$\hbar\omega$ transitions between p-h states around the Fermi edge E_F, i.e. between bound p-h states. There are, however, also transitions possible between bound hole and unbound particle states in the continuum. Those latter transitions should give rise to a rather broad E2 GR or equivalently a "background" state on which sharper "resonant" structures might be superimposed which are due to transitions between the bound p-h states. The total E2 strength is then

$$B(E2,0^+ \to 2^+ GR) = \left| \sum_{\alpha=1}^{M} c_\alpha <\phi_\alpha | r^2 Y_2 | 0> + \sum_{\beta=1}^{\Lambda} c_\beta <\phi_\beta | r^2 Y_2 | 0> \right|^2 \qquad (9)$$

where ϕ denotes the 1p-1h wave functions of M bound (α) and Λ unbound (β) configurations (see fig. 11) and the c's are their respective amplitudes. An estimate[24] of the two terms on the r.h.s. of eq. (9) yields $|42.6|e$ fm^2 and $|46.2|e$ fm^2, so that the square of their sum yields a total E2 strength of roughly 7800e^2fm^4. This estimate is of course very qualitative since it would have definitely to be checked in a continuum shell model calculation[25]. Nevertheless it is a guideline for the experimentalist to search carefully for E2 strength hidden in the "background" of the measured spectra besides the one in the resonant structures of smaller width.

To see how well these expectations are actually fulfilled let us turn over to the experiments. An example of one of several ^{208}Pb(e,e') high resolution spectra is shown in fig. 12. Despite the fact that we are looking at an excitation energy region way above the neutron emission threshold ($Q(\gamma,n) = 7.37$ MeV) we are still observing a large number of fine structure peaks. After background subtraction (the boundary conditions for this procedure are essentially the correct reproduction of the radiation tail in the region of bound states below $E_x \approx 8$ MeV and of the strength of the E1 and E0 giant resonances around $E_x \approx 14$ MeV) those spectra were decomposed by a line shape fitting computer program taking the line shape of the line due to elastically scattered electrons as a reference line. The result of this procedure is also illustrated in fig. 12. Note, that for a line to be properly identified it has to occur in all measured spectra. This is - in some way - a further consistency check for the background subtraction.

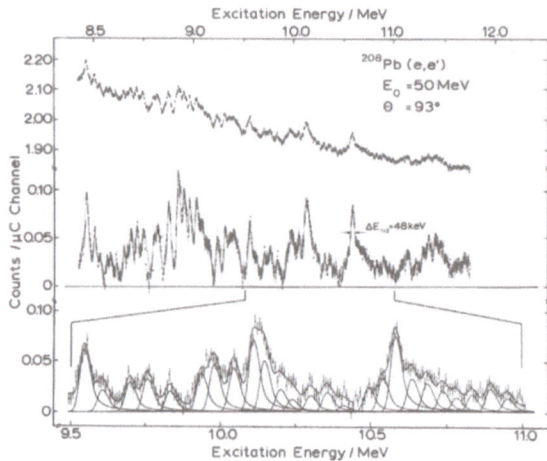

Fig. 12

Original ^{208}Pb(e,e') spectrum at E_O = 50 MeV and Θ = 93° between E_x ≈ 8-12 MeV (upper part). The same spectrum but with the background subtracted is shown in the middle part and an excerpt of it displaying the part between E_x = 9.5-11 MeV decomposed by a computer fit into various lines in the lower part of this figure.

In order to determine multipolarities and transition strengths for all those lines (altogether about 60) the experimental angular distributions and excitation functions have been compared to DWBA predictions using a standard code[26] and transition densities from the RPA-MSI model[23]. Examples of such a comparison are shown in fig. 13 for a few experimental states ($J^\pi=2^+$ at E_x=10.593 MeV; $J^\pi=1^-$ at E_x=10.152 MeV and $J^\pi=2^-$ at E_x=7.932 MeV). The lowest DWBA curve in this figure is for M1 transitions but none have been found in ^{208}Pb(e,e'), the detection limit for M1 transitions between E_x ≈ 8-12 MeV being as low as $0.5\mu_K^2$ for an individual state. The implications of this result will be discussed in a later part of these lectures. Despite the fact that - as remarked in sect. I - in such a heavy nucleus like ^{208}Pb a model independent analysis

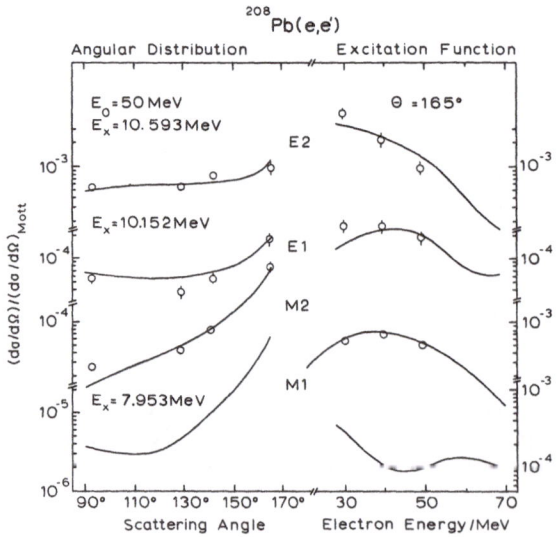

Fig. 13

Examples of ^{208}Pb(e,e') angular distributions and excitation functions. Compared to the experimental points (open circles) are DWBA predictions (solid lines) using RPA transition densities. Note that no M1 strength (theoretical predictions lowest curves) has been found in ^{208}Pb(e,e') between E_x ≈ 8-12 MeV (detection limit ≈ $0.5\mu_K^2$ for an individual state).

of the multipolarity and strength of a transition is not possible
anymore, fig. 13 should really illustrate that angular distributions
and excitation functions for transitions of different multipolarities are also distinctly different. A mapping of both with data
points therefore ensures that assignments can be made with fairly high
confidence. The so obtained rather detailed E1 and E2 strength distributions are given in fig. 14. There is indeed considerable E1
strength which previous medium energy resolution ^{208}Pb(e,e') experiments identified wrongly as E2. Infact, this E1 strength is in
fair agreement with what is known from photon scattering[21] and
(n,γ) and (γ,n) experiments (for a summary, see[27]) and with what
is predicted by Knüpfer in the frame of the MSI model. The $J^{\pi} = 1^{-}$
states displayed in fig. 14 carry about 8% of the EWSR for electric
dipole transitions (remember that most of the remaining strength
rests in the GDR at $E_x \approx 13.5$ MeV). There is also considerable E2
strength left between $E_x \approx 8-12$ MeV, but much less than originally
thought[16]. The EWSR for isoscalar E2 transitions is now only exhausted by about 20-40%. The uncertainty of this number is due to
the uncertainty in the background subtraction of the spectra. We
therefore have the remarkable result that the E2 strength found in
(e,e') is much smaller than that in hadron scattering experiments

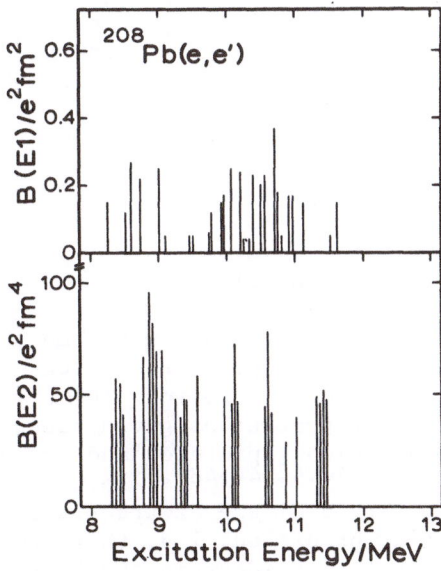

Fig. 14   E1 and E2 strength distribution between $E_x \approx 8-12$ MeV
          from ^{208}Pb(e,e').

and of the theory. We furthermore have also no evidence for a broad
"background" E2 continuum as expected from our arguments quoted
above in connection with fig. 11 and eq. (9).

Let us finally compare in fig. 15 our medium and high resolu-
tion ^{208}Pb(e,e') spectra with a ^{208}Pb(α,α') spectrum[17] and a
^{208}Pb(α,α'nγ) spectrum[28]. There is not very much resemblance between the (e,e') and (α,α') spectra, except perhaps for the case
of the ^{208}Pb(α,α'nγ) coincidence experiment, but for this the spectrum would require an overall energy shift of about 300 keV. Only
a short time ago - when we compared the medium energy resolution
(e,e') with medium energy resolution hadron scattering experiments -
did we find the two in fair agreement. It will certainly be a necessary challenge for us in the future to understand the physical significance behind this difference in the E2 excitation strengths.
The cause might well be the different reaction mechanism.

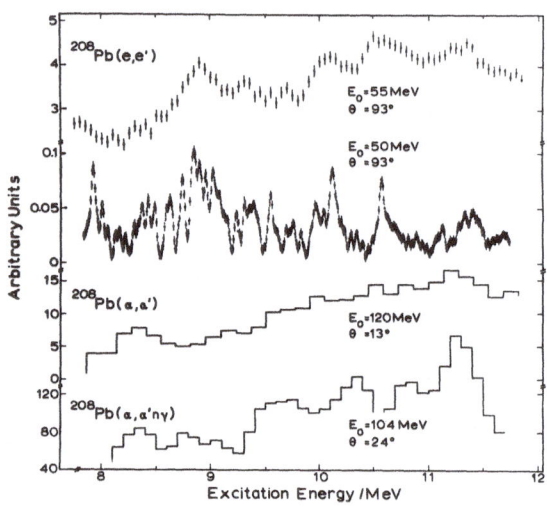

Fig. 15

Comparison of medium and
high energy resolution
(e,e') spectra from Darm-
stadt with (α,α') and
(α,α'nγ) taken at Groningen[17] and Karlsruhe[28], respectively.

IV. EXAMPLES OF MAGNETIC TRANSITIONS

In this second lecture I will discuss some phenomena of magnetic dipole and magnetic quadrupole transitions studied with high
resolution inelastic electron scattering. Thereby I also follow
closely comprehensive representations given recently[3,4,29] but enlarge it with a few educational pictures and our newest experimental results on ^{16,18}O and ^{40,42,44,48}Ca. In proceeding from light
to heavy nuclei the observed M1 and M2 transition strengths become
smaller as theoretically predicted, i.e. they are quenched. This
pattern of quenching is also evident in the behaviour of the magnetic moments and we therefore will try at the end of the lectures
to interprete this quantitatively - though with a finite theoreti-

cal effort - as a mesonic renormalization of the nuclear spin current.

Why are we interested to study magnetic transitions? Let us
recall a few facts which - by way of selected examples from our work -
we later on will give a serious consideration:

(i)   If magnetic moments deviate from simple shell model predic-
       tions, two effects - core polarization and mesonic contribu-
       tions - are usually made responsible for that. Those two ef-
       fects, however, cannot easily be disentangled from each other
       by just looking at magnetic moments alone.

(ii)   It is therefore very important to study other nuclear pro-
       cesses involving spin-dependent interactions[30], like the ex-
       citation energy of 1^+, 2^-, 3^+, ... unnatural parity states
       reached from a 0^+ ground state by an M1, M2, M3, ... transi-
       tion, respectively. In the common ansatz[31] for an effective
       spin-spin and spin-isospin particle-hole interaction

$$F_{p-h} = g_o \underset{\sim}{\sigma}\cdot\underset{\sim}{\sigma} + g_o' \underset{\sim}{\sigma}\cdot\underset{\sim}{\sigma}\ \underset{\sim}{\tau}\cdot\underset{\sim}{\tau} + \pi\text{-exchange term} + \rho\text{-exchange} + .. \quad (10)$$

       the Landau-Migdal parameters g_o and g_o' influence the proper-
       ties (excitation energies and transition strengths) of those
       unnatural parity states in a very distinct way.

(iii) Since it is expected that the pion field in the nucleus coup-
       les strongly to those p-h states which carry the quantum num-
       ber of bound pions, i.e. the 1^+, 2^-, 3^+,.... states, the stu-
       dy of those pion like states might give us a hint about the
       possible existence of a pion condensate in the nuclear ground
       state[32]. In the case that the pion in nuclear matter is more
       bound than its rest mass, real pions can be created without
       needing additional energy. We know that the parameter g_o' in
       the Migdal force (c.f. eq. (10)) is positive and momentum
       independent. The p-h interaction of the zero-range part of
       this force is therefore repulsive. Consequently pion conden-
       sation is not expected. But there is in eq. (10) due to pion
       exchange a strongly momentum dependent attractive tensor term
       of the form $-q^2/(m_\pi^2+q^2)$ as well as a repulsive term $q^2/(m_\rho^2+q^2)$
       due to rho meson exchange (see also the lectures of Wolfram
       Weise[2] at this school). Consequently, the parameter g_o' be-
       comes momentum dependent, however, its momentum dependence
       is not well known at present. Remember, that in our (e,e')
       experiments at the DALINAC where $q \lesssim 0.7$ fm^{-1} we determine
       g_o' close to its static value.

(iv)   Do magnetic giant resonances exist as the well known electric
       giant resonances and what is from the distribution of B(M1),
       B(M2), ... strength to be learned with respect to correla-
       tions in the nuclear ground state ?

(v)   Finally, since magnetic transitions involve the spin operator
       they are related to analogous processes like Gamow-Teller
       ß-decay and reactions of the type (π^+,π^0), (π^-,γ), (μ^-,γ),
       (p,n),....

       Let us ask next, what is the mechanism for existing M1 tran-
sitions. This is best illustrated in the frame of the simple IPM
(see fig. 16). The transition strength is given by

$$B(M1) \sim |<j_f| e^{\,i\underset{\sim}{q}\underset{\sim}{r}}(g_\ell\underset{\sim}{\ell} + g_s\underset{\sim}{s})|j_i>|^2, \qquad (11)$$

where the g's are the gyromagnetic factors for the orbital and spin
motion of the nucleon, respectively.

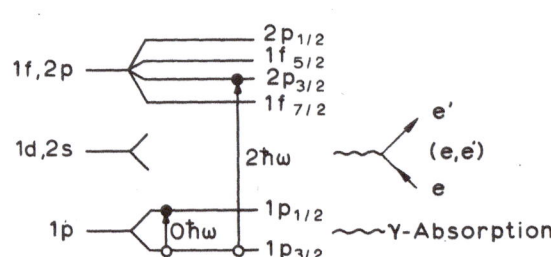

Fig. 16

Schematic illustration
of $0\hbar\omega$ and $2\hbar\omega$ M1 transitions in the indepen-
dent particle shell mo-
del. In the absorption
of real photons only
$0\hbar\omega$ transitions are expected, while in (e,e')
the excitation of $2\hbar\omega$
transition becomes pos-
sible due to the radial
dependence in eq. (11).

The r.h.s. of eq. (11) reflects the following properties:

(a) If there is no radial dependence of the transition operator
    (essentially like in the absorption of real photons or in low
    energy inelastic electron scattering) then $\ell_f = \ell_i$, and we are
    dealing with $0\hbar\omega$ "spin-flip" p-h transitions between $j_i = j_> =$
    $(\ell+1/2)$ and $j_f = j_< = (\ell-1/2)$. Those transitions should be
    strongest - or even collective - in nuclei where the $j_>$ shell
    is full and the $j_<$ shell is empty. In a doubly closed shell
    nucleus like ^{16}O and ^{40}Ca $0\hbar\omega$ transitions are therefore not
    expected in the pure IPM.

        A 1p-1h RPA-MSI calculation of Knüpfer[3] employing in
    eq. (11) the g factors of bare nucleons yields for a few select-
    ed nuclei the strength distribution shown in fig. 17.
    Note especially that the excitation energy of the $J^\pi = 1^+$ states
    is expected to follow the relation $E_x \approx 40A^{-1/3}$ MeV. We will
    mainly discuss such $0\hbar\omega$ transitions in these lectures.

(b) There are "ℓ-forbidden" M1 transitions, like $1d_{3/2} \to 2s_{1/2}$ in
    ^{39,41}K, which can also be investigated in (e,e') but which we
    will not be going to evaluate here, see, however, ref.[33].

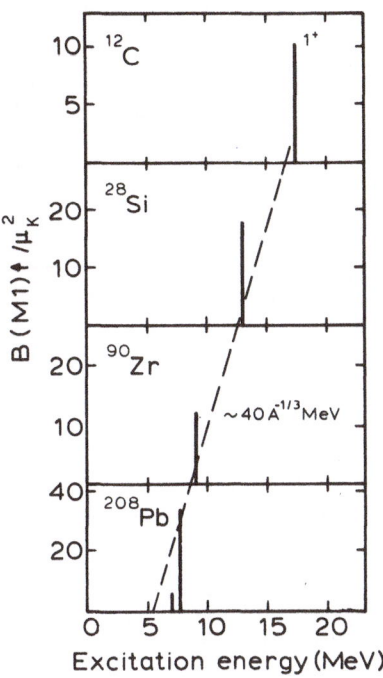

Fig. 17   RPA-MSI predictions[3] using bare g factors for a few
selected nuclei. Note the $40A^{-1/3}$ MeV dependence of
the excitation energy.

(c)   The radial dependence of (e,e'), i.e. the exp(iqr) factor in
eq. (11), might possibly lead to very collective $2\hbar\omega$ M1 transitions[34]. The search for such transitions noticeably in ^{208}Pb
has so far been negative (see also the results from ref.[35]).
Those 1[+] states lie in the continuum and should be investigated[36] at a slightly higher q than what is available at our present linac.

The excitation mechanism for M2 transitions is by no means
as transparent as the one for M1 transitions (because of the greater
number of possible p-h excitations). Figure 25 summarizes a sample
calculation of the strength distribution of $J^{\pi} = 2^-$ states in various nuclei ranging from ^{12}C to ^{208}Pb. Those strength distributions
were obtained[37] in the frame of a (1-3) $\hbar\omega$ RPA shell model calcula-

tion (1p-1h,g_{bare}). Inspecting fig. 18 we note, that the M2 strength is (even in the 1p-1h approximation) much more fragmented than the corresponding M1 strength (see fig. 17) and that the center of gravity of the excitation energy follows rather closely an $E_x \approx 44A^{-1/3}$MeV law, i.e. E_x(M2) $\approx E_x$(M1). This effectively means that where we

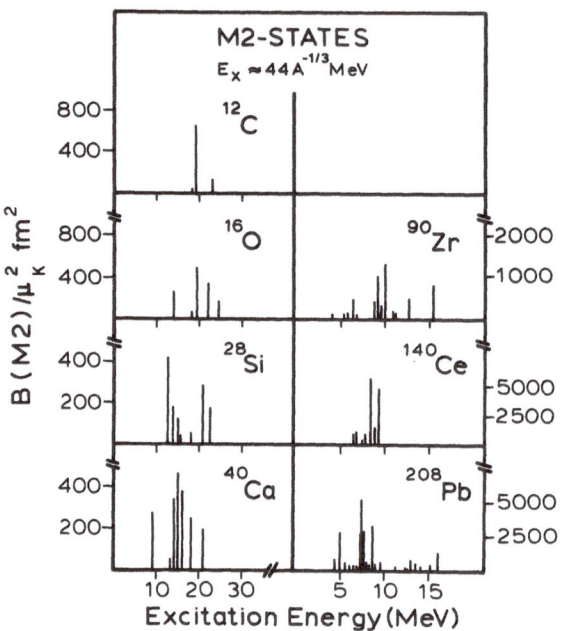

Fig. 18   (1-3)$\hbar\omega$ RPA shell model prediction[37] of M2 excitations
           (1p-1h,g_{bare}) for various light and heavy nuclei.

search for M1 transitions we find most likely also M2 transitions and vice versa and we definitely have to use the different behaviour of the respective form factors as a function of momentum transfer for the assignment of multipolarities to a particular transition.

     The B(M2) strength is given by

$$B(M2) \sim |<j_f|e^{i\underset{\sim}{q}\underset{\sim}{r}}(g_\ell\underset{\sim}{\ell}+g_s\underset{\sim}{s})\underset{\sim}{\nabla}(r^2 Y_2(\hat{\underset{\sim}{r}}))|j_i>|^2, \qquad (12)$$

     We are now in a position to illustrate the subject of M1 and M2 transitions by a few examples from low q inelastic electron scattering.

Example 1: M1 ground state transitions in ^{16}O

     Since the doubly closed shell model nucleus ^{16}O is composed

of filled levels of both spin-orbit partners up to the Fermi-edge,
0ħω single particle M1 transitions are not expected in the pure IPM.
As is shown schematically in fig. 19, they become, however, possible through 2p-2h components in the ground state of ^{16}O. A detection
of M1 transitions in N=Z nuclei with filled shells of both spin-
orbit partners is therefore a direct measure of such ground state
correlations. We will return in more detail to the effect of ground

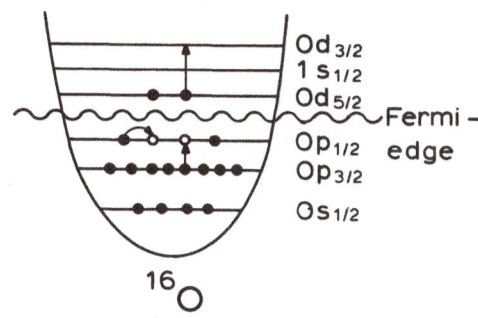

Fig. 19 Schematic illu-
stration of pos-
sible M1 transi-
tions in ^{16}O due
to ground state
correlations,
i.e. the presen-
ce of 2p-2h com-
ponents in the
ground state.

state correlations on M1 and M2 transitions in ^{28}Si and ^{40}Ca in a
later part of this lecture.

    The first experimental evidence for the existence of an M1
ground state transition in ^{16}O stems from an ^{16}O(e,e') experiment
performed[38] at our accelerator in 1970 with medium energy resolution (see fig. 20). At that time a state at E_x = 16.22 MeV had already been identified as a J^{π} = 1$^+$ state. A later experiment[39]
using polarized proton capture on ^{15}N verified this assignment and
furnished additional 1$^+$ states at E_x = 17.13 and possibly 18.82 MeV.

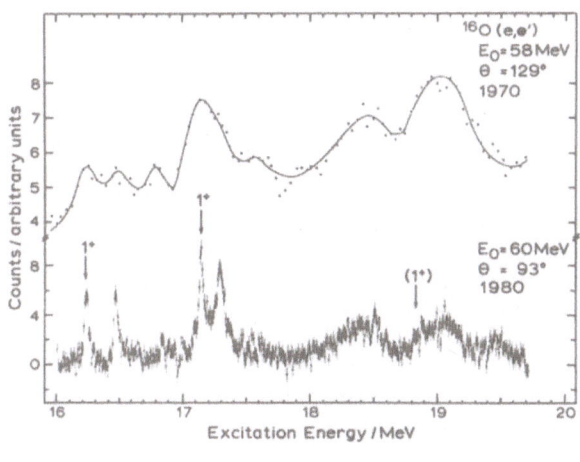

Fig. 20 Comparison of
two ^{16}O(e,e')
spectra - both
measured at Darm-
stadt - but with
medium and high
resolution in
1970 and 1980,
respectively.

In the meantime those M1 transitions have also been seen in a
high resolution $^{16}O(e,e')$ experiment. One of the spectra is shown
in the lower part of fig. 20 where it is compared with one of the
old spectra from the DALINAC. The enormous progress which inelastic
electron scattering has made in recent years is obvious.

Before I discuss the (e,e') and $(\vec{p},\gamma)$ results in more detail,
I should like to spend a few moments explaining for the young students of this school how such an improvement in energy resolution
has been achieved. Spectra like the one in the lower part of fig. 20
are obtained with so called energy loss spectrometers. A simple
sketch of the spectrometer operated in the energy loss mode as compared to the conventional mode is given in fig. 21. In the conventional mode, all electrons are focussed onto one point of the target. The scattered electrons appear then in the focal plane of the

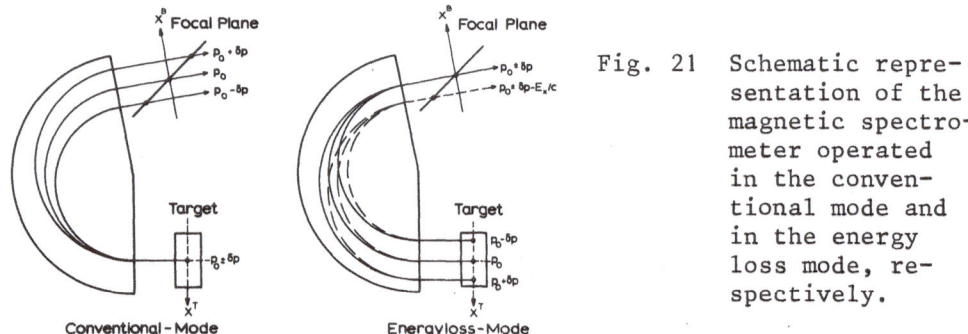

Fig. 21   Schematic repre-
          sentation of the
          magnetic spectro-
          meter operated
          in the conven-
          tional mode and
          in the energy
          loss mode, re-
          spectively.

spectrometer at momenta p_0 and $p_0 \pm \delta p$. In order to keep δp small
(or equivalently the width of the detected line) the spread in momentum of the initial beam has to be small. This of course can be
achieved but only at the expense of reducing the beam intensity by
narrowing some slits in the beam energy analyzing magnet system. This
disadvantage might be overcome when the spectrometer is operated in
the energy loss mode. This mode (see fig. 21) is essentially a mir-
ror image of the conventional mode. If the beam electrons are positioned at the target according to their deviation δp from their
nominal value p_0 of the momentum, then all electrons are focussed
onto the same point in the focal plane independent of their primary momentum. The position of this image point depends only on the
energy loss E_x of the scattered electron in the target. Therefore,
the dispersion of the beam has to be changed with the excitation
energy E_x of a nuclear level to be investigated. The size of the
beam is e.g. 1 mm x 10 mm when we at the DALINAC investigate levels
around $E_x \approx 10$ MeV. Our energy loss system is described in detail
in refs. [40-43].

We return now to a comparison of a $^{15}N(p,\gamma)^{16}O$ excitation function[39] at $\Theta = 90°$ with an $^{16}O(e,e')$ spectrum at $E_0 = 30$ MeV and

$\Theta = 165^O$, where the transverse magnetic transitions are particularly distinct (see fig. 22). There is a correspondence between the excitation function and the spectrum with respect to the lowest two
1^+ states. This correspondence can be understood theoretically by

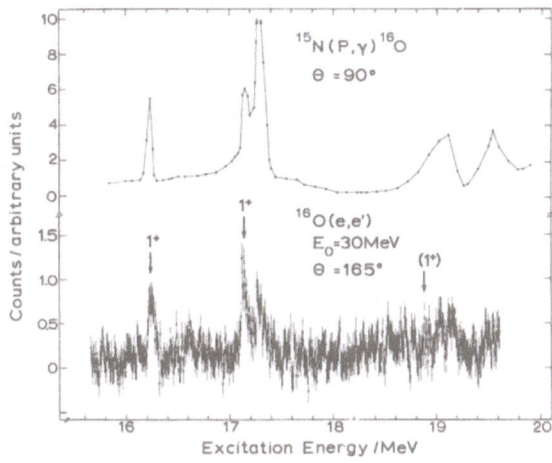

Fig. 22

Comparison of a radiative
proton capture excitation
function on ^{15}N (ref.[39])
and a $^{16}O(e,e')$ high resolution spectrum. Note,
that the strongly excited
1^- state seen in (p,γ) on
the high energy side of
the 1^+ state at 17.13 MeV
is almost absent in the
backward angle (e,e') spectrum.

relating the inclusive (e,e') cross section to the exclusive (p,γ)
cross section (for details, see my lectures at the Alushta school[4]).
As is evident from fig. 22 strength of electrical origin seen in
proton capture (note the strong 1^- state around $E_x \approx 17.3$ MeV and
the clustering of strength around $E_x \approx 19$ MeV) is almost absent in
the backward angle electron scattering spectrum.

The summed M1 strength in ^{16}O determined in polarized radia-
tive proton capture[39] is $\Sigma B(M1)\uparrow \geq 0.69\mu_K^2$, the value from (e,e')
$0.65 \pm 0.12\mu_K^2$. The general features of this unexpectedly large M1
ground state transition strength in ^{16}O can be explained[44] at least
qualitatively in terms of the shell model with 2p-2h components in
the g.s. wave function of ^{16}O.

Example 2: M1 and M2 ground state transitions in ^{18}O and their
relation to radiative pion capture

The nearest even-even open shell nucleus to ^{16}O which can be
studied with electron scattering is ^{18}O in which we can investigate
the effect of adding two neutrons to the ^{16}O core. Little is known
with respect to the structure of high lying states and the properties of electric and magnetic multipole strength distributions.
Additional interest in a high resolution $^{18}O(e,e')$ experiment is
raised through the possibility to make a comparison with the above
mentioned radiative pion capture process $^{18}O(\pi^-,\gamma)^{18}N$ which has been
studied recently[45]. The experiment[46] which I will discuss here briefly is the first high resolution (e,e') study of nuclear states at
excitation energies $E_x > 15$ MeV using a gas target.

The spectra of fig. 23 clearly demonstrate the selectivity of
the high resolution, low momentum transfer (e,e') experiment with respect to certain low multipolarity transitions from the ground state of
^{18}O. In an excitation energy region of high level density, only two
peaks at $E_x = 16.38 \pm 0.01$ and 18.86 ± 0.01 MeV stand out clearly
on an almost continuous background, which has fine structure indicating the high density of underlying ^{18}O levels. The $\Theta = 105^{\circ}$
spectra show that there are longitudinal transitions in addition
to the two sharp transverse ones. This is expected since the excitation energy region displayed overlaps the low-energy side of the

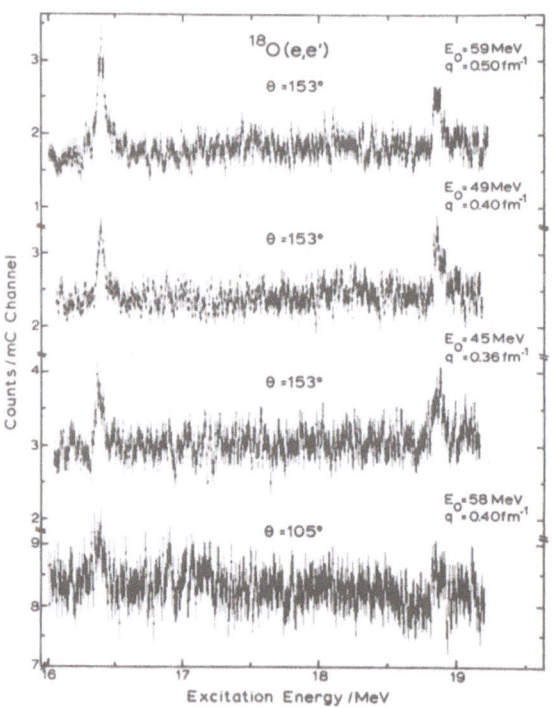

Fig. 23   Spectra of inelastically scattered electrons at three
          different momentum transfers. Notice the supressed zero
          of the scales on both axes.

giant dipole resonance (GDR). In the lower part of fig. 24 to be
discussed below, the 16 to 27 MeV part of the spectrum obtained at
54.3 MeV (153°) is plotted as a histogram over 50 keV energy bins.
In addition to the two narrow peaks, the spectrum shows broad but
distinct maxima. The strongest one, centered at $E_x \approx 23.7$ MeV, coincides with the position of the GDR peak that dominates the photo-
absorption cross sections[47].

The multipolarity of the transition and the transition strength have been determined with the help of the formalism given in the first lecture (see sect. II) and $J^\pi = 2^-$ and $B(M2)\uparrow = 58 \pm 8\mu_K^2$ fm^2 is found for the state at $E_x = 16.38$ MeV, likewise $J^\pi = 1^+$ and $B(M1)\uparrow = 0.28 \pm 0.04\mu_K^2$ for the one at $E_x = 18.86$ MeV. Their core states in ^{16}O are the $J^\pi = 2^-$ at $E_x = 12.97$ MeV and the $J^\pi = 1^+$ state at $E_x = 16.22$ MeV just discussed, i.e. adding the 2 neutrons to the ^{16}O core causes an energy shift of roughly 3 MeV. Furthermore, those tow sharp states (see fig. 23) have isospin $T = 2$, i.e. they are isobaric analogue states of states in ^{18}N. In fact, the present ^{18}O(e,e') experiment provides for the first time conclusive evidence that the g.s. of ^{18}N has $J^\pi = 2^-$. The Coulomb displacement energy to the 16.38 MeV state in ^{18}O is $\Delta E_c = 3110 \pm 10$ keV.

Next we will try to relate the ^{18}O(e,e') results to the ones

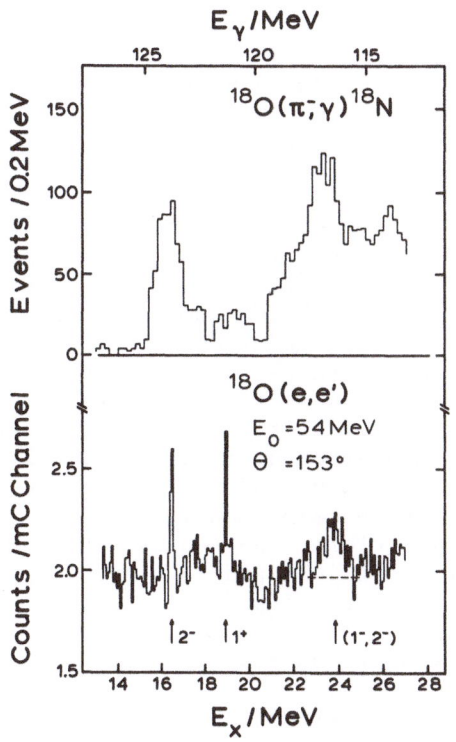

Fig. 24

A comparison of the photon spectrum from the (π^-,γ) experiment (top, adapted from ref.[45]) with the inelastic (e,e') spectrum at 54.3 MeV and 153° (bottom). The area above the dashed line has been used in the cross section estimates discussed in ref.[46].

from the $^{18}O(\pi^-,\gamma)^{18}N$ reaction. Remember that there is an analogy between various axial vector matrix elements as is usually shown in impulse approximation. To facilitate a comparison fig. 24 has been prepared. The (e,e') and (π^-,γ) spectra appear to be quite similar if the difference in energy resolution and momentum transfer between the two experiments is considered. Obviously the 16.2 MeV state populated in the (π^-,γ) spectrum corresponds to the transition to the 2^- ground state of ^{18}N as suggested by the authors of ref.[45]. Corresponding to the (e,e') peak at 18.86 MeV there is a weak state at 19.2 ± 0.2 MeV in the (π^-,γ) spectrum with a ratio $R_\gamma(19.2)/R_\gamma(16.2)$ ≈ 0.11. This is consistent with the M1 nature of the 18.86 MeV peak. An approximate extrapolation of our data to $q = 0.62$ fm^{-1} (the momentum transfer in (π^-,γ)) yields for the respective squares of the (e,e') formfactors

$$F^2(18.86)/F^2(16.38) = F^2(M1)/F^2(M2) \le 0.2.$$

The peak at 23.2 MeV in the (π^-,γ) spectrum, being broader than a similar resonance in ^{16}O, has been interpreted as being due to the existence of perhaps another $J^\pi = 2^-$ level in addition to the $J^\pi = 1^-$ GDR. The (e,e') data show that those excitations are intrinsically broad. This may be understood on the basis of neutron decay channels opening at 19.1 MeV for $J^\pi = 1^-$ states and at 20.5 MeV for $J^\pi = 2^-$, T = 2 states.

Finally, the results of excitation energy and strengths of the states found in the present experiment may be compared with theoretical predictions using a core excitation model[48]. This model predicts $J^\pi = 2^-$ states at $E_x = 15.1$ and 16.6 MeV, whose configurations are predominantly $0d_{5/2}(0p_{1/2})^{-1}$ and $0d_{3/2}(0p_{1/2})^{-1}$, respectively. The summed strength from the g.s. into those states is $\Sigma B(M2)\uparrow \approx 60\mu_K^2$fm^2 in fair agreement with what is found experimentally to the state at 16.38 MeV $(B(M2) = 58\mu_K^2$fm$^2)$. The model predicts further a clustering of $J^\pi = 2^-$ states with dominant members at 21.6 and 25.6 MeV. The center of gravity and the spread in energy of the calculated states agrees well with the broad maxima observed in the (e,e') and (π^-,γ) data. Moreover the total predicted strength of $\Sigma B(M2)\uparrow \approx 150\mu_K^2$fm^2 for the two states is in fair agreement with the M2 strength in the (e,e') spectrum if at least 50% of the experimental strength is M2, as discussed above. The upper two strong states in the calculation have major configurations $0d_{5/2}(0p_{3/2})^{-1}$ and $0d_{3/2}(0p_{3/2})^{-1}$. Therefore, the separation of M2 strength in ^{18}O in two parts, centered at 16.4 and 23.5 MeV is a consequence of the spin-orbit splitting in the p-shell.

Example 3: M1 transitions in ^{28}Si and ground state correlations

We shall now specify what we mean by ground state correlations and their effect on the M1 strength in nuclei. In a search for M1 strength in ^{28}Si states with $J^\pi = 1^+$ have been found[49]at E_x=10.594,

10.725, 10.901, 11.445 and 12.331 MeV. Those are shown in fig. 25 together with three more 1^+ states at E_x = 14.030, 15.146 and 15.500 MeV which we have recently discovered. As is seen in fig. 25, the M1 strength is very fragmented but we feel that we have detected all the expected M1 strength. The reason will become obvious below. The total measured M1 strength is now $\Sigma B(M1)\uparrow = 6.78 \pm 0.32\mu_K^2$. Before this result is compared to nuclear model calculations let us recall Kurath's[50] EWSR prediction for isovector M1 transitions in selfconjugate nuclei:

$$\sum_n (E_n - E_o) B(M1, 0 \to n) = \frac{3}{4\pi} a (\mu_p - \mu_n + \frac{1}{2})^2 <0|\Sigma \underset{\sim}{\ell} \cdot \underset{\sim}{s}|0> \qquad (13)$$

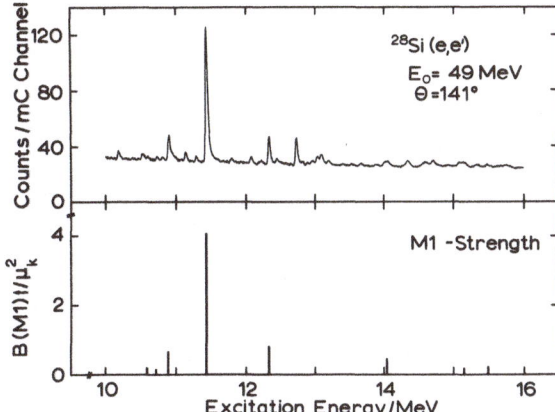

Fig. 25

Spectrum from ^{28}Si(e,e') at Θ = 141° and E_o = 49 MeV (upper part) and the M1 strength distribution derived from several of such spectra (lower part).

The left-hand side is determined rather accurately from experiment. By comparison with the right-hand side (a is the spin-orbit splitting parameter) the necessity of introducing effective magnetic charges (μ_p and μ_n stand for the magnetic moments of the proton and neutron, respectively) can be studied as well as the effect of ground state correlations via the expectation value of the $\underset{\sim}{\ell} \cdot \underset{\sim}{s}$ operator in the nuclear ground state. Assuming first bare magnetic moments we will demonstrate the effect of ground state correlations with the help of fig. 26. Since the M1 operator has only matrix elements between single particle states with the same orbital angular momentum and principal quantum number, $0\hbar\omega$ M1 transitions occur only between spin-orbit partners (see the beginning of sect. IV). In the pure independent particle model (IPM) the average occupation number $<n_j>$ in the $d_{5/2}$ shell is 12 while the $d_{3/2}$ shell is empty or equivalently, if the normalized quantity $<n_j>/2(2j+1)$ is plotted as a function of energy, we observe a sharp break from unity at the Fermi edge E_F. The calculated total strength $\Sigma B(M1)\uparrow = 20.30\mu_K^2$ is far too large compared to the experimental number. In the RPA (which contains some amount of ground state correlations) the Fermi surface becomes smoother and $<n_{d5/2}> = 11$ and $<n_{d3/2}> = 0.5$ lead to a strength $\Sigma B(M1)\uparrow = 17.34\mu_K^2$ still too large.

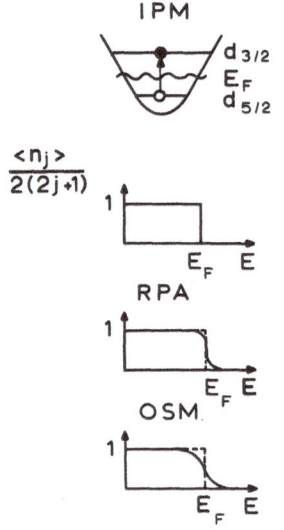

Fig. 26 Schematic representation of
the effect of ground state correlations on the occupation
number $\langle n_j \rangle$ of a shell with
given j. The various shell mo-
dels are abbreviated by IPM
(Independent Particle Model),
RPA (Random Phase Approximation model) and OSM (Open Shell
Model).

Only an open shell model[51] (OSM) calculation (with an ^{16}O core and
12 active particles around) produces the necessary change in occu-
pation numbers ($\langle n_{d5/2} \rangle$ = 6.8 and $\langle n_{d3/2} \rangle$ = 2.5) or a soft enough
Fermi edge (see fig. 26) to produce a total strength of
$\Sigma B(M1)$ = 6.74μ_K^2 in almost perfect agreement with the experimental
number quoted above. Of course, in the OSM calculation effective
magnetic charges were used. Ground state correlations can hence sensitively be detected by a determination of ground state M1 transitions.

It is interesting to compare (see table I below) the mean occupation numbers of the d-shell in ^{28}Si from various experiments
(the proton occupation number measured in the (e,e'p) and (d,^{3}He)
experiments have been multiplied by two). The uncertainty of $\langle n_d \rangle$
is about 10% in the (e,e') reaction. There is a rather good agreement between the three different methods.

Table I: Mean occupation numbers of the d-shell in ^{28}Si determined in various experiments. The (e,e'p) result is
given in[11] and the (d,^{3}He) result in[52].

	(e,e')	(e,e'p)	(d,^{3}He)
$\langle n_d \rangle$	9.3	$11\,{}^{+\,1}_{-\,4}$	10.5

Finally, we note that ground state correlations (as in the case of ^{28}Si) lead in general to a weakening of the strength (the $j_>$ shell is partly emptied and the $j_<$ shell partly filled and hence blocks particle jumping). However, N = Z closed shell nuclei like ^{16}O and ^{40}Ca are the prime exceptions to this rule: M1 ground state transitions become just possible for the very reason of ground state correlations. We turn now to a further discussion of them using (e,e') on the stable even-even Ca isotopes.

Example 4: M1 and M2 transitions in ^{40,42,44,48}Ca and the relation of M1 transitions to Gamow-Teller strength distributions

Like the doubly closed shell nucleus ^{16}O, the nucleus ^{40}Ca is also composed of filled $j_> = \ell-1/2$ shells. Magnetic dipole transitions of $0\hbar\omega$ character become therefore only possible in the presence of ground state correlations of the sort as shown schematically in fig. 27. Due to the presence of two holes in the $0d_{3/2}$ shell and two particles in the $0f_{7/2}$ shell we might expect p-h jumps of $d_{5/2} \to d_{3/2}$ and $f_{7/2} \to f_{5/2}$ character. If this picture of ground state correlations is correct, the dominant p-h M2 transition in the IPM, i.e. $0d_{3/2} \to 0f_{7/2}$, should be blocked since two particles are occupying already the $0f_{7/2}$ shell. With a continuous filling of the $0f_{7/2}$ shell with neutrons in ^{42,44,48}Ca the M2 strength is then expected to decrease more and more while the M1 transitions should become stronger and stronger. Are those features of M1 and M2 transitions really born out by the experiments? .

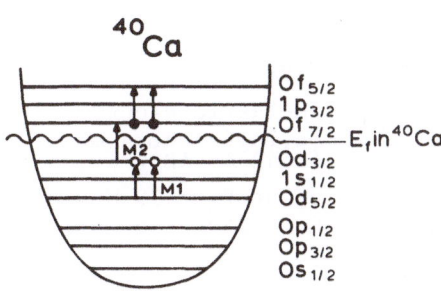

Fig. 27 Schematic illustration of two possible M1 transitions in ^{40}Ca due to 2p-2h excitations in the ground state. The excitation mode of the dominant M2 transitions is also indicated.

Let us discuss ^{40}Ca first. We have compelling experimental evidence[53,54] for a strong M1 transition of a J^π; T = 1^+; 1 state at E_x = 10.319 ± 0.005 MeV (see fig. 28) to the ground state of ^{40}Ca.

The M1 transition strength is B(M1)↑ = 1.12 ± 0.07μ_K^2 (or, when measured in terms of the ground state radiative width Γ_γ^0 = 4.74 ± 0.30 eV). We are hence dealing here with an unusually

strong M1 transition (of about half the strength of the famous M1
giant resonance in ^{12}C at E_x = 15.11 MeV).

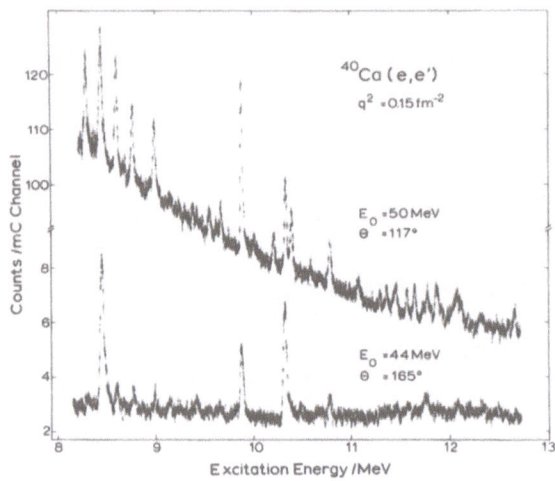

Fig. 28

Spectra from ^{40}Ca(e,e')
at the same q. By compar-
ing the spectrum at
Θ = 117° with the one at
Θ = 165° the selectivity
of backward angle data
with respect to trans-
verse excitations is ob-
vious. The state at
E_x = 10.319 MeV has J^{π}=1^{+},
the one at E_x = 8.428 MeV
J^{π} = 2^{-}.

A strong M2 transition from the ground state to a state located at E_x = 8.428 ± 0.005 MeV (see fig. 28) is also detected.

In fig. 29 in addition to ^{40}Ca, inelastic electron scattering
spectra all measured[55] at the same energy E_0 = 39 MeV and scattering
angle Θ = 165° are shown for ^{42,44,48}Ca. This picture is rather exciting. While we observe a J^{π} = 1^{+} state at E_x = 11.235 MeV in ^{42}Ca
(B(M1)↑ = 0.6μ_K^2), which is the only strong state in the spectrum,

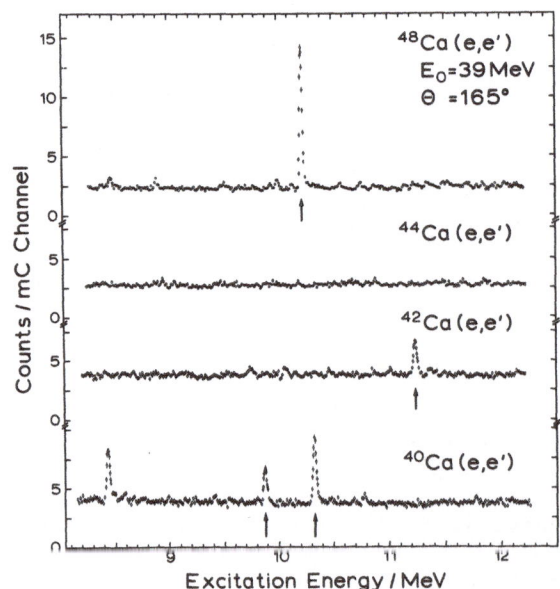

Fig. 29

Inelastic electron scat-
tering spectra on
40,42,44,48Ca all measur-
ed at E_0 = 39 MeV and
Θ = 165°. The arrow
points to J^{π} = 1^{+} states
(the line at 9.87 MeV is
probably a doublett con-
taining a weakly excited
1^{+} state). Note the absence of strong M1 tran-
sitions in ^{44}Ca and the
strongly excited 1^{+} states
in ^{40}Ca and ^{48}Ca at virtually the same excita-
tion energy.

is the spectrum of ^{44}Ca essentially flat with at most very weak
lines signaling an extreme fragmentation of M1 strength due to the
very deformed nature or softness of the ^{44}Ca ground state. But in
^{48}Ca we discovered[55] – in agreement with the expectation from the
IPM – an isolated strongly excited 1$^+$ state at E_x = 10.227 MeV
(B(M1)↑ = 4μ_K^2). Note the closeness of excitation energies for the
1$^+$ states in ^{40}Ca and ^{48}Ca, i.e. E_x ≈ 10.32 vs. 10.23 MeV. Furthermore, there is an indication of M2 strength in ^{48}Ca between
E_x ≈ 8 - 9 MeV, which is however diluted as compared to the M2
strength associated with the $J^\pi = 2^-$ state at 8.43 MeV in ^{40}Ca, i.e.
the expectation from the IPM stated above is realized.

An excitation function of the prominent M1 transition in ^{48}Ca
is displayed in fig. 30. We are presently extending those measurements to higher bombarding energies at the electron accelerator of
Mainz[56]. The formfactor of the $J^\pi = 1^+$ state in ^{48}Ca, i.e. in a
nucleus where the simple shell model works best, should then in
some way be a better example for the test of the so-called precritical phenomena[2,57,58] and pion condensation than ^{12}C which is not a
good nucleus from the point of the simple IPM.

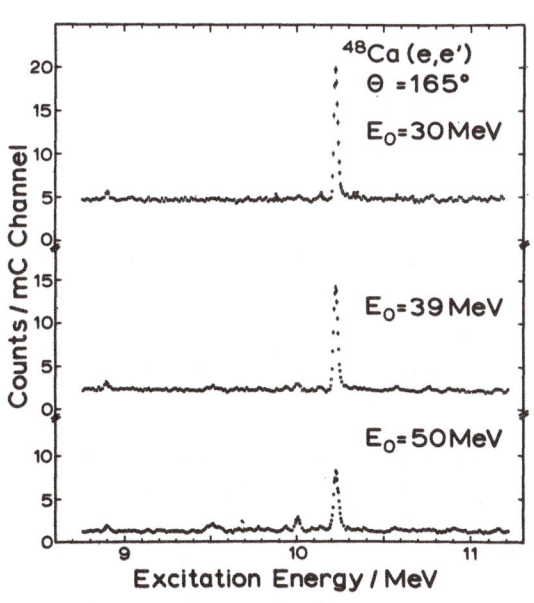

Fig. 30   Three ^{48}Ca(e,e')
spectra all at
Θ = 165° but dif-
ferent bombarding
energies (exci-
tation function)
displaying the
prominent M1
transition.

We can derive some very straightforward conclusions about the
strong M1 transitions in ^{40}Ca and ^{48}Ca:

(i)   In order to explain this transition in ^{40}Ca we again employ
      the concept of ground state correlations introduced in the
      discussion of the M1 strength in ^{28}Si. The prominent M1 tran-
      sition is expected to stem from an open shell ground state,
      i.e. a correlated many-particle many-hole state. There is

overwhelming experimental evidence for that and (for details
see ref.[53]) a reasonable percentage of p-h components in the
^{40}Ca ground state is e.g. the following: Op-Oh (39%), 2p-2h
(31%), 4p-4h (24.5%), 6p-6h (5%) and 8p-8h (0.5%). These numbers lead to an average occupation number of the $d_{3/2}$ shell
$\langle n_{d3/2} \rangle = 6.06$ compared with n = 8 in the pure IPM picture of
an uncorrelated ground state. Most likely the two missing particles are raised into the $f_{7/2}$ shell. As an immediate consequence two types of M1 spin-flip transitions can be realized
(see fig. 27 above) from the now partially occupied $f_{7/2}$ into
the still empty $f_{5/2}$ and from the full $d_{5/2}$ into the partially empty $d_{3/2}$ shell. We therefore can construct a simple quantum mechanical two-state model in which the M1 strength is
given by the coherent superposition of two matrix-elements

$$B(M1) = |\alpha\langle f_{5/2}|M1|f_{7/2}\rangle + \beta\langle d_{3/2}|M1|d_{5/2}\rangle|^2. \qquad (14)$$

Taking the unperturbed energies of the two spin-flip excitations as 7.2 and 6.2 MeV, respectively, and interaction matrix
elements calculated within the frame of the MSI model[23] we obtain two states at $E_x = 7.65$ MeV ($B(M1)\uparrow = 0.24\mu_K^2$) and at
$E_x = 10.05$ MeV ($B(M1)\uparrow = 1.47\mu_K^2$), the latter being in fairly
good agreement with the experimental number quoted above.

It should be noted in this connection that the strong
magnetic dipole spin polarizability, detected in ^{40}Ca supports
naturally the introduction[51] of effective g factors for the
description of magnetic moments in the upper part of the sd
shell.

(ii) The observed M1 strength in ^{48}Ca is most likely entirely due
to the dominant $f_{7/2} \rightarrow f_{5/2}$ spin flip transition. Taking bare
g factors, the IPM prediction is $B(M1)\uparrow \approx 12\mu_K^2$. However, the
experimental value is only $B(M1)\uparrow = 4.0 \pm 0.3\mu_K^2$. We make two
mechanisms responsible for this fact, renormalized g factors
and ground state correlations.

The main effect for the reduction of strength results
from the g_s factor, which is considerably quenched with respect to its bare value (see also the discussion at the end
of these lectures). The comparison between experimentally determined and theoretically calculated M2 strength distributions[59] and also M1 strength distributions and magnetic moments and their behaviour from light to heavy nuclei[60] indi-
cate for ^{48}Ca that $g_s = \gamma \cdot g_s^{free}$ with $\gamma = 0.79$ for protons and
neutrons alike. If we take this value for g_s and $g_\ell^n = 0$ we obtain the upper full curve shown in fig. 31. In this figure the
expected dependence of M1 strength in ^{48}Ca on the mean occupation number of neutrons in the $f_{7/2}$ shell (assuming a closed proton shell, which is probably a very good assumption[61])

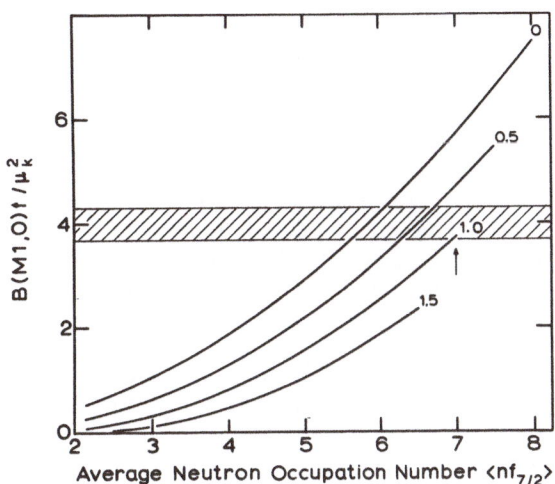

Fig. 31

Dependence of the M1
strength in ^{48}Ca on the
average occupation num-
ber $\langle nf_{7/2}\rangle$. The numbers
written on the right side
of the solid lines denote
the average occupation
number $\langle nf_{5/2}\rangle$. The experimentally determined
transition strength and
its uncertainty is plott-
ed in form of a band. If
there is one neutron missing in the closed $f_{7/2}$
shell (indicated by the
arrow) the experiment
tells that between 1 and
0.75 neutrons might be
sitting in the $f_{5/2}$ shell.

for different mean occupation numbers in the $f_{5/2}$ shell is
plotted. Those curves are calculated from the squared expec-
tation value of the $\underset{\sim}{\ell}\cdot\underset{\sim}{s}$ operator in the ^{48}Ca ground state
which yields

$$B(M1)\uparrow = (\frac{3\langle nf_{7/2}\rangle - 4\langle nf_{5/2}\rangle}{24})^2 \; B(M1; \; f_{7/2}\rightarrow f_{5/2})\uparrow. \qquad (15)$$

For $B(M1; \; f_{7/2} \rightarrow f_{5/2})\uparrow$ we have the IPM value of $12\mu_K^2$ reduced
to $7.5\mu_K^2$ due to the effective g factors quoted above. Obvious-
ly, if there are 8 neutrons in the $f_{7/2}$ shell, none can be
in the $f_{5/2}$ shell and the first term on the r.h.s. is unity.
This factor is therefore a measure for the distribution of
neutrons among the $f_{7/2}$ and $f_{5/2}$ shells in ^{48}Ca, i.e. of
ground state correlations. Drawn in fig.31 in form of a horizontal band is the experimentally measured value
$B(M1)\uparrow = 4.0 \pm 0.3\mu_K^2$. Together with $\langle nf_{7/2}\rangle = 7$ as experimentally determined from pick up reactions[62] a mean occupa-
tion number $\langle nf_{5/2}\rangle$ of neutrons between 0.75 and 1 is obtained and the difference is likely to be distributed among other
shells. Those numbers, which confirm that ^{48}Ca is a good
shell model nucleus should of course be viewed within the
$\pm$ 20% uncertainty of the quenched g_s factor. In passing we
note that such a simple estimate as performed for ^{48}Ca cannot be made at present for neither ^{42}Ca nor ^{44}Ca since the
proton shell is not well closed[61].

(iii) Finally, there is an interesting corroboration of our discovery of the $J^\pi = 1^+$ state in ^{48}Ca. As fig. 32 from ref.[63]
shows, the neutron spectra from the ^{48}Ca(p,n)^{48}Sc charge exchange reaction at 160 MeV exhibit a prominent narrow peak at
$E_x = 16.81$ MeV. This peak is the isobaric analogue state of
the $J^\pi = 1^+$ state at $E_x = 10.23$ MeV and carries a significant
fraction of the $T_>$ Gamow-Teller strength. Remember, that in
single charge exchange reactions with hadrons like (p,n),
(^{3}He,t), (^{6}Li, ^{6}He),..... or pions like (π^+,π^o) the Gamow-
Teller operator $\sum_i \vec{\tau}_i \cdot \vec{\sigma}_i$ operating on a target ground state
with N > Z leads to states with isospin $T_>$ and $T_<$ in the fi-
nal nucleus. Both components (see upper part of fig. 32) are
most likely identified in the charge exchange reaction.

The relation between our ^{48}Ca(e,e') spectra and the
spectra from ^{48}Ca(p,n)^{48}Sc might be explained in terms of the
shell model (see lower part of fig. 32). The $J^\pi = 1^+$ state at
$E_x \approx 10.3$ MeV in ^{48}Ca reached from the ^{48}Ca ground state with
an $(f_{7/2})^8$ neutron configuration corresponds to a neutron particle-hole state of the structure $(\nu f_{5/2} \nu f_{7/2}^{-1})_{1^+}$. Now we look
at the (p,n) charge exchange reaction. The low lying states

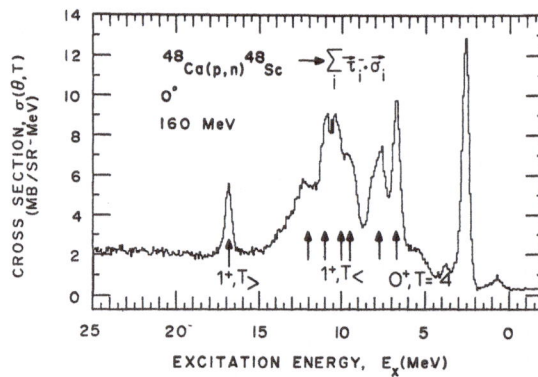

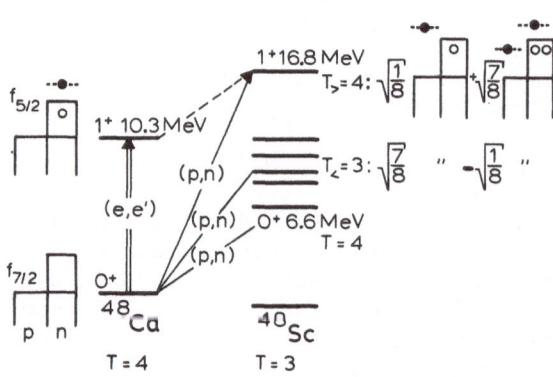

Fig. 32

In the upper part, a neu-
tron spectrum from the
^{48}Ca(p,n)^{48}Sc charge exchange reaction[63] is shown.
Note the narrow 1^+, $T_> = 4$
state at 16.81 MeV and
other possible 1^+, $T_< = 3$
states around $E_x \approx 10$ MeV
as well as the 0^+, $T = 4$
isobaric analogue state
of the ^{48}Ca ground state
at $E_x \approx 6.6$ MeV in ^{48}Sc.
The 1^+, $T_> = 4$ state at
16.8 MeV is the isobaric
analogue state of the 1^+,
$T = 4$ state seen at
$E_x \approx 10.3$ MeV in ^{48}Ca
(e,e') and an explanation
of this in terms of the
shell model is given in
the lower part of this
figure.

in ^{48}Sc belong to the $(\nu f_{7/2}^{-1}\pi f_{7/2})$ multiplet[64]. The strongest
state at $E_x = 2.52$ MeV in the spectrum of fig. 32 is the J^{π};
$T = 1^+$; 3 state belonging to this multiplet, as well as another
one, the J^{π}; $T = 0^+$; 4 state at $E_x = 6.67$ MeV which is the
isobaric analogue state of the ^{48}Ca ground state. The isobaric
analogue state of the $J^{\pi} = 1^+$ state at $E_x \approx 10.3$ MeV in ^{48}Ca
which with a Coulomb energy difference between ^{48}Ca and ^{48}Sc
(ref.[65]) of $\Delta E_C = 7.18$ MeV falls almost exactly into the place
where it is observed in the (p,n) reaction, has the shell mo-
del structure $\sqrt{1/8}(\pi f_{5/2}\nu f_{7/2}^{-1}) + \sqrt{7/8}(\pi f_{7/2}\nu f_{7/2}^{-2}\nu f_{5/2})$ where
the coefficients are due to the usual isospin Clebsch-Gordan
coefficient $(T_c T_{cz} 1/2 \pm 1/2 | T\, T_z)$ with $T_c = 7/2$. There is an
orthogonal component to that $T_> = 4$ state, the $T_< = 3$ anti-
analogue state with $\sqrt{7/8}(\pi f_{5/2}\nu f_{7/2}^{-1}) - \sqrt{1/8}(\pi f_{7/2}\nu f_{7/2}^{-2}\nu f_{5/2})$.
In fact, the 1^+ model space is much larger because of single
nucleon excitations also in the p shell, but this is not so
important for the present discussion; see also ref.[66]. Important is that the (p,n) single charge exchange reaction at
high bombarding energies is through its reaction mechanism
predominantly exciting the 1p-1h component of nuclear states.
We therefore see immediately that from the respective isospin
Clebsch-Gordan coefficients of the nuclear wave function the
Gamow-Teller excitation strength of the $T_>$ state must be much
lower than of the $T_<$ state, i.e. roughly by 1/8 over 7/8 for
the specific shell model configurations given in fig. 32.
This is indeed born out by the (p,n) experiment. With increasing neutron excess in the target ground state this ratio will
favour more and more the excitation of $T_<$ Gamow-Teller strength
in charge exchange reactions. There are experimental hints
for that from charge exchange reactions performed for a
search of such strength in heavier nuclei like ^{90}Zr and ^{208}Pb.

We are in fact dealing here with an example of isovector
collective vibrations with charge exchange components predicted already some time ago[67,68]. Further examples of the different isospin modes of the isovector M1 transitions are given
in refs.[3,69].

Example 5: M1 transitions in ^{58}Ni

This nucleus for which in the simple IPM three $J^{\pi} = 1^+$ states
of the structure $|1^+\rangle = [(f_{7/2})^{-1}(f_{5/2})]_{p,n} + [(p_{3/2})^{-1}(p_{1/2})]_n$
are expected to be excited around $E_x \approx 11$ MeV, is the heaviest nucleus in which strong M1 transitions are seen in (e,e'). Most of
them line up rather closely with states predicted by a refined
3p-1h shell model calculation[70] using the surface delta interaction
(fig. 33). They constitute the $T_> = 2$ part of the isovector M1
strength. The $T_< = 1$ part of this strength is predicted to lie
around $E_x \approx 6$ MeV but the only 1^+ state we observe - in agreement
with resonance scattering experiments[71] - is at $E_x \approx 7.7$ MeV. There
is also no M1 strength seen at $E_x \approx 12 - 15$ MeV.

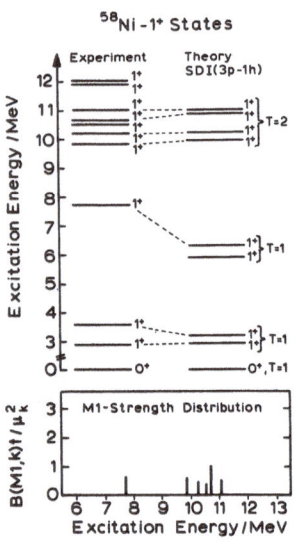

Fig. 33   Comparison between the ex-
           perimentally determined
           $J^\pi = 1^+$ states in ^{58}Ni and
           a shell model prediction[70].
           The M1 strength distribution
           derived from the ^{58}Ni(e,e')
           high resolution data is also
           shown.

All theoretical calculations[70,72,73] yield approximately 60%
more M1 strength than found experimentally. Note, that the $T_> = 2$
states are the analogue states of low lying 1^+ states in ^{58}Co studied in the charge exchange reactions ^{58}Fe(p,n)^{58}Co (ref.[74]) and
^{58}Ni(t,^{3}He)^{58}Co (ref.[75]).   A detailed comparison of the various
strength distributions measured in the different reactions and of
shell model calculations is given in [74].

Example 6:   Distribution of M1 and M2 strength in ^{90}Zr, the Landau-
             Migdal parameter g_0' and pion condensation

     In the search for M1 strength in heavy nuclei we have found
in ^{58}Ni only about 40% of the total predicted isovector M1 strength
and in ^{90}Zr about 10% of the RPA prediciton[9]. Three $J^\pi = 1^+$ states
have been identified at E_x = 8.233, 9.000 and 9.371 MeV. There is
some indication of further very fragmented dipole strength and an
upper limit for the total M1 strength in the investigated energy
region is $\Sigma B(M1) \leq 2.5\mu_K^2$. This is surprising, since we expect a
sizable M1 strength due to a pure neutron $g_{9/2} \rightarrow g_{7/2}$ spin-flip
transition. Instead, the spectra in fig. 34 (the high level density in the investigated energy region calls for an even higher resolution than the one presently available) yield strong evidence
for many 2^- states with a center of gravity at $E_x \approx 44A^{-1/3}$ MeV.
This clustering of strength may be loosely called an M2 giant resonance which is a common feature of other nuclei[59]. Altogether 34
candidates for 2^- states have been experimentally identified and
their detailed strength distribution is in fig. 35 compared to RPA

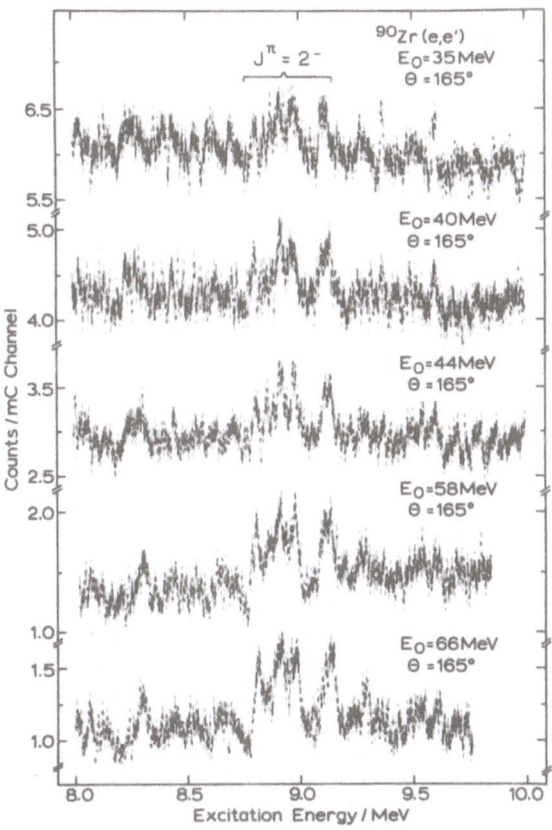

Fig. 34   Various ^{90}Zr(e,e') spectra measured at different bombard-
ding energies E_O but all at Θ = 165°.

preditions within the MSI model. Other theoretical predictions[76]
also show that the M2 strength distribution is concentrated in the
excitation energy region of 6 - 12 MeV.

Three interesting facts emerge immediately from the distribution in fig. 35. First, less than 20% of the predicted M2 strength
is found experimentally. Hence, like in the case of the M1 excitations, the experimental strength seems to be quenched. Second, the
large body of 34 2⁻ level spacings is more consistent with a Wig-
ner-distribution[77] than an experimental ("random") distribution.
Furthermore, the B(M2) values are in good agreement with a Porter-
Thomas distribution[77]. Third, the location of the 2⁻ states around
E_x = 9 MeV yields a Landau-Migdal parameter $g_O' \approx 0.7$ for the effective p-h force, in rather good agreement with a nuclear structure
calculation[78] taking π- and ρ meson exchange into account. This
has the remarkable consequence with respect to the existence of
pion condensates in the nuclear ground state that the threshold
density for pion condensates in infinite systems is larger than

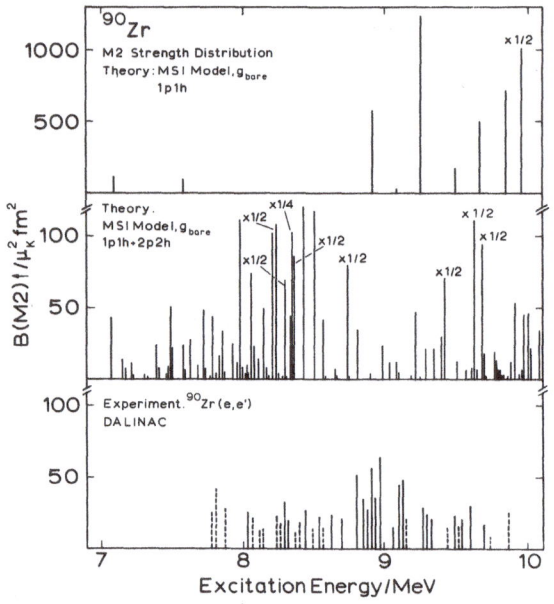

Fig. 35 Comparison of theoretically predicted and experimentally measured M2 strength distributions in ^{90}Zr.

twice the normal nuclear density. This conclusion has been derived[79] within the Brown-Weise[80] model of pion condensation. Such a large value of g_0' most likely also excludes the occurence of pion conden- sation in relativistic heavy ion collisions. We indeed have here another fine example where a careful measurement at a low energy electron accelerator helps to elucidate questions in connection with the modern trend of relativistic heavy ion research.

V.   SUMMARY ON M1 AND M2 STRENGTH DISTRIBUTIONS AND THEIR QUENCH-
     ING, ANOMALOUS MAGNETIC MOMENTS AND THE POSSIBLE RENORMALIZA-
     TION OF THE NUCLEAR SPIN CURRENT

     In these lectures I have tried to show how low multipolarity electric and magnetic transitions are studied. Our concrete achieve- ments at the present time based very much on the availability of high resolution electron beams, however, demonstrate clearly that even those very elementary nuclear excitations discussed are still not yet fully understood theoretically. One of the most striking results is the strong retardation both of M1 and M2 strength in the heavy nuclei as compared to the theoretical predictions. In addition to the examples we have discussed I should note that from the high resolution experiments ^{140}Ce(e,e') and ^{208}Pb we have no indication for the occurence of M1 transitions in these nuclei (see ref.[29]). The M2 strength in these nuclei is also very much retard- ed. On the l.h.s. of fig. 36, the measured M2 strengths in (e,e') on ^{28}Si, ^{90}Zr and ^{208}Pb are collected. They might be directly com- pared to the theoretical predictions of the MSI model on the r.h.s. of this figure. In all cases much less M2 strength is experimental-

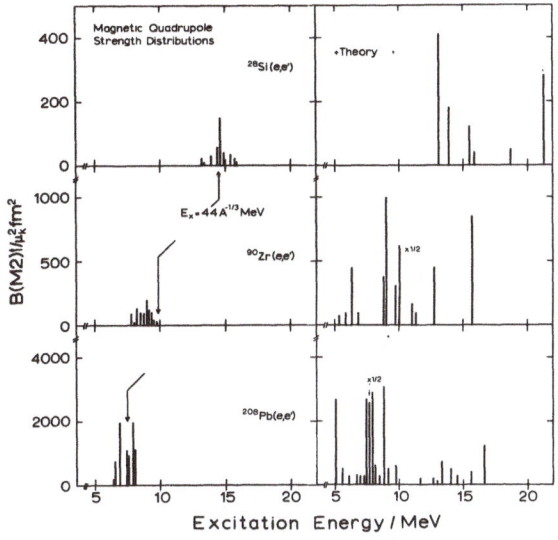

Fig. 36 Comparison of the experimentally measured M2 strength distribution on ^{28}Si, ^{90}Zr and ^{208}Pb (l.h.s.) with theoretical predictions from the MSI model (r.h.s.).

ly observed as is theoretically predicted. This statement is fairly invariant against a specific theoretical model prediction[4].

The observed inhibition of the M2 strength in nuclei can also be demonstrated through a comparison of the measured experimental M2 strength with the corresponding theoretical sum rule estimates. We will hence express the fact of the missing strength in terms of a sum rule. Magnetic sum rules (for multipolarities $\lambda \geq 2$) have only recently achieved some attention[10,81,82].

The rather small percentage of at most 20% by which the EWSR for M2 transitions is exhausted and that none or very little M1 strength is seen in heavy nuclei points strongly to a non-collective nature of magnetic as compared to the collective nature of electric transitions. We consequently have no magnetic giant resonances in heavy nuclei. There is here neither time nor space to discuss this phenomenon (see, however ref.[3]), which can be understood best in terms of the schematic model of Brown and Bolsterli[83].

What are the possible reasons for such an inhibition of magnetic strength — which is after all observed also for magnetic transitions of higher multipolarity and in charge exchange reactions measuring Gamow-Teller strength - in heavy nuclei? Here are four reasons:

(i)   If by a particular physical mechanism the strength would be shifted to higher excitation energy than where we searched it would have escaped the measurement. Such a mechanism invoked in [84] has prompted us to look for M1 strength in ^{208}Pb at $E_x > 8$ MeV, so far we have only a negative result. We are also extending presently our ^{90}Zr(e,e') experiments into the region of the higher excitation energies.

(ii) In case of a very strongly fragmented strength[59,85], it is the question if our experimental detection limit is low enough in order to detect it. A natural limit to what can be done in (e,e'), represents the treatment of the background (mainly due to the radiation tail) in the spectra and we are working on that problem also.

(iii) Ground state correlations, which in the course of these lectures we have shown to be very important in the light nuclei, have a strong tendency to damp magnetic transitions. We find for heavy nuclei typically $B(M2)_{IPM}/B(M2)_{RPA} \approx 1/0.8$ and know that the ground state correlations are generally too weak in the RPA.

(iv) It is clear that the magnetic charge of the nucleus has to be renormalized due to mesonic effects and core polarization, i.e. $g^{free} \rightarrow g^{eff}$ and part of the observed retardation of the magnetic strength has definitely to be attributed to this effect[86].

Let us finally assemble a few arguments for this last reason (iv). Remember from eqs. (11) and (12) discussed above, that always $B(M1, M2,...) \sim |<|..(g_\ell \ell + g_s s)..|>|^2$, While g_ℓ is fairly well known from measurements of magnetic moments of high spin states[87], $g_\ell^p = 1.119$ and $g_\ell^n = -0.031$, is this not so in the case of g_s. We therefore make here the hypothesis that g_s might be renormalized such that $g_s^{eff} = \gamma \cdot g_s^{free}$ both for the proton and the neutron. This quenching factor γ is then determined by the amount of which the with g_s^{free} theoretically calculated $B(M1,M2)$ distribution have to be adjusted by a variation of γ such that the respective experimentally measured strengths are reproduced. The so obtained mass dependent quenching factor γ is shown in the upper and middle part of fig. 37. If the same procedure is also performed such that the overall trend of the measured magnetic moments especially for A > 60 nuclei near closed shells is reproduced, i.e. $\mu_{exp}=\mu_{the}(\gamma)$, the same general behaviour of the quenching factor is observed as for M1 and M2 transitions (see lower part of fig. 37). Inspecting the results in fig. 37 we feel that there is a very strong hint that the quenching of the magnetic strength distribution and of the magnetic moments might be due to the same mechanism. In light nuclei, where $\gamma \approx 1$, strong M1 transitions are observed and the so called Arima-Horie core polarization effect[89] is an important correction to the magnetic moments. In fact, shell model calculations[51,90] reproduce the observed magnetic moments of sd-shell nuclei with only a small quenching of g_s. In heavy nuclei, M1 excitations are inhibited and consequently the Arima-Horie effect alone is not capable to predict the trend of the measured moments without invoking quenched g_s-factors ($\gamma \approx 0.7$).

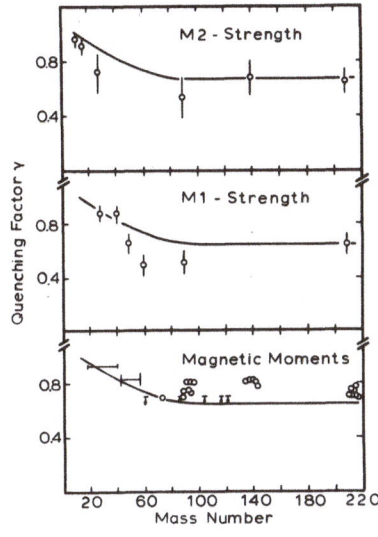

Fig. 37 Mass number A dependent
quenching factor γ of the
g_s factor. Upper part: derived from the observed M2
strength in (e,e'). Middle
part: derived from the observed M1 strength in (e,e'),
except for ^{208}Pb where photonuclear data were used[88].
Lower part: derived from the
magnetic moments of which a
complete list of references
is given in [60].

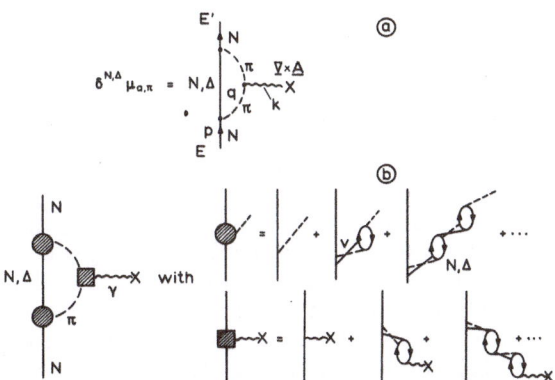

Fig. 38 Upper part (a): Pionic contribution to the anomalous nucleonic magnetic moment. The involved energies and momenta are indicated.
Lower part (b): Renormalization of the pion and photon
vertices in the diagram of the anomalous nucleonic magnetic moment by the coupling to intermediate particle hole
states. The particle can be either a nucleon or a Δ isobar. The dashed line represents the pion, the weavy line
a photon, the combined dashed-solid line (V) an e.g.
$\omega-\pi$ or $\rho-\pi$ exchange (short range correlations).

We[60] have started to think about a possible mechanism which explains the observed amount of quenching over and above the core polarization contribution. We consider as the main source the pionic contribution to the anomalous magnetic moment of the free nucleon in fig. 38. Remember that $g_s \sim (f_{\pi NN})^2$. We have investigated in the frame of an infinite matter approximation the renormalization effect of the nuclear medium on this pionic contribution. This is essentially equivalent to the mechanism known as the Lorentz-Lorenz effect[91] which plays an important rôle in the derivation of the pion-nucleus optical potential[92] and for the renormalization of the axial current in nuclei[93-95]. Of course, the infinite matter calculation is thougt to be only a first step towards the final solution of the problem. A totally realistic treatment would have to include in fig. 38 the momentum distribution of the baryons and the state dependent propagator renormalization, i.e. the proper angular momentum conservation of the p-h bubbles[95,96]. All details of the calculation can be found in ref.[60] and we quote here only that the quenching factor is calculated to be $\gamma \approx 0.7$ in heavy nuclei where the infinite matter approximation of fig. 37(b) is most likely appropriate. A calculation of the quenching effect in finite nuclei (the lower graph in fig. 38(b) has been considered) yields a somewhat smaller quenching factor γ which is furthermore depending on the multipolarity of the transition[97].

With those remarks on interesting many body effects which can even be studied with the help of such very elementary magnetic excitations in inelastic electron scattering on nuclei discussed here I close these lectures.

ACKNOWLEDGEMENTS

Most of the material presented here is still unpublished and is the subject of theses of students working at the DALINAC. I therefore would like to thank my experimental and theoretical colleagues at Darmstadt for the extremely pleasant collaboration in preparing these lectures. I thank D. Bender, R. Benz, G. Bayer, G. Eulenberg, A. Friebel, H. Genz, H.-D. Gräf, W. Gross, T. Grundey, V. Heil, D.H.H. Hoffmann, P. Ickelsheimer, U. Krämer, G. Küchler, G. Kühner, P. Manakos, D. Meuer, S. Müller, G. Schrieder, K. Seegebarth, E. Spamer, W. Steffen, W. Stock and O. Titze. For many years we have had a very intense collaboration with W. Knüpfer (Erlangen), whom I am particularly indebted to. He has shown a continuous effort concerning the theoretical implications of our work. Finally, I am very grateful to Mrs. R. Lamatsch for her efficient typing of these lecture notes and to Mrs. I. Kairies for her help with the references.

REFERENCES

1. J. McGrory, Lectures on the shell model given at this school.
2. W. Weise, Lectures on pion-nuclear many body problems given at this school.
3. A. Richter and W. Knüpfer, Collective Excitations, Giant Resonances and Sum Rules, Lectures given at the "International School of Intermediate Energy Nuclear Physics", Ariccia (Rome), June 18-29, 1979, Harwood Publ. Co., London, in press.
4. A. Richter, Electric and Magnetic Giant Resonances of Low Multipolarity Studied with Inelastic Electron Scattering, Lectures given at the "International School on Nuclear Structure", Alushta (USSR), April 14-25, 1980, to be published.
5. G. R. Satchler, Phys. Rep. 14C (1974) 97.
6. A. van der Woude, Lectures on giant resonances given at this school.
7. T. W. Donnelly and J. D. Walecka, Ann. Rev. Nucl. Sci. 25 (1975) 329.
8. H. Theissen, Springer Tracts in Modern Physics 65 (1972) 1.
9. D. Meuer, R. Frey, D. H. H. Hoffmann, A. Richter, E. Spamer, O. Titze and W. Knüpfer, Nucl. Phys. A, in press.
10. W. D. Teeters and D. Kurath, Nucl. Phys. A283 (1977) 1.
11. J. Mougey, M. Bernheim, A. Bussière, A. Gillebert, P. X. Hô, M. Priou, D. Royer, I. Sick and G. J. Wagner, Nucl. Phys. A262 (1976) 461; see also J. Mougey, J. Phys. Soc. Japan 44, Suppl. (1978) 420.
12. H. Morinaga, private communication.
13. P. Ickelsheimer, A. Richter, E. Spamer, W. Stock and O. Titze, Phys. Lett. B, submitted.
14. R. Frey, A. Richter, A. Schwierczinski, E. Spamer, O. Titze and W. Knüpfer, Phys. Lett. 74B (1978) 45.
15. F. R. Buskirk, H.-D. Gräf, R. Pitthan, H. Theissen, O. Titze and Th. Walcher, Phys. Lett. 42B (1972) 194.
16. A. Schwierczinski, R. Frey, A. Richter, E. Spamer, H. Theissen, O. Titze, Th. Walcher, S. Krewald and R. Rosenfelder, Phys. Rev. Lett. 35 (1975) 1244.
17. M. N. Harakeh, K. van der Borg, T. Ishimatsu, H. P. Morsch, A. van der Woude and F. E. Bertrand, Phys. Rev. Lett. 38 (1977) 676.
18. D. H. Youngblood, C. M. Rozsa, J. M. Moss, D. R. Brown and J. D. Bronson, Phys. Rev. Lett. 39 (1977) 1188.
19. F. E. Bertrand, G. R. Satchler, D. J. Horen and A. van der Woude, Phys. Lett. 80B (1979) 198.
20. H. P. Morsch, P. Decowski and W. Benenson, Nucl. Phys. A297 (1978) 317.

21.   P. Axel, Lecture Notes in Physics 108 (Springer, Berlin, 1979), p. 256.

22.   E. Grecksch, W. Knüpfer and M. G. Huber, Lett. Nuovo Cim. 14 (1975) 505.

23.   W. Knüpfer and M. G. Huber, Phys. Rev. C14 (1976) 2254.

24.   W. Knüpfer, private communication.

25.   For references, see C. Mahaux and H. A. Weidenmüller, Shell-Model Approach to Nuclear Reactions (North-Holland, Amsterdam, 1969).

26.   S. T. Tuan, L. E. Wright and D. S. Onley, Nucl. Instr. Meth. 60 (1968) 70.

27.   S. Raman, in Neutron Capture Gamma-Ray Spectroscopy (Plenum Press, New York, 1979), p. 193.

28.   W. Eyrich, A. Hofmann, U. Scheib, S. Schneider, F. Vogler and H. Rebel, Phys. Rev. Lett. 43 (1979) 1369.

29.   A. Richter, Lecture Notes in Physics 108 (Springer, Berlin, 1979), p. 19.

30.   J. Speth, E. Werner and W. Wild, Phys. Rep. 33C (1977) 127.

31.   A. B. Migdal, Theory of Finite Fermi Systems and Applications to Atomic Nuclei (Wiley, Intersci. Publ., New York, 1967).

32.   S.-O. Bäckman and W. Weise, Phys. Lett. 55B (1975) 1.

33.   Th. Grundey, A. Richter, G. Schrieder, E. Spamer and W. Stock, Nucl. Phys. A, submitted.

34.   J. Speth, J. Wambach, V. Klemt and S. Krewald, Phys. Lett. 63B (1976) 257.

35.   C. Woodward and G. A. Peterson, Phys. Rev. C20 (1979) 2437.

36.   E. Grecksch, W. Knüpfer and M. G. Huber, Phys. Rev. C15 (1977) 837.

37.   W. Knüpfer, in Proc. of the Sendai Conf. on Electro- and Photoexcitations, 1977, Suppl. to Research Report of the Laboratory of Nuclear Science, Tohoku University, Tomizawa, Sendai, Japan, Vol. 10 (1977) 123.

38.   M. Stroetzel and A. Goldmann, Z. Physik 233 (1970) 245.

39.   K. A. Snover, P. G. Ikossi and T. A. Trainor, Phys. Rev. Lett. 43 (1979) 117.

40.   H.-D. Gräf, H. Miska, E. Spamer, O. Titze and Th. Walcher, Nucl. Instr. Meth. 153 (1978) 9.

41.   Th. Walcher, R. Frey, H.-D. Gräf, E. Spamer and H. Theissen, Nucl. Instr. Meth. 153 (1978) 17.

42.   D. Schüll, J. Foh, H.-D. Gräf, H. Miska, R. Schneider, E. Spamer, H. Theissen, O. Titze and Th. Walcher, Nucl. Instr. Meth. 153 (1978) 29.

43.   J. Foh, R. Frey, R. Schneider, D. Schüll, A. Schwierczinski, H. Theissen and O. Titze, Nucl. Instr. Meth. 153 (1978) 43.

44.   A. Arima and D. Strottman, Los Alamos Scientific Laboratory Report, LA-UR 78-2969.

45.   G. Strassner, P. Truöl, J. C. Alder, B. Gabioud, C. Joseph,
      J. F. Loude, N. Morel, A. Perrenoud, J. P. Perroud,
      M. T. Tran, E. Winkelmann, W. Dahme, H. Panke, D. Renker
      and H. A. Medicus, Phys. Rev. C20 (1979) 248.
46.   E. J. Ansaldo, C. Rangacharyulu, D. Bender, U. Krämer,
      A. Richter, E. Spamer and W. Knüpfer, Phys. Lett. 95B
      (1980) 31.
47.   J. G. Woodworth, K. G. McNeill, J. W. Jury, R. A. Alvarez,
      B. L. Berman, D. D. Faul and P. Meyer, Phys. Rev. C19
      (1979) 1667.
48.   W. Knüpfer, K. Knauss and M. G. Huber, Phys. Lett. 66B
      (1977) 305.
49.   R. Schneider, A. Richter, A. Schwierczinski, E. Spamer,
      O. Titze and W. Knüpfer, Nucl. Phys. A323 (1979) 13.
50.   D. Kurath, Phys. Rev. 130 (1963) 1525.
51.   B. H. Wildenthal, in: Proc. of the International School of
      Physics "Enrico Fermi", Course LXIX, Varenna 1976, eds.
      A. Bohr and R. A. Broglia (North-Holland, Amsterdam, 1977)
      p. 383.
52.   H. Mackh, G. Mairle and G. J. Wagner, Z. Physik 269 (1974)
      353.
53.   W. Gross, D. Meuer, A. Richter, E. Spamer, O. Titze and
      W. Knüpfer, Phys. Lett. 84B (1979) 296.
54.   P. Burt, L. Fagg, H. Crannell, D. Sober, J. O'Brien, X. Ma-
      ruyama, J. Lightbody, R. Lindgren and K. Seth, Contr.
      3.5 to the Int. Conf. on Nucl. Phys. with Electromagnetic
      Interactions, Mainz, June 5-9, 1979 (unpublished).
55.   W. Steffen, H.-D. Gräf, W. Gross, D. Meuer, A. Richter,
      E. Spamer, O. Titze and W. Knüpfer, Phys. Lett. 95B
      (1980) 23.
56.   Collaboration with R. Neuhausen from the Mainz electron
      accelerator.
57.   J. Delorme, A. Figureau and N. Giraud, Phys. Lett. 91B
      (1980) 328.
58.   H. Toki and W. Weise, Phys. Lett. 92B (1980) 265.
59.   W. Knüpfer, R. Frey, A. Friebel, W. Mettner, D. Meuer,
      A. Richter, E. Spamer and O. Titze, Phys. Lett. 77B
      (1978) 367.
60.   W. Knüpfer, M. Dillig and A. Richter, Phys. Lett. 95B
      (1980) 349.
61.   P. Doll, G. J. Wagner, K. T. Knöpfle and G. Mairle, Nucl.
      Phys. A263 (1976) 210.
62.   J. L. Yntema, Phys. Rev. 186 (1969) 1144.
63.   B. D. Anderson, J. N. Knudson, P. C. Tandy, J. W. Watson,
      R. Madey and C. C. Foster, Phys. Rev. Lett. 45 (1980) 699.
64.   A. Richter, J. R. Comfort, N. Anantaraman and J. P. Schiffer,
      Phys. Rev. C5 (1972) 821.

65.   J. A. Nolen, Jr. and J. P. Schiffer, Ann. Rev. Nucl. Sci. 19
       (1969) 471.
66.   C. Gaarde, J. S. Larsen, M. N. Harakeh, S. Y. van der Werf,
       M. Igarashi and A. Müller-Arnke, Nucl. Phys. A334 (1980)
       248.
67.   D. F. Petersen and C. J. Veje, Phys. Lett. 24B (1967) 449.
68.   E. R. Flynn, J. Sherman and N. Stein, Phys. Rev. Lett. 32
       (1974) 846.
69.   A. Bohr and B. R. Mottelson, Nuclear Structure Vol. II
       (Benjamin, Reading, 1975).
70.   P. W. M. Glaudemans and A. G. M. van Hees, private communi-
       cation.
71.   K. Ackermann, K. Bangert, G. Junghans, R. Schneider, R. Stock
       and K. Wienhard, Lecture Notes in Physics 92 (Springer,
       Berlin, 1979) p. 447.
72.   D. Strottman, private communication.
73.   V. Yu. Ponomarev, Ch. Stoyanov, A. I. Vdovin and V. V. Voronov,
       JINR-preprint E4-12093, Dubna (1979).
74.   C. Wong, S. D. Bloom, S. M. Grimes, R. F. Hausman, Jr. and
       V. A. Madsen, Phys. Rev. C18 (1978) 2052.
75.   E. R. Flynn and J. D. Garrett, Phys. Rev. Lett. 29 (1972) 1748.
76.   V. Yu. Ponomarev, V. G. Soloviev, Ch. Stoyanov and A. I.
       Vdovin, Nucl. Phys. A323 (1979) 446.
77.   See e.g. M. L. Mehta, Random Matrices and the Statistical
       Theory of Energy Levels (Academic Press, New York, 1967).
78.   J. Speth, V. Klemt, J. Wambach and G. E. Brown, Nucl. Phys.
       A343 (1980) 382.
79.   J. Meyer-ter-Vehn, Proc. of the Symposium on Relativistic
       Heavy Ion Research, GSI Darmstadt, 1978 (Report No. GSI-
       P-5-78), Vol. 1, p. 302.
80.   G. E. Brown and W. Weise, Phys. Rep. 27C (1976) 1.
81.   T. Suzuki, Phys. Lett. 83B (1979) 147.
82.   M. Traini, Phys. Rev. Lett. 41 (1978) 1535.
83.   See G. E. Brown, Unified Theory of Nuclear Models and Forces
       (North-Holland, Amsterdam, 1967).
84.   G. E. Brown, J. S. Dehesa and J. Speth, Nucl. Phys. A330
       (1979) 290.
85.   J. S. Dehesa, J. Speth and A. Faessler, Phys. Rev. Lett. 38
       (1977) 208.
86.   See also, though in a slightly different context, H. Ejiri
       and J.I. Fujita, Phys. Rep. 38C (1978) 85.
87.   See e.g. T. Yamazaki, in Mesons in Nuclei, eds. M. Rho and
       D. H. Wilkinson, (North-Holland, Amsterdam, 1979), Vol. II.,
       p. 651.
88.   D. J. Horen, J. A. Harvey and N. W. Hill, Phys. Rev. Lett. 38
       (1977) 1344.
89.   A. Arima and H. Horie, Progr. Theor. Phys. 12 (1954) 623.
90.   B. H. Wildenthal and W. Chung, in Mesons in Nuclei, eds.
       M. Rho and D. H. Wilkinson (North-Holland, Amsterdam, 1979),
       Vol. II, p. 721.

91.   M. Ericson and T. E. O. Ericson, Ann. Phys. 36 (1966) 323.
92.   S. Barshay, G. E. Brown and M. Rho, Phys. Rev. Lett. 32
         (1974) 787.
93.   M. Rho, Nucl. Phys. A231 (1974) 493.
94.   J. Delorme, M. Ericson, A. Figureau and C. Thévenet, Ann.
         Phys. 102 (1976) 273.
95.   E. Oset and M. Rho, Phys. Rev. Lett. 42 (1979) 47.
96.   H. Toki and W. Weise, Z. Physik A292 (1979) 389.
97.   H. Toki and W. Weise, Phys. Lett. B, submitted.

EXOTIC NUCLEI

P.G. Hansen

Institute of Physics
University of Aarhus
DK-8000 Aarhus

and

The ISOLDE Collaboration
CERN
CH-1211 Geneva 23

INTRODUCTION

These lectures deal with nuclei with abnormally large or small neutron-to-proton ratios. In comparison with many other activities in nuclear physics, this is a relatively small field but at the same time one that has links to many of the traditional specialities in nuclear-structure physics. In these lectures I shall given some examples of recent results in the study of exotic nuclei. A somewhat broader picture of the many activities in this area can be obtained from the proceedings of specialized conferences[1,2] and from reviews[3-5].

The problem of techniques for producing and detecting the far-unstable nuclei is a central one to the field, but not one that I shall have time to discuss. Let me just in passing mention the important role played by on-line mass separation and refer to the review by Ravn[6] and to the forthcoming Proceedings of the 1980 Zinal Conference[7].

NEW RADIOACTIVE DECAY MODES: THE BETA-DELAYED EMISSION
OF TWO AND THREE NEUTRONS

It is clear from the mass formulas that beta-delayed emission
of more than one neutron or proton is expected to become possible
far from the line of beta stability. The first evidence for such
a process was obtained a little over a year ago from the energy
spectrum of neutrons emitted from 8.6 ms ^{11}Li produced by on-line
mass separation. The spectrum clearly showed a broad distribution
that was interpreted[8] as 2n emission from the 8.84 MeV state[9] in
^{11}Be (Fig. 1).

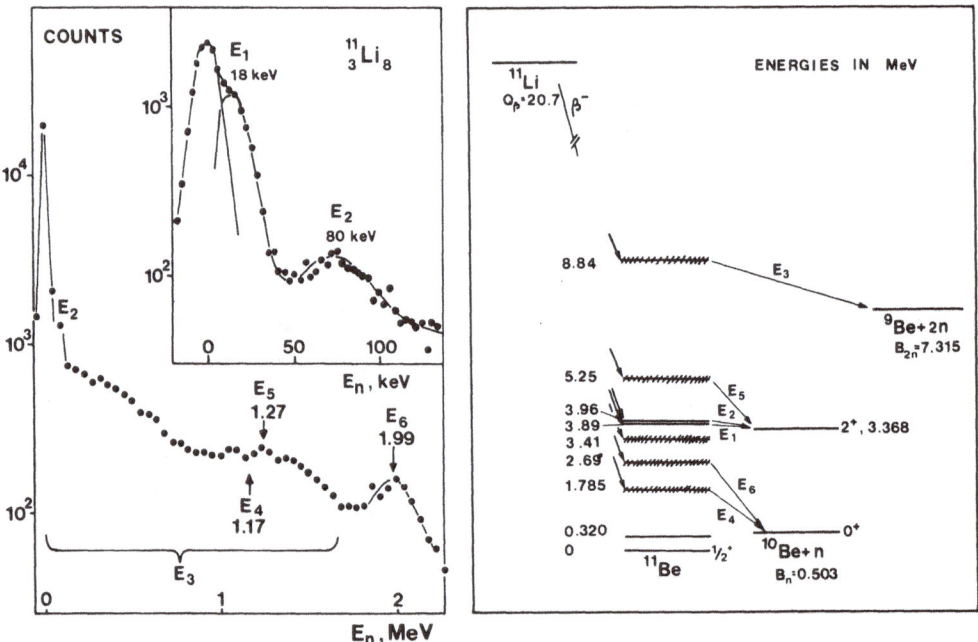

Fig. 1 Left: the energy spectrum of β-delayed neutrons from ^{11}Li.
The expected positions of peaks corresponding to known resonances
in ^{11}B are indicated by arrows. Right: the corresponding level
scheme. The 8.84 MeV level has a natural width of 200 ± 50 keV;
as it is strongly populated in the (t,p) reaction it must have a
large overlap with ^{9}Be + two neutrons, and it is most likely the
origin of the two-neutron emission observed in the present work.
(from Ref. 8).

This interpretation was confirmed in a time correlation experiment[8] in which a beam of ^{11}Li was directed to the centre of a paraffin-filled 4π counter. The measurement clearly showed pairs of neutron counts separated by the characteristic residence time for a neutron in the detector.

In a new experiment[10] the time correlation technique was extended to allow a search for three-neutron emission from ^{11}Li, which is ideally suited for this purpose since the Q_β value is 20.7 MeV while the threshold for break-up into $(3n + 2\alpha)$ lies at 8.888 MeV only. The detector had been improved so that the neutron detection efficiency now was 20% and the average residence time 89 ± 1 ms. In order to detect multiple events, the arrival times of individual neutrons were recorded by a "flying clock" and stored on magnetic tape. The strategy chosen for detecting multiple correlations was to examine a time interval θ following an initial event. If $(q - 1)$ neutrons had been registered during that time, the event was classified as a q-fold event. The calculation of the true multiplicities from the event rates, however, require a very detailed analysis, which takes into account also random events; thus random triples will occur as combinations of doubles and singles. The accuracy of the analysis was verified by varying the data rates in the experiment and also by varying the correlation time θ. Examples of the results are shown in Fig. 2 and Table 1. The p_n value for ^{11}Li has recently been revised[11,12] to $95 \pm 8\%$ from which one calculates the absolute value $p_{3n} = 1.8 \pm 0.2\%$.

A second series of experiments[13] studied two-neutron emission from isotopes of sodium. In addition to the time correlation technique described above, this work used gamma spectroscopy on mass-separated samples to detect gamma rays from lower mass chains than the one collected. These gamma rays could be shown to be due to one- and two-neutron decay processes and gave intensities in good agreement with the time-correlation measurements. The adopted p_{2n} values are ^{30}Na ($1.2 \pm 0.2\%$), ^{31}Na ($0.70 \pm 0.25\%$) and ^{32}Na ($5.1 \pm \pm 1.8\%$). These branching ratios agree well[13] with those calculated in a statistical model based on the gross theory of beta decay[14] and furnish a sensitive test on the calculations of beta strength at high energies (Fig. 3). The experiments show that beta-delayed two-neutron emission is not a rare process far from stability; in fact, for ^{32}Na it gives rise to half of the neutrons emitted. This mechanism therefore has to be taken into account in calculations of yields in the astrophysical r-process[15].

THE BETA STRENGTH FUNCTION

The high Q-values in the decay of far-unstable nuclei lead to extremely complex decays and it is usually convenient to discuss the beta decay pattern in terms of a strength function $S_\beta(E)$ defined as a reciprocal <u>ft</u>-value calculated per MeV of final levels

Table 1

Relative two- and three-neutron branching ratios[*)]

Singles rate c/s	$10^2 \times P_{2n}/P_{1n}$	$10^2 \times P_{3n}/P_{1n}$
7.9 [a)]	5.2 ± 0.6	2.2 ± 0.2
22.9 [a)]	4.5 ± 0.3	2.45 ± 0.09
45.5 [a)]	4.6 ± 0.5	2.35 ± 0.2
202.8 [b)]	5.0 ± 1.2	2.0 ± 0.4
Correlation time θ μs [c)]		
79.8 [c)]	5.2 ± 1.0	2.2 ± 0.3
114 [c)]	5.9 ± 0.6	2.0 ± 0.2
228 [c)]	5.2 ± 0.6	2.2 ± 0.2
456 [c)]	5.0 ± 0.6	2.39 ± 0.16

[*)] The quantities p_{in} denote the probabilities per beta decay for emitting i neutrons. Consequently, we have $p_n = \Sigma_i i p_{in}$.

[a)] $\theta = 228$ μs.

[b)] $\theta = 114$ μs.

[c)] Singles rate R = 7.9 c/s. The results shown correspond to the analysis of the same set of data for four different values of the correlation time θ.

$$S_\beta(E) = b(E)/\left[\underline{f}(Z, Q - E)\ T_{1/2}\right],$$

where b(E) is the absolute beta intensity per MeV of final levels in the daughter with atomic number Z, Q the total energy available, and $\underline{f}$ the usual statistical rate function. The concept of a strength function is most useful if the variations with energy are slow.

It is well known (see, for example, Ref. 16) that the beta strength function is dominated by collective states at high energy, and that the pattern for allowed beta decay can be understood in terms of a partially conserved Wigner supermultiplet symmetry. It

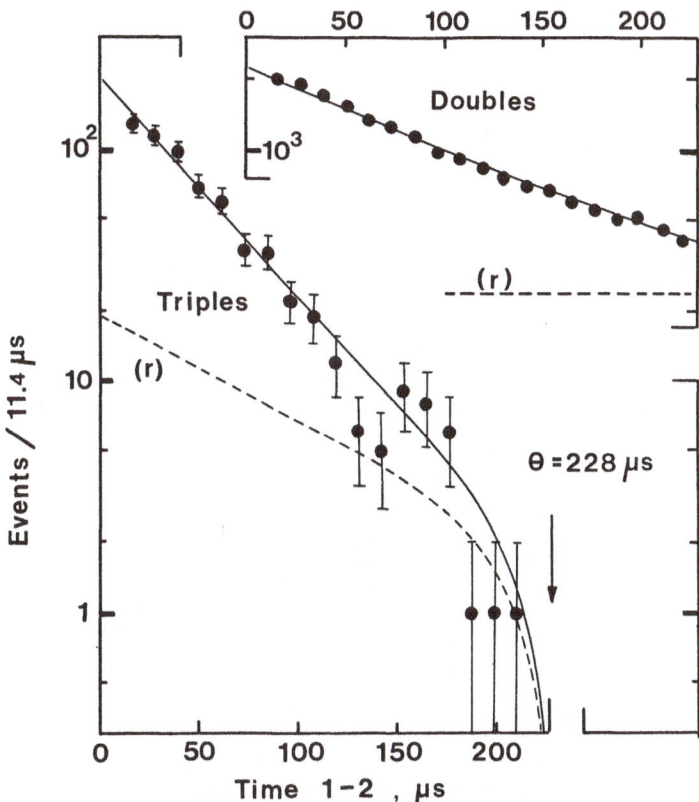

Fig. 2 Distribution of the time interval between the first and the second neutrons for events registered as doubles and triples with a correlation time θ = 228 μs. The data are from a 12 h run with an average neutron count rate of 22.9 c/s. The theoretical curves showing the total number of events and the contribution from random events (r) do not represent a fit: they have been calculated on an absolute scale from the results of the analysis. Note that the triples events clearly show the $e^{-2\lambda t}$ dependence expected theoretically, while the doubles vary as $e^{-\lambda t}$.

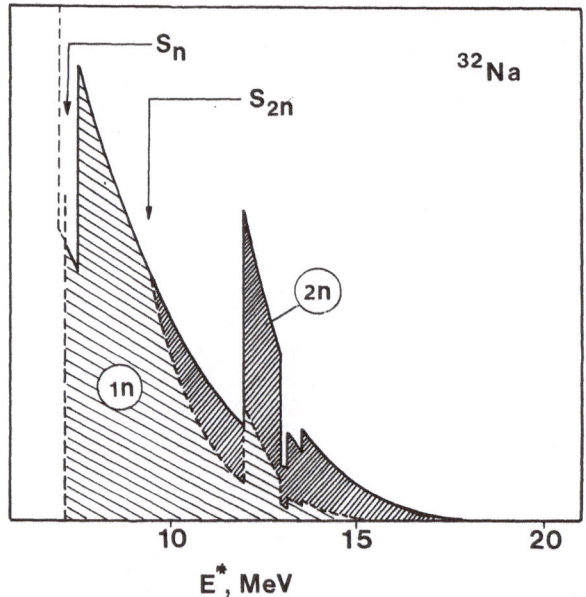

Fig. 3 Calculated ^{32}Na beta intensity[13] as a function of excitation energy in the ^{32}Mg daughter. The calculated one- and two-neutron contributions are 13% and 4.5%, respectively, in excellent agreement with the experimental values of 10 ± 4% and 5.1 ± 1.8%. Note that the discontinuities in the spectrum arise from the schematic treatment of nuclear pairing in the gross theory of beta decay, and note also that the calculation is based entirely on the parameters used in Ref. 14, so that no adjustment has taken place. The neutron-emission experiments clearly furnish a sensitive test on the beta-decay theory.

is a consequence of this picture that the collective states can be
attained in beta decay of nuclei with Z > N only. The main experimental evidence, so far, comes from the family of nuclei with $T_Z =$
$= -\frac{3}{2}$ (discussed by J.C. Hardy in Ref. 1), but it is an encouraging
sign that it has recently been possible to produce and detect[17,18] the
first members of the $T_Z = -2$ family (Fig. 4). Owing to its small
spreading width the isobaric analogue state corresponding to a collective isospin mode is clearly detectable, although it accounts for
only a small fraction of the decay. In neither of the two spectra
shown in Fig. 4, nor in the spectra of ^{24}Si and ^{36}Ca, recently discovered by Cerny and co-workers, does one observe directly the collective Gamow-Teller strength corresponding to a spin-isospin mode.
It is clear that much interest is attached to improving the intensities in which light proton-rich nuclei can be produced.

It is clearly not possible to produce heavier nuclei with Z > N,
and direct observation in beta decay of the collective Fermi and
Gamow-Teller strength is excluded. Fortunately, help has once again
come from the study of charge-exchange reactions. It turns out[19,20]
that the (p,n) reaction at high energy (120-160 MeV) and in forward
angles strongly favours the spin-isospin collective states so that
a direct observation of the Gamow-Teller giant resonance has been
possible in reactions on ^{90}Zr and ^{208}Pb. The results for the latter
are shown in Fig. 5. An analysis by Goodman et al.[21] finds a definite
correlation between medium-energy (p,n) cross-sections at 0° and
allowed beta decay rates. We may thus hope that reaction studies
may considerably add to our understanding of the beta strength function.

The best direct method for studying the beta strength function
is the measurement of beta-delayed particle spectra (see Figs. 3 and
6). The analysis of these, however, depends not only on the beta
strength function but is also very sensitive to assumptions about
transmission coefficients, level densities and gamma widths. Therefore it is important that the particle X-ray coincidence technique,
developed by Hardy and collaborators, in certain cases allows the
measurement of 10^{-16} s lifetimes and thus indirectly the γ-widths.
and level density. (In the following section we point to yet another
possibility for determining the level density.) Figure 6 shows the
results of an experiment[22] on the decay of ^{73}Kr.

We shall not attempt here to review the theory of beta strength.
Only one general model exists, the "gross theory" due to Yamada[23]
and developed by him and his collaborators[14,24]. It combines in a
schematic way a supermultiplet picture of allowed beta decay with a
Fermi-gas picture of the nucleus. In spite of being essentially a
one-parameter global theory, it has considerable predictive powers
as can be seen from the survey of beta-decay half-lives[14] and also
from the example given in Fig. 3.

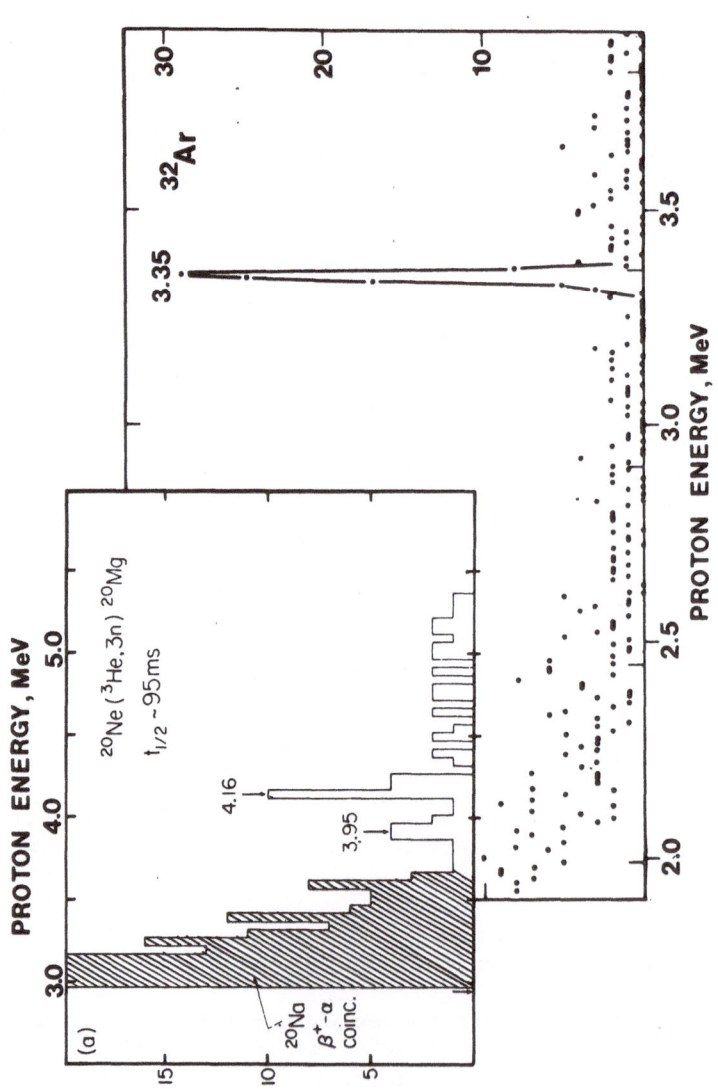

Fig. 4 Beta-delayed proton spectra[17,18] from the decay of the $(T,T_Z) = (2,-2)$ nuclei ^{32}Ar and ^{20}Mg. The proton peaks arise from the decay of the $T,T_Z = 2,-1$ isobaric analogue states, which can be estimated to be populated in beta decay with an intensity of $\sim 17\%$ (^{32}Ar) and 3% (^{20}Mg).

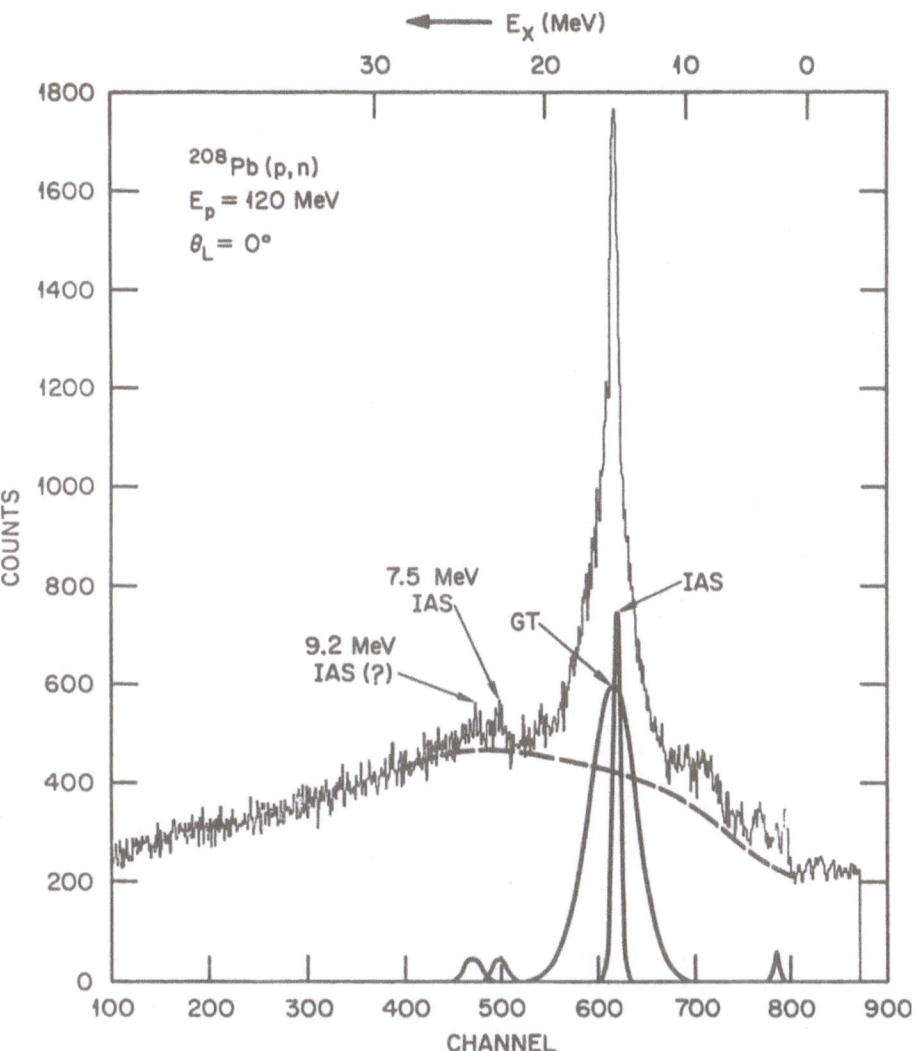

Fig. 5 Time-of-flight neutron spectrum for ^{208}Pb(p,n)^{208}Bi at 0°
for E_p = 120 MeV (from Ref. 20).

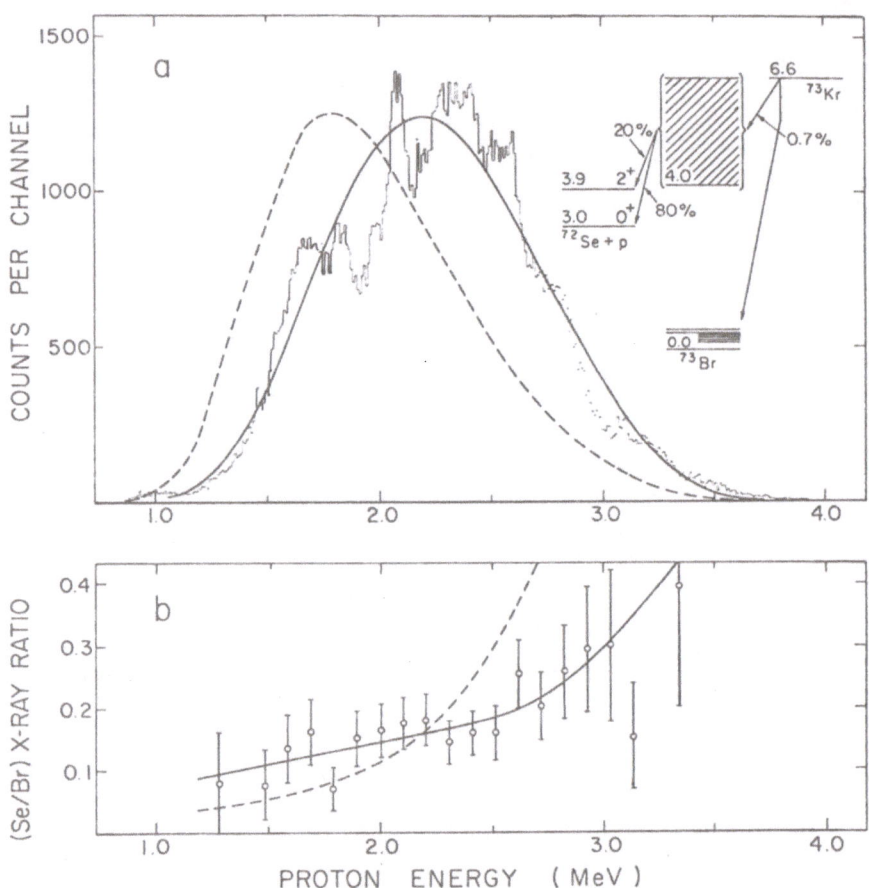

Fig. 6   a) Spectrum of protons observed following the decay of ^{73}Kr. In the simplified decay scheme, which is inset, all energies are given in MeV relative to the ^{73}Br ground state. b) Ratio of X-rays from Se relative to those from Br, plotted as a function of coincident proton energy. The curves are the results of calculations with different parameter sets (from Ref. 22).

Evidence of how nuclear shell structure modifies the strength distribution is beginning to accumulate especially near the closed shells[5], but there is a clear need for very complete experiments that carefully measure a number of parameters: spectral shapes and intensities, spins, Q-values, competing decay channels, level densities and average lifetimes. One of the difficulties encountered in the past has been that nuclear fluctuation phenomena sometimes were mistaken for structural effects. Since the statistical properties of nuclei at high energy play a significant role in spectroscopy away from stability, this subject is briefly reviewed in the following section.

THE PROPERTIES OF NUCLEAR NOISE

The experimentally measured strength function will fluctuate owing to the finite number of levels contributing to the average. To study these fluctuations is of interest for two reasons: (i) they give rise to a "noise level", which may make the observation of real structural effects difficult, and (ii) they provide a new way of measuring the nuclear level density.

The theory of nuclear statistical phenomena was developed in connection with neutron physics (see, for example, Lynn[25]) and applications to nuclei away from stability have been discussed a number of times[16,26-28]. The fluctuations arise primarily from fluctuations in the transition probabilities. These are governed by the Porter-Thomas law, which for a single reaction channel can be written

$$p(x) = (2\pi x)^{-\frac{1}{2}} \exp(-x/2) ,$$

where $p(x)$ is the probability density for observing a reduced width x. Here, x is measured in units of its average, so that $\langle x \rangle \equiv 1$. The variance on x is 2. This distribution is strongly asymmetric.

From the Porter-Thomas law it is possible to derive distributions governing many other experimental situations[26]. Assume that the experiment observes a weak particle branch with width y from a state populated in beta decay with width x. If the *total* width of the intermediate states is approximately constant, as if often the case, the observed intensities of y will be proportional to the product $v = x \cdot y$. The variable v obeys the product distribution law

$$p(v) = \pi^{-1} v^{-\frac{1}{2}} K_0(v^{\frac{1}{2}}) ,$$

where K_0 denotes the modified Bessel function. This distribution has a mean value of one and a variance of eight!

Finally, consider the case of a group of states decaying by two competing particle channels x and y, which are Porter-Thomas distributed. The average branching ratio for particle x can be expressed in terms of the average widths $\langle \Gamma_x \rangle$ and $\langle \Gamma_y \rangle$, where, defining $\langle \Gamma_y \rangle / \langle \Gamma_x \rangle = \alpha$

$$\langle \Gamma_x / (\Gamma_x + \Gamma_y) \rangle = \left(1 + \alpha^{1/2} \right)^{-1} ,$$

which appears strongly to favour the weak branch. It is interesting to note that the decay of the *group* will not be exponential: it is easily shown that the decay law is approximately t^{-2} for large values of the time t.

In most studies of far-unstable nuclei, one deals with unresolved levels so that one has $\sigma \gg D \gg \Gamma$, where σ denotes the Gaussian resolution parameter of the detector, D the average level spacing, and Γ the average natural width of the levels. Mathematical techniques for analysing fluctuations in unresolved nuclear spectra can be adapted from those used for interpreting noise in electrical circuits[29]. For the simple case of a spectrum g(x) with mean intensity unity, one intermediate spin, constant level spacing D, and experimental resolution parameter σ ($\gg$ D), one finds for the autocorrelation function

$$\psi_g(\tau) = \langle g(x) g(x + \tau) \rangle$$

$$= 1 + \frac{\alpha D}{2\pi^{1/2} \sigma} \exp \left(- \frac{\tau^2}{4\sigma^2} \right) ,$$

where α is the variance of the intensity of a line with average intensity unity. This expression permits the determination of the quantity $\alpha D / \sigma$ and (independently) of the experimental resolution. A related result[26] is that for a detector with resolution $W_{1/2}$ (full width at half maximum) the average spacing between "peaks" will be 2.2 $W_{1/2}$. (It is interesting to note how many spectra reported in the spectroscopic literature approach the theoretical peak density of noise ... .)

As an example of the application of the techniques discussed here, Fig. 7 shows the measured fine structure[30] of the ^{99}Cd delayed-proton spectrum, from which the level density in ^{99}In is derived. To determine the fluctuation amplitudes from such a spectrum introduces the delicate question that these, in principle, should be measured from the "true" value, which is, of course, both unknown and operationally undefined. This problem can be solved by a mathematical filtering technique, to which an electrical analogue would be the determination of a high-frequency noise component in the presence of a background of fluctuations of lower frequency in the d.c. level.

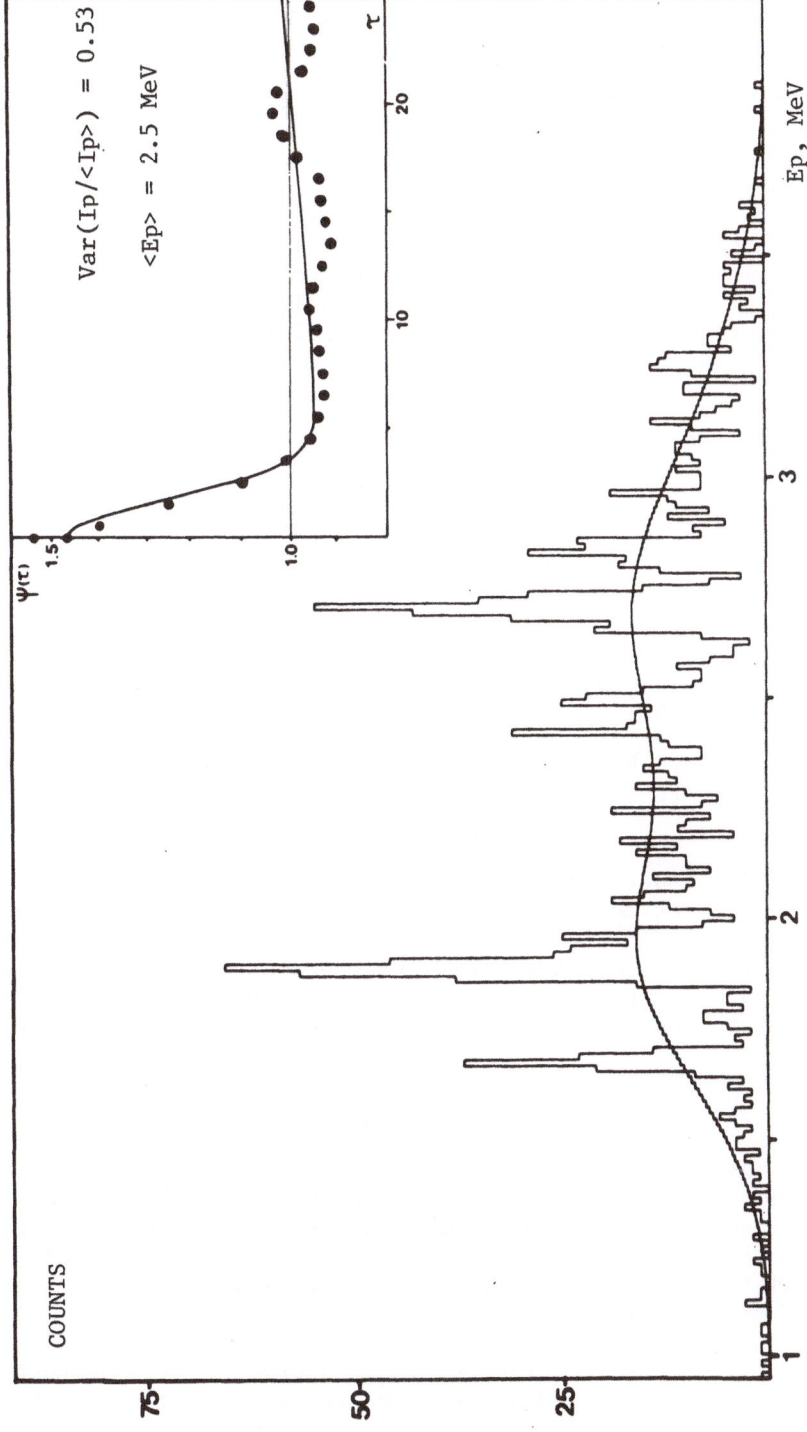

Fig. 7 The measured proton energy spectrum of 16 s ^{99}Cd and a smoothed spectrum obtained with a folding function of second order. The inset shows the autocorrelation function $\psi(\tau)$ for a spectrum formed as the quotient of the two spectra. The solid curve shows a theoretical fit to the autocorrelation function, from which a level density parameter a = 8.5 was obtained. The low value of a is evidence for the magicity of the nearby 50-50 doubly closed shell (from Ref. 30).

Several investigations[27,28] have resorted to Monte Carlo techniques in order to simulate the fine structure of delayed-particle spectra. The primary aim of these calculations has been pedagogical: even if the initial strength functions are structureless, a considerable fine structure may appear. Therefore, claims that a "prominent line structure" is evidence for a "selective beta population" are unjustified. Hardy et al.[27], named this structureless element "pandemonium" and pointed out that nuclear structure effects extracted from real data must be obtained along lines that would give the correct (structureless) picture of pandemonium from its "data". A second and probably more important application of pandemonium-type calculations is to derive level-density parameters in the cases in which suitable analytical expressions are not readily available[31]. Increased precision could probably be obtained by comparing autocorrelation functions of data and pseudodata.

THE NUCLEUS AT LOW EXCITATION ENERGY

Until this point we have mainly been regarding the exotic aspects of far-unstable nuclei. We now briefly indicate some of the many ways in which studies of nuclei far from stability contribute to our understanding of the systematics of nuclear structure.

A number of techniques from atomic physics have played a key role in these studies, as discussed in the reviews by Otten[32] and by Klapisch and Jacquinot[4]. Most productive of them all has probably been the classical atomic-beam magnetic resonance technique, which has been applied by Ekström and collaborators for determining spins and magnetic moments in very many cases, most recently to alkalies[33] and halogens[34].

The tunable laser has been an important tool in many applications[4,32]. The most recent development is the so-called collinear laser spectroscopy[35,36] in which an ion beam from an isotope separator is deflected into a laser beam. Following this, the ions are neutralized in a charge-exchange cell. Since the spread in the kinetic energy of the beam remains unchanged under electrostatic acceleration one has

$$\delta E = \delta(\tfrac{1}{2} \, mv^2) = mv\delta v = \text{constant} \, ,$$

which means that the Doppler width, which is determined by δv, is greatly reduced by acceleration. This new technique offers essentially Doppler-free spectroscopy and eliminates a number of chemical problems associated with the use of resonance cells. It has recently been applied[37] to the elements barium and ytterbium, and shows great promise for many other cases.

We finally give a few examples of studies of nuclear structure. The rubidium isotopes (Z = 37) have been studied by several techniques.

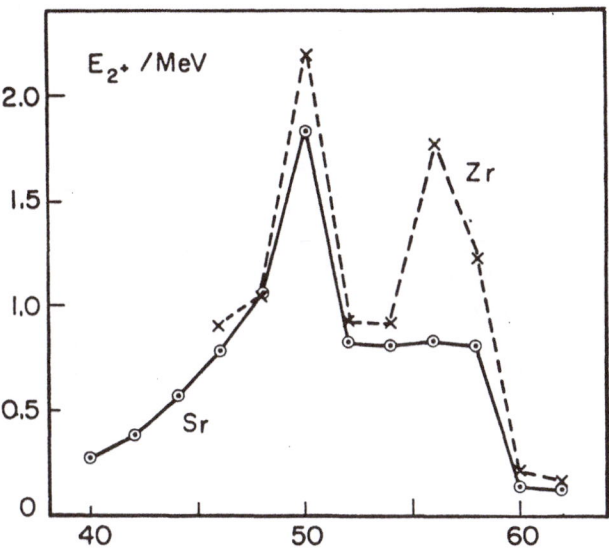

Fig. 8 The systematics of the energies of the first 2$^+$ state for
the elements strontium (Z = 38) and zirconium (Z = 40) (from Ref. 3).

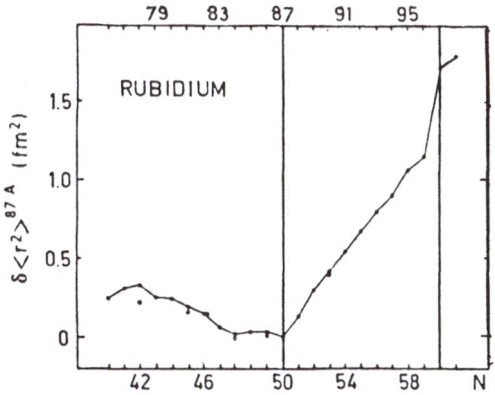

Fig. 9 The charge radii of the rubidium isotopes relative to ^{87}Rb
as determined by laser spectroscopy. Note the decrease in $\langle r^2 \rangle$ towards N = 50 and also the sudden onset of deformation at N = 60
(from Klapisch et al.[40]).

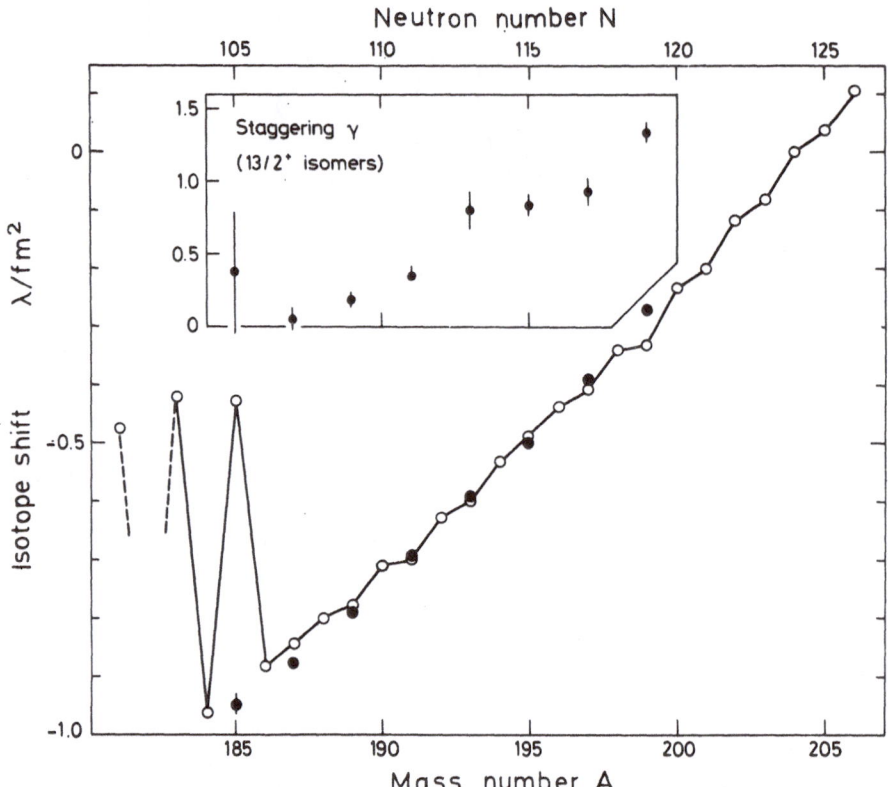

Fig. 10   The charge radii of the mercury isotopes relative to ^{204}Hg.
The size parameter is defined as $\lambda = \delta\langle r^2 \rangle = 1.1 \times 10^{-3}\ \delta\langle r^4 \rangle +$
$+$ higher terms.   Open and full circles indicate ground states and
isomers.   With the exception of ^{185m}Hg the experimental error is
smaller than the diameter of the circle.   The line connects ground
states of neighbouring isotopes.   The inset shows for the $^{13}/_2{}^+$ isomers the even-odd staggering parameter defined as $\gamma = 2(\langle r^2 \rangle_{N+1} -$
$- \langle r^2 \rangle_N)/(\langle r^2 \rangle_{N+2} - \langle r^2 \rangle_N)$, for even neutron number N (from Refs.
32, 41 and 42).

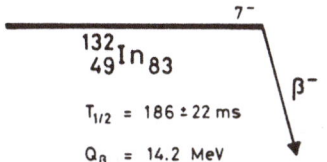

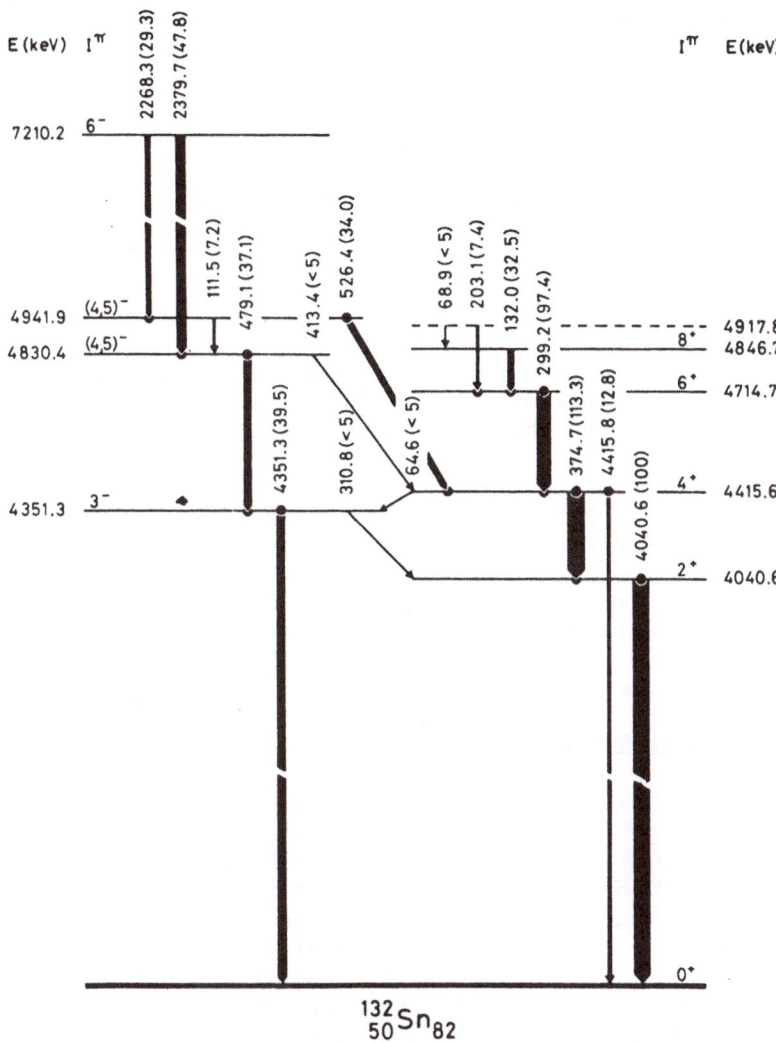

Fig. 11 The level scheme of ^{132}Sn obtained from the beta decay of 186 ms ^{132}In produced in mass separation and from the decay of the 1.7 μs isomer produced directly in fission. The ^{132}In ground state is presumably $(\nu f_{7/2} \, \pi g_{9/2}^{-1})_{7^-}$ decaying to the state at 7.2 MeV, presumably $(\nu f_{7/2} \, \nu g_{7/2}^{-1})_{6^-}$. The even-parity states (to the right) are $(\nu f_{7/2} \, \nu h_{11/2}^{-1})_{2,4,6,8^+}$ (From Bjørnstad et al.[43]).

The masses were measured by direct mass spectroscopy[38] and the systematics of one- and two-neutron separation energies clearly indicate the role of the magic and sub-magic numbers N = 50 and N = 56 (the closure of the $d_{5/2}$ shell) and also that strong deformations set in at N = 60. The systematics of the first-excited 2^+ energies of the elements strontium and zirconium given in Fig. 8 clearly bring out the same structural points. Further, the half-life of the (surprisingly low-lying) 129.2 keV level in ^{100}Sr has been measured[39] to be 5.15 ± 0.20 ns corresponding to a deformation parameter ε of 0.29. Finally, a new series of optical experiments by Klapisch and collaborators[40] has determined the charge radii of the rubidium isotopes (Fig. 9). The onset of deformation at N = 60 is again very clear.

The results of a long series of experiments dealing with the mercury isotopes are shown in Fig. 10, which represents the results of a 10-year concentrated effort by E.W. Otten and collaborators. A discussion of the individual data points would in itself provide an illuminating illustration of the applications of atomic-physics techniques to nuclear structure within the last decade. The results have been discussed in many papers, but the theoretical picture is still not completely clear. Note in particular the difference of 0.5 fm^2 between the two ^{185}Hg isomers and also the regular behaviour of the even-odd staggering parameter for the $^{13/2^+}$ isomers.

The examples of the rubidium-strontium region and of the mercury isotopes discussed above illustrate the interplay between spherical and deformed structures in a predominantly spherical region. One of the pure shell-model cases away from stability is the doubly magic ^{132}Sn, the level scheme of which is shown in Fig. 11, based on results obtained at JOSEF (Jülich) and at ISOLDE (CERN).

CONCLUDING REMARKS

The study of exotic nuclei contributes in several ways to nuclear physics. First of all it allows the exploitation of decay modes that are totally unknown for the species near stability. Second, the experiments with long mass chains bring in what we may call the "N-Z degree of freedom" and allow the observation of patterns that would not emerge from a few cases measured near stability. A good example of this is provided by the rubidium and mercury radii (Figs. 9 and 10), which, incidentally, both are very far from the $A^{1/3}$ behaviour in which the textbooks would have us believe. Third, there are certain individual nuclei away from stability that we would like to know about. One example would be the doubly magic systems such as ^{132}Sn, which are fundamental for our understanding of the shell model, and another example would be neutron-rich nuclei along the r-process path, which although exotic under terrestrial conditions must have been very common when the heavy elements were formed.

REFERENCES

1. Proc. Int. Conf. on Nuclei Far From Stability, Cargèse, 1976
       (CERN Report 76-13, Geneva, 1976), 608 pp.
2. Proc. Int. Symposium on Future Directions in Studies of Nuclei
       Far From Stability, Nashville, 1979, North-Holland, Amsterdam
       (1980).
3. P.G. Hansen, Ann. Rev. Nucl. Part. Sci. 29:69 (1979).
4. R. Klapisch and P. Jacquinot, Rep. Progr. Phys. 42:773 (1979).
5. B. Jonson, in Proc. Int. Conf. on Nucl. Phys., Berkeley, August
       1980, to be published in Nuclear Physics.
6. H.L. Ravn, Phys. Reports 54:201 (1979).
7. Proc. Int. Conf. on Electromagnetic Isotope Separation, Zinal,
       1980, to be published in Nucl. Instrum. Methods.
8. R.E. Azuma, L.C. Carraz, P.G. Hansen, B. Jonson, K.-L. Kratz,
       S. Mattsson, G. Nyman, H. Ohm, H.L. Ravn, A. Schröder and
       W. Ziegert, Phys. Rev. Lett. 43:1652 (1979).
9. F. Ajzenberg-Selove and C. Langnell-Busch, Nucl. Phys. A336:4
       (1980).
10. R.E. Azuma, T. Björnstad, H.Å. Gustafsson, P.G. Hansen,
       B. Jonson, S. Mattsson, G. Nyman, A.M. Poskanzer and
       H.L. Ravn, Beta-delayed three-neutron radioactivity of ^{11}Li,
       to be published.
11. The ISOLDE Collaboration, p_n values of ^{9}Li and ^{11}Li, to be pub-
       lished.
12. C. Détraz, D. Guillemaud, M. Langevin, F. Naulin, M. Epherre,
       R. Klapisch, M. de Saint-Simon, C. Thibault and F. Touchard,
       Journal de Physique Lettres (in press).
13. C. Détraz, M. Epherre, D. Guillemaud, P.G. Hansen, B. Jonson,
       R. Klapisch, M. Langevin, S. Mattsson, F. Naulin, G. Nyman,
       A.M. Poskanzer, H.L. Ravn, M. de Saint-Simon, K. Takahashi,
       C. Thibault and F. Touchard, Beta-delayed two-neutron emission
       from ^{30,31,32}Na, Physics Letters B (in press).
14. K. Takahashi, M. Yamada and T. Kondoh, Atomic Data and Nuclear
       Data Tables 12:101 (1973).
15. T. Kodama and K. Takahashi, Phys. Lett. 43B:167 (1973).
       J.B. Blake and D.N. Schramm, Astrophysical Letters 14:207 (1973).
       J.W. Truran, J.J. Cowan and A.G.W. Cameron, Astrophys. J. 222:
       L63 (1978).
16. P.G. Hansen, Adv. Nucl. Phys. 7:159 (1973).
17. E. Hagberg, P.G. Hansen, J.C. Hardy, A. Huck, B. Jonson,
       S. Mattsson, H.L. Ravn, P. Tidemand-Petersson and G. Walter,
       Phys. Rev. Lett. 39:792 (1977).
18. D.M. Moltz, J. Äystö, M.D. Cable, R.D. von Dincklage, R.F. Parry,
       J.M. Wouters and J. Cerny, Phys. Rev. Lett. 42:43 (1979).
19. D.E. Bainum, J. Rapaport, C.D. Goodman, D.J. Horen, C.C. Foster,
       M.B. Greenfield and C.A. Goulding, Phys. Rev. Lett. 44:1751
       (1980).

20. D.J. Horen, C.D. Goodman, C.C. Foster, C.A. Goulding,
        M.B. Greenfield, J. Rapaport, D.E. Bainum, E. Sugarbaker,
        T.G. Masterson, F. Petrovich and W.G. Love, to be published
        (1980).

21. C.D. Goodman, C.A. Goulding, M.B. Greenfield, J. Rapaport,
        D.E. Bainum, C.C. Foster, W.G. Love and F. Petrovich, Phys.
        Rev. Lett. 44:1755 (1980).

22. P. Asboe-Hansen, E. Hagberg, P.G. Hansen, J.C. Hardy, B. Jonson
        and S. Mattsson, Average excited state lifetimes in ^{73}Br,
        Nuclear Physics (in press).

23. M. Yamada, Bull. Sci. Eng. Research Lab. (Waseda Univ., Tokyo),
        No. 31-32 (1965).

24. K. Takahashi and M. Yamada, Progr. Theor. Phys. 41:1470 (1969).

25. J.E. Lynn, The Theory of Neutron Resonance Reactions, Clarendon,
        Oxford (1968), 503 pp.

26. B. Jonson, E. Hagberg, P.G. Hansen, P. Hornshøj and P. Tidemand-
        Petersson, *in* Ref. 1, p. 277.

27. J.C. Hardy, L.-C. Carraz, B. Jonson and P.G. Hansen, Phys. Lett.
        71B:307 (1977).

28. J.C. Hardy, B. Jonson and P.G. Hansen, Nucl. Phys. A305:15 (1978).

29. S.O. Rice, Bell Syst. Technical Journal 23,24:166 pp (1944-45).

30. T. Elmroth, E. Hagberg, P.G. Hansen, J.C. Hardy, B. Jonson,
        H.L. Ravn and P. Tidemand-Petersson, Nucl. Phys. A304:493
        (1978).

31. O.K. Gjøtterud, P. Hoff and A.C. Pappas, Nucl. Phys. A303:295
        (1978).

32. E.W. Otten, *in* Ref. 2.

33. C. Ekström et al., Nucl. Phys. A311:269 (1979); A292:144 (1977);
        Phys. Lett. 76B:565 (1978).

34. C. Ekström and L. Robertson, to be published in Physica Scripta.

35. S.L. Kaufman, Opt. Comm. 17:309 (1976).

36. K.R. Anton, S.L. Kaufman, W. Klempt, G. Moruzzi, R. Neugart,
        E.W. Otten and B. Schinzler, Phys. Rev. Lett. 40:642 (1978).

37. F. Buchinger, W. Klempt, A.C. Müller, E.W. Otten, C. Ekström
        and R. Neugart, to be published (1980).

38. M. Epherre, G. Audi, C. Thibault, R. Klapisch, G. Huber,
        F. Touchard and H. Wollnik, Phys. Rev. C. 19:1504 (1979).

39. R.E. Azuma, G.L. Borchert, L.-C. Carraz, P.G. Hansen, B. Jonson,
        S. Mattsson, O.B. Nielsen, G. Nyman, I. Ragnarsson and
        H.L. Ravn, Phys. Lett. 86B:5 (1979).

40. R. Klapisch et al., to be published (1980).

41. P. Dabkiewicz, F. Buchinger, H. Fischer, H.-J. Kluge,
        H. Kremmling, T. Kühl, A.C. Müller and H.A. Schüssler, Phys.
        Lett. 82B:199 (1979).

42. T. Kühl, P. Dabkiewicz, C. Duke, H. Fischer, H.-J. Kremmling
        and E.W. Otten, Phys. Rev. Lett. 39:180 (1977).

43. T. Bjørnstad, J. Blomqvist, G.T. Ewan, P.G. Hansen, B. Jonson,
        K. Kawade, A. Kerek, S. Mattsson and K. Sistemich, Contribu-
        tion to the Int. Conf. on Nuclear Physics, Berkeley, August
        1980.

LOW ENERGY NUCLEAR FISSION

H. Nifenecker

Département de Recherche Fondamentale, C.P.N.
Centre d'Etudes Nucléaires de Grenoble
85 X - 38041 Grenoble Cedex, France

Introduction

Fifteen years after the discovery of the fission process[1-3] a very important step was made towards the synthesis of the shell[7] and liquid drop model[3-6]. This was the unified model of Bohr and Mottelson,[9] and Mottelson and Nilsson,[10]. In this model the nuclear potential is allowed to be deformed. This deformation induces a deformation of the nuclear matter density which should be consistent with the potential itself. This was the socalled self consistent approach. In this approach the self consistency was not required in "detail" but on the second moments of the potential and density distributions. The potential was therefore related to smoothed average density. It was also recognized by Swiatecki et al.,[11] that all systems with "leptodermous" (thin-skin) density distribution should behave as liquid drops, (as far as their potential energy goes). Was the liquid drop model related to the average density responsible for the self consistent average potential? The answer to this question would have still to wait more than ten years. The Nilsson model was able to account very satisfactorily for ground state deformations but failed at large deformations and predicted much too strong stabilities of nuclei towards fission. The real break through was made in 1966 by Myers and Swiatecki,[12] and Strutinsky,[13]. Myers and Swiatecki related the small level density of closed shell nuclei to an increased binding energy leading to a shell correction to the liquid drop mass formula. They also predicted the possible existence of superheavy nuclei. Independently Strutinsky had found this relationship between level densities and nuclear masses. This was the basis for an averaging and renormalization procedure by which the liquid drop model was used to provide average nuclear energies on top of which shell corrections obtained from the Nilsson model were added.

Strutinsky made the very unexpected finding that the fission barrier was split in two. This double-humped fission barrier provided
a very natural explanation of fission isomerism. It predicted and
explained a whole wealth of phenomena such as certain structures in
fission cross sections. Once again fission studies came to the fore-
ground in low energy nuclear physics until the upsurge of heave ions
physics put them back on a more modest footing.

Bloch and Balian,[14]   showed that the deformed shells responsible
for the secondary minimum in the fission barriers could be understood from very general geometrical considerations.   Finally the
Hartree Fock theory of nuclear deformation energies[15]   gave a
microscopic basis to the Strutinsky procedure and illuminated the
relationship between density and energy averagings. The mass asy-
metry was qualitatively understood on the basis of deformed shell
structures in the nascent fragments and quasi stationnary scission-
point models met some quantitative successes. However the dynamics
of the fission process remains largely ununderstood. A cross-fer-
tilization between low energy fission and heavy ion reactions is
already taking place and may help understanding nuclear dynamics.

In these lectures we shall start with the liquid drop model of
fission and compare some of its predictions with experiment. The
liquid drop analogy allows to define in a rather simple and intui-
tive way a number of useful concepts and possible observables.

We will first of all show how a synthesis of the liquid drop
model and of the shell model can be made using the Strutinsky shell
averaging procedure. Some experimental data related to the existence of shape isomers are presented and discussed. We conclude by
discussing some aspects, both experimental and theoretical, of
fission dynamics.

1. Fission of a charged liquid drop.
1.1. Generalities.

We first consider an uncharged liquid drop. In a liquid,
molecules only interact with their nearest neighbors. It is therefore possible to define a local energy density which is a functio-
nal of the matter density

$$E (\rho) = E (\rho(\vec{r})) = E (\vec{r})$$

As long as the density is constant within the drop the energy density is also constant and the total energy of a definite
volume of the liquid is proportional to the volume itself. Molecules close to the surface are less bound than those which are in
the bulk of the liquid. Therefore the total internal energy of the
drop has a negative (for attractive forces) volume term and a positive surface term which accounts for the deficit in binding

$$E = - A_V V + A_S \varepsilon$$

For an incompressible liquid with a total number of molecules N, V is proportional to N and S to N 2/3 thus

$$E = - a_V N + a_s N^{2/3}$$

If the drop is charged one has to take into account a Coulomb repulsion term

$$E_c = \frac{Z^2}{V^2} \int \frac{e^2}{/r-r'/} \; d\vec{r} \; d\vec{r}'$$

where $\frac{Z}{V}$ is in fact the charge density of the drop.

For a sphere of radius R one obtains the known formula

$$E_c = \frac{3}{5} \frac{Z^2 e^2}{R}$$

It is clear that, due to the $\frac{1}{/r-r'/}$ factor, the Coulomb term will be less for stretched liquid drops than for spherical ones while the reverse is true for the surface energy term. Therefore when the drop deforms there will be an antagonism between the Coulomb and Surface terms. To be more specific it is necessary to describe, in some way, the shape of the liquid-drop. This is a rather tricky task when one wants to describe all the shapes which might take place during the fission in two parts of an originally spherical drop. One possibility is to describe the surface in terms of a complete set of spherical harmonics. This description is certainly very good for small or moderate deformation but fails for separated fragments. Due to the large amount of work which has been done using spherical harmonics we give some details on this approach. We use the definitions of Wilets (16). The surface is defined by its distance to the origin

$$R(\theta,\Phi) = R_\alpha \left\{ 1 + \sum_{\lambda=1}^{\infty} \sum_{\mu=-\lambda}^{\lambda} \alpha_{\lambda\mu} \; Y_{\lambda\mu} \; (\theta,\Phi) \right\}$$

with $\alpha_{\lambda\mu} = (-)^{\mu} \alpha_{\lambda-\mu}$

In most cases one assumes an axial symmetry of the drop. Thus R depends only on θ and

$$R(\theta) = R_\alpha \left\{ 1 + \sum_{\lambda=1}^{\infty} a_\lambda P_\lambda \; (\cos\theta) \right\}$$

R_α is a normalization constant which should keep the volume of the drop constant.

The term with $\lambda=1$ corresponds to a motion of the center-of-mass for small values of a and it is usually assumed that $a_1=0$.

At least a third parameter is needed to describe completely enough
the fission of the drop. This parameter is called the mass asymmetry parameter and corresponds, for example to the $\lambda=3$ term of the
Legendre polynomial expansion, or to some combination of $\lambda=3$ and
$\lambda=5$. These terms are not reflection symmetric and they may therefore
describe an increase of the mass of one fragment at the expense of
the other.

It has recently appeared that a full description of fission had
to take into account some departure from axial symmetry which
means the inclusion of the terms with $m \neq 0$ of the second order
spherical harmonics. It can be shown, however, that this only includes an additional parameter namely the parameter "γ" defined
as

$$\alpha_{20} = \beta \cos \gamma$$

$$\alpha_{22} = \alpha_{2-2} = \frac{1}{\sqrt{2}} \beta \sin \gamma$$

$$\alpha_{21} = \alpha_{2-1} = 0$$

In summary the fissionning liquid drop is usually defined by four
parameters
- stretching
- necking
- mass asymmetry
- axial asymmetry

Some additional parameters have been introduced especially for the
treatment of the dynamics of the drop by R. Nix for example[17].

1.2. Main results.
a)- Stability with respect to fission.

It can be shown that the quadrupole $\lambda=2$ mode is the softest.
For this mode the deformation energy for small deformations is
given by [16]

$$E_S = E_{SO} \left\{ 1 + \frac{a_2^2}{2\pi} \right\}$$

The Coulomb energy is given by

$$E_C = E_{CO} \left\{ 1 - \frac{a_2^2}{4\pi} \right\}$$

where E_{SO} and E_{CO} refer to the spherical configuration. Thus

$$E_S + E_C - (E_{SO} + E_{CO}) = \frac{a_2^2}{2\pi} \left(E_{SO} - \frac{E_{CO}}{2} \right)$$

This difference is positive and thus the spherical shape stable with respect to fission if

$$x = \frac{E_{CO}}{2E_{SO}} < 1$$

x is called the fissility parameter

Since $E_{SO} \sim A^{2/3}$

and $\quad E_{CO} \sim \dfrac{Z^2}{A^{1/3}}$

$$x \quad \sim \frac{Z^2}{A}$$

more precisely for nuclei

$$x \quad \simeq \frac{Z^2}{50A}$$

Fission is exothermic for $x > 0.35$ ($\frac{Z^2}{A} \simeq 17$)

Therefore for all nuclei with $0.35 < x < 1$ one may state that the spherical ground state is only a metastable state with respect to fission. To undergo fission the nucleus has to overcome a fission barrier very much similar to an activation barrier in chemistry. The shape which corresponds to the minimum fission barrier (minimax) is called the saddle point shape and corresponds to an unstable equilibrium shape with respect to fission. Some "saddle point" shapes are shown in Fig. 1. It can be seen that a transition occurs for fissility parameters between 0.6 and 0.7. For $x < 0.6$ the saddle point shapes may be approximated by two spheroidal nuclei in contact, a definite neck has formed. For $x > 0.7$ the saddle shapes are more cylindrical_like and do not show any necking. It is important to note that for $x < 0.6$ the saddle and scission shapes are rather similar so that the dynamical effects taking place during the motion of the drop between those two configurations should not have much importance. This might not be true for heavy systems with $x > 0.7$.

b)- Mass distributions.

Let us first clarify the terminology. Of course, since in most cases fission gives rise to only two fragments, mass distributions should be symmetric with respect to approximately half the mass of the fissioning nucleus. However, the mass split is asymmetrical except when both fragments have precisely the same mass. By extension we say that the mass distribution is symmetric when the most probable mass split is symmetrical and that it is asymmetric when the most probable mass split is not symmetrical.

For x > 0.38 the liquid-drop model always produces symmetric mass distributions. This result is by no means trivial since it involves not only the knowledge of shape dependance of the potential energy of the drop but also the full treatment of the dynamics of the drop motion. Such a calculation has been made, for example, by R. Nix,[17].

It is, however, instructive to consider the case of two drops in contact and study their potential energy as a function of their mass ratio. A simplified calculation where deformations are not taken into account suffices to show the trends of the variations of this energy. The energy of the double drop system is computed by adding the ground state masses of the drops to their mutual Coulomb repulsion which, limited to the monopole-monopole contribution writes

$$C = \frac{Z_1 Z_2 e^2}{r_o (A_1^{1/3} + A_2^{1/3})}$$

The set of curves shown in Fig. 2 illustrates the curves obtained for different values of the fissility parameter. It is seen that very asymmetric splits always present a relative minimum. This means that the nuclear drop is stable with respect to particle emission (neutrons or protons for example). Symmetric mass splits appear for x > 0.38 which is called the Businaro Gallone point. If two equally sized drops are put into contact for x < $x_{Businaro\,Gallone}$ one of the drops will tend to digest the other. On the contrary for x > x_{BG} the two drops may coexist peacefully ! This finding has profound implications for our understanding of heavy ion reactions.

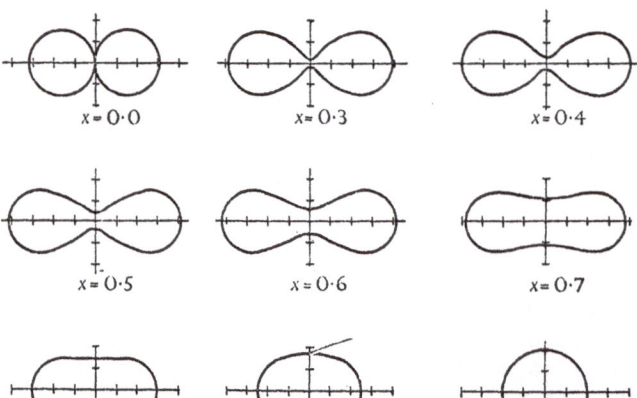

Fig. 1.

Saddle point shapes for various values of x according to Cohen and Swiatecki, UCRL 10450 (1962).

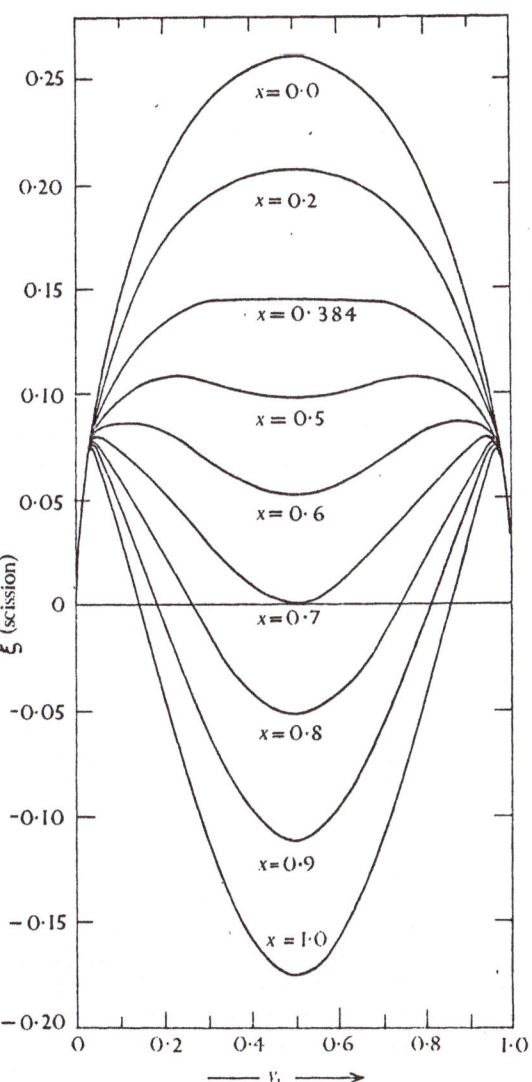

Figure 2. Energy of deformation at scission, in units of E_{so} as a
function of $C_1 y_1 = \dfrac{A_1}{A} = \dfrac{Z_1}{Z}$ for various values of the
fissility parameter x. From L. Wilets (Ref. 16).

2. Some significant experimental results and their relation to the liquid drop model predictions.

We now present some experimental results which cannot be explained in the frame of the liquid drop model of fission. Many other results may be seen in the proceedings of the four international symposia on the Physics an Chemistry of Fission which were held by the International Atomic Energy Agency at Salzburg (1965), Vienna (1969), Rochester (1973) and Julich (1979).

2.1. Mass distributions.

Fig. 3 shows some low energy fission mass distributions. It can be seen that for nuclei heavier than Thorium the mass distributions are clearly asymmetric. This situation persists up to the Fermium isotopes where a transition back to symmetric distributions occurs in the region of $Fm^{257,258}$. Around Radium triple humped distributions appear, leading to symmetric distributions for lighter systems. It can also be seen that above Uranium the light edge of the heavy fragments distribution remains more or less invariant in the region corresponding to the shells with 50 protons and 82 neutrons. Below Uranium it is the light edge of the light group which seems to stabilize around the magic neutron number N = 50.

While the liquid drop model is unable to explain these mass distributions, the latest remarks strongly suggest that shell effects are at work in determining the mass distributions.

When rising the excitation energy of the fissionning nucleus it appears that the mass distributions become more and more symmetrical, which again agrees with the well known tendency of shell effects to wash out with excitation energy.

2.2. Fission barriers.

Fission barriers are obtained from the study of fission excitation functions. Before discussing the results we shortly recall the principle of measurement of fission barriers. One has to distinguish the high excitation energy technique and the low excitation energy one. At high energy fission is only in significant competition with the neutron emission. One assumes that all available channels for fission and neutron emission are fully opened. Therefore the relative probability of fission to neutron emission is simply proportional to the number of available channels in each case. Thus

$$\frac{\Gamma_F}{\Gamma_n} \propto \frac{\rho_S (A,Z,E^* - B_F)}{\rho (A-1,Z,E^* - B_n)}$$

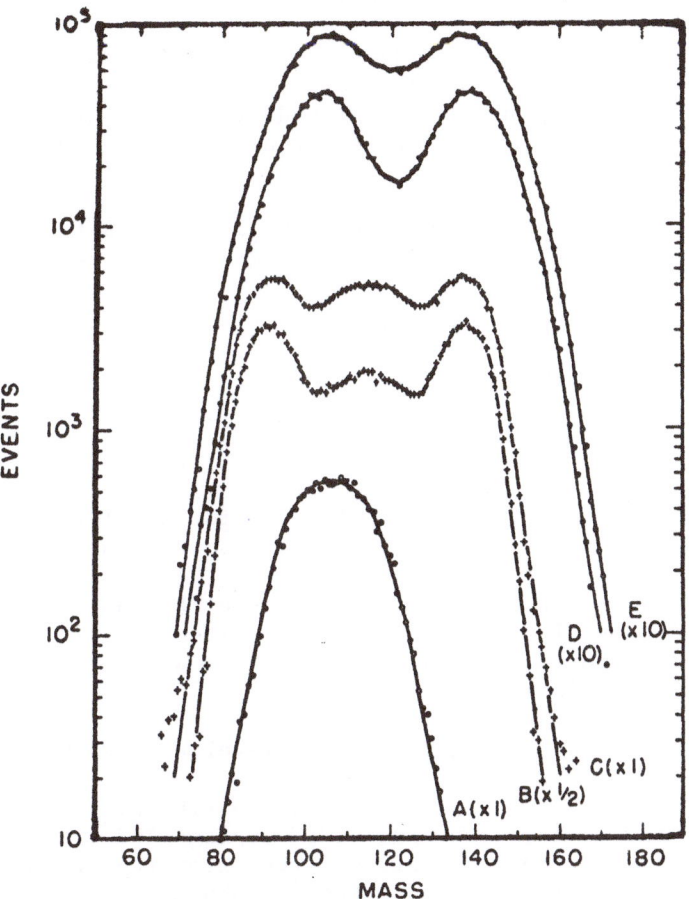

Fig. 3. Primary fission fragment mass yield distributions for helium-
         ion induced fission:
         (A) Bi^{209} (42 MeV He^4,f)
         (B) Ra^{226} (30.8 MeV He^4,f)
         (C) Ra^{226} (38.7 MeV He^4,f)
         (D) U^{238} (29.4 MeV He^4,f)
         (E) U^{238} (42.0 MeV He^4,f)
         Taken from Ref. 19.

where $\rho_S(A,Z,E^* - B_F)$ is the level density in the saddle point con-figuration for the fissionning species A,Z, the excitation energy at saddle being $E^* - B_F$. Similarly $\rho(A-1, Z, E^* - B_n)$ is the level density in the daughter nucleus obtained by neutron evaporation and $E^* - B_n$ the available excitation energy in this nucleus.

In principle it is necessary to evaluate carefully the level densities taking into account shell structure effects which are different at saddle and in the ground state configuration. This has been done by various authors and realistic fission barriers are obtained this way [18].

For simplicity we assume that the level densities have the same exponential form at saddle point and in the daughter nucleus. Thus

$$\frac{\Gamma_F}{\Gamma_n} \simeq e^{(B_n - B_F)/T}$$

$$\text{with} \quad T \simeq \sqrt{\frac{8(E - B_F)}{A}}$$

In light systems $B_F \gg B_n$ and $\Gamma_F \ll \Gamma_n$ for small temperatures. It is therefore very difficult to observe fission near the threshold.
It is also wrong, in this case, to assume the same temperature at saddle point and in the daughter nucleus because of the rather large difference in the available excitation energies in both configurations. In fact one usually fits the energy dependance of $\frac{\Gamma_F}{\Gamma_n}$ to obtain B_F knowing B_n and the level density parameters.
See for example Ref.19 for more detailed treatment. For heavy nuclei B_F and B_n have the same order or magnitude. The fission barrier may be, then, directly observed from the steep rise of the fission cross section close to the threshold.

Fig.4 shows the neutron induced fission cross section of U^{238} as a function of neutron energy. The fission barrier corresponds approximately to the first rise in the cross section. The following rises are due to fission after one or some neutrons are emitted. In fact fission cross sections near the theshold are very seldom as simple as that displayed on Fig.4 and show structures which we shall come back to.

On Fig.5 we show the comparison between the experimental fission barriers and the predictions of the liquid drop model. It is seen that the model cannot account for the very small variation of the barriers in the actinide region.

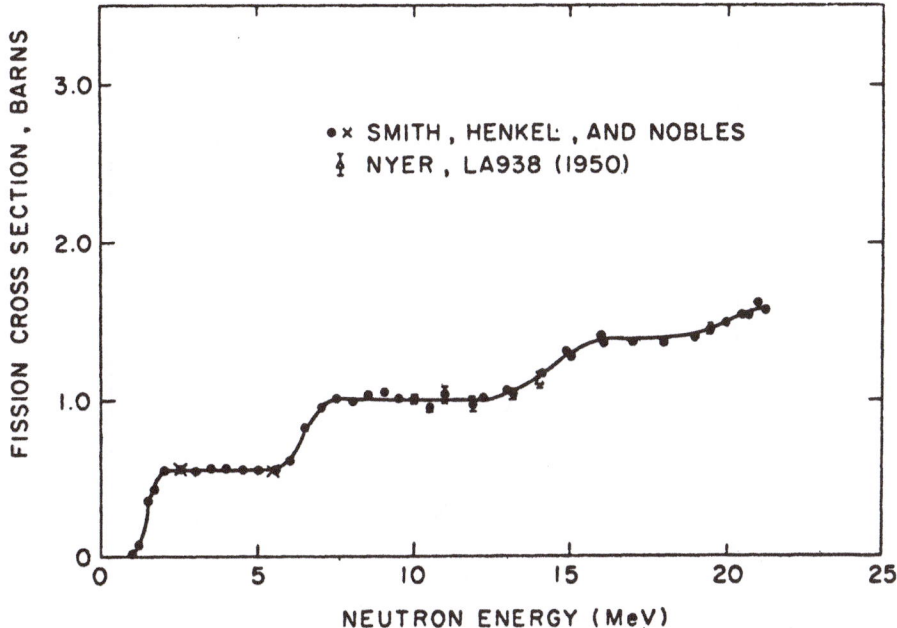

Fig. 4.   The neutron-induced fission cross section of U^{238} is plotted
as a function of neutron energy. From Ref. 19.

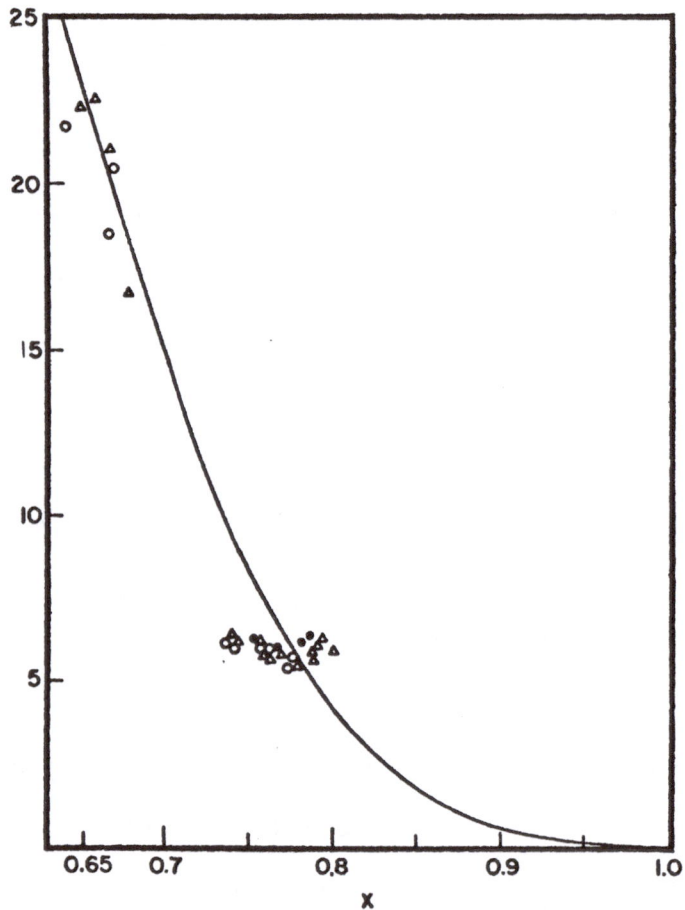

Fig. 5. The simple liquid-drop model prediction of the fission bar-
         rier height as a function of the fissility parameter x (Cohen
         and Swiatecki, 1963).
         o Even-even,
         Δ odd-A,
         o odd-odd.
         From Ref. 19.

2.3. Fission Isomers.

In their ground state nuclei may experience spontaneous fission by a quantum mechanical tunnelling through the fission barrier. Accounting for this tunnelling is a very involved task since it requires the treatment of tunnelling through a multidimensional barrier and further the knowledge of the inertia parameter for the motion. A step towards such a treatment was made by H.C.Pauli[20].

Nevertheless it has been possible to show systematic trends in spontaneous fission half lives and relate them to the fissility parameter. This is shown in Fig.6 together with the calculation of Pauli and Ledergerber [20]. The plot is shown for even-even isotopes. Odd and Odd-Odd isotopes show increased lifetimes. It was a big surprise when Polikanov et al.,[8] discovered a 14 ms half life fissionning isomer of Am^{242}. Many other such fission isomers were discovered later on. The half life of some of them are shown in Fig. 7. The isomeric lifetimes may be as much as 10^{-25} times shorter than the ground state lifetimes. At the time of their discovery these isomers could not be understood even taking into account the possibility of spin isomerism.

2.4. Structures in fission cross sections.

As stated above fission cross sections close to the barriers may have rather complicated behavior. This is examplified in Fig. 8 which shows some fission probabilities for even-even nuclei. The structures observed in Fig. 8 are seen with rather modest resolutions of several tens of keV. We may call them gross structures. Much higher resolutions (a fraction of eV) are those obtained with the neutron time of flight technique. With this technique it is possible to resolve individual compound states. In fissile nuclides which were first studied by this method it was found that the ratio $\frac{\sigma_F}{\sigma_T}$ of the fission cross section to the total cross section fluctuated from resonance to resonance in a purely statistical manner. An example of typical total and fission cross section is shown in Fig. 9(a) for $U^{235}(n,f)$. It was a surprise when the study of below threshold resonance fission showed (21,22) a very different behavior which is examplified in Fig. 9(b) in the case of $Pu^{240}(n,f)$. It is seen that resonances which fission cluster around definite energies. Some type of intermediate structure in fission cross section is at work. It should be noted that this structure is in the exit fission channel while most of the familiar ones in nuclear physics are present in the entrance channel. By 1968 all these structures in fission cross sections were hardly understood.

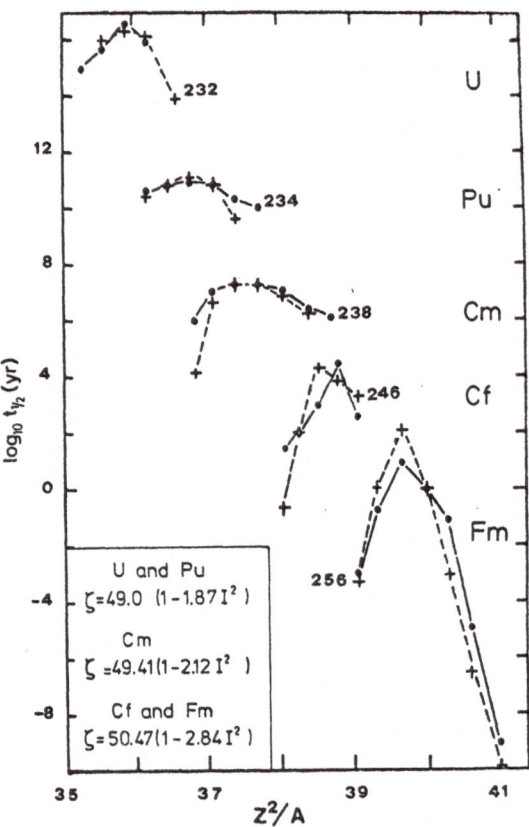

Fig. 6.   Comparison of calculated (dots) and measured (crosses) life
times for spontaneous fission. From Ref. 20.

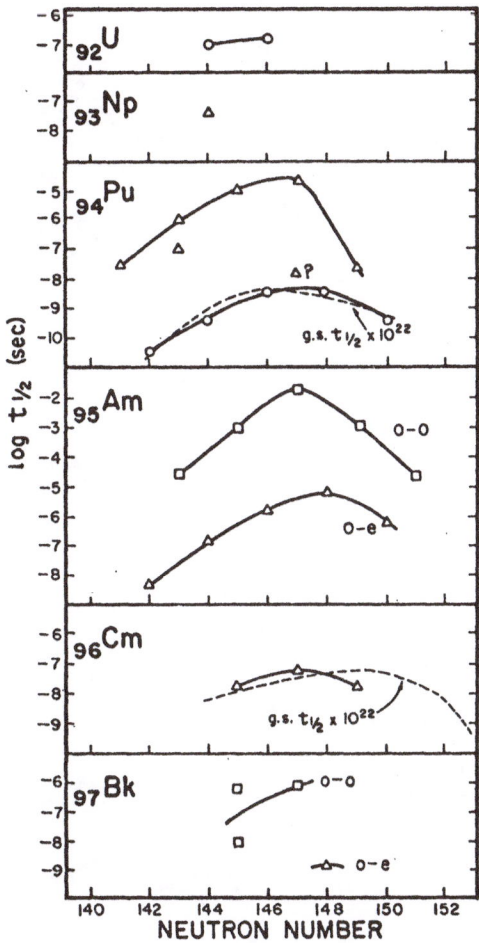

Fig. 7. Spontaneously fissioning isomer half-lives as a function of neutron number. Circles, triangles and squares represent values for even-even, odd-A, and odd-odd nuclei, respectively. From R. Vandenbosh, 3rd Symposium on Physics and Chemistry of Fission, IAEA, Rochester (1973) 252.

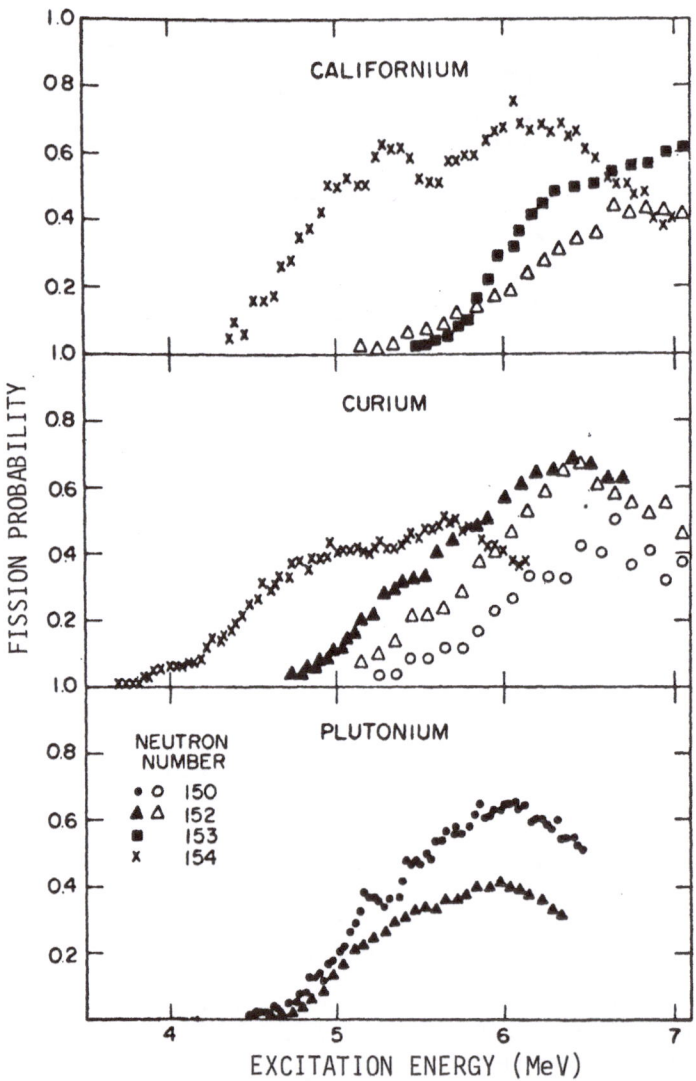

Fig. 8.   Fission probabilities for plutonium, curium and californium
          isotopes. From Ref. 33.

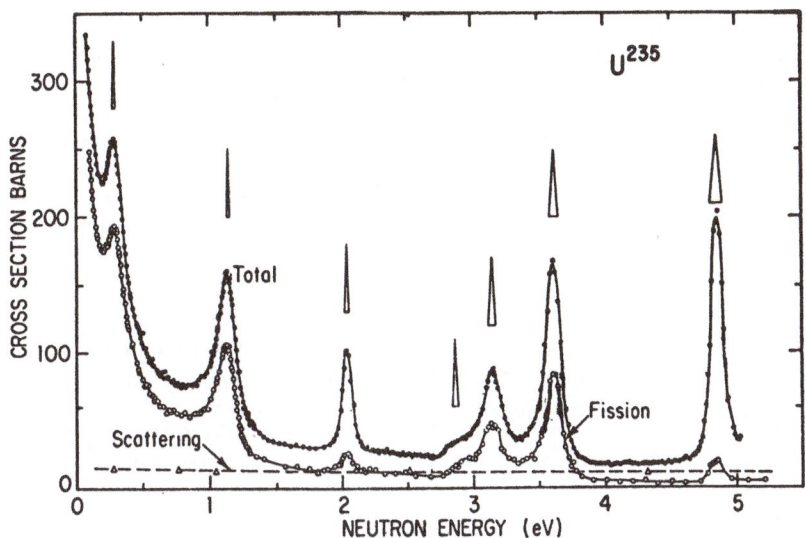

Fig. 9(a).   Resonance structure in the interaction of neutrons with
             U^{235} in the energy range from 0.1 to 5 eV. The observed
             total cross section, fission cross section, and scatter-
             ing cross section are displayed.

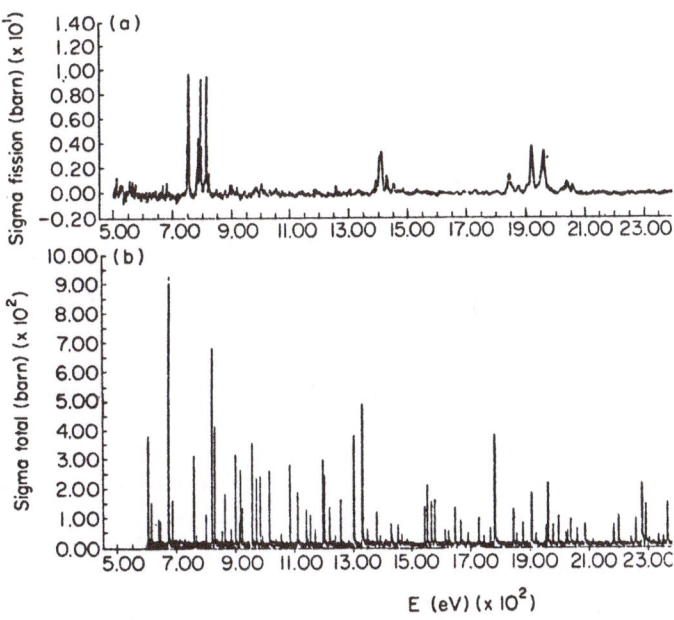

Fig. 9(b).   The neutron fission cross section of ^{240}Pu is compared
             with the total neutron cross section to demonstrate the
             grouping of fission resonances. From Ref. 22.

2.5. Even-odd effects on charge distributions.

It is only recently that it has become possible to obtain charge distributions of fission fragments. These charge distributions reflect the gross features of the corresponding mass distributions since, due to the isospin symmetry energy the ratio of charge to mass of the fragments is close to that of the fissioning nucleus. However it appeared that, besides these gross features, in some cases, fragments with even proton numbers where produced in much larger quantities thant fragments with odd-charges. The enhancement may reach a factor of 2. Furthermore this enhancement is a very sensitive function of the initial state of the fissionning nucleus. For example, while even odd effects are clearly present and strong in the thermal neutron induced fission of U^{235} they almost disappear for 3 MeV neutrons[23].

It was also shown[24] that the even-odd effects are almost absent in the thermal neutron induced fission of Pu^{239}. It seems clear, that these effects are related to the pairing interaction in nuclei. It is speculated that they offer[25] a powerful tool for studying the dynamics of fission.

3. The synthesis between the liquid drop and shell models of nuclei

Because of the inadequacies of the liquid drop model it was natural to try to apply the shell model to the study of fission. The Nilsson modified oscillator model[10] allowed the study of level schemes in deformed nuclei. We recall its form[26] which allows studies of large deformations

$$V = \frac{1}{2}\hbar\,\omega_o\rho^2\left\{1 - \alpha_2\sqrt{\frac{4\pi}{5}}\left[\cos\gamma\,Y_{20} - \frac{\sin\gamma}{\sqrt{2}}(Y_{22} + Y_{2-2})\right]\right.$$

$$\left. - 2\alpha_4\sqrt{\frac{4\pi}{9}}\,Y_{40} + \sum_{i\neq 2,4}\alpha_i\,Y_{io}\right\}$$

$$- K\,\hbar\,\overset{o}{\omega}_o\left\{2\,\ell.s + \mu\,(\ell^2 - <\ell^2>_N)\right\}$$

with $\rho^2 = \frac{M}{\hbar}(\omega_x X^2 + \omega_y Y^2 + \omega_z Z^2)$

In the last parenthesis one finds the usual $\bar{\ell}.\bar{s}$ coupling term and the ℓ^2 term which tends to simulate sharper edges for the potential than has the harmonic oscillator itself. For $\gamma = 0$ one obtains an axially symmetric potential which, if one considers only the α_2 term, reduces to the standard harmonic oscillator.

The total energy is then given by adding up the single particle energies so that

$$E(\alpha_2,\alpha_4) = \sum_{i=1}^{N\ \text{neutrons}} \varepsilon_i\,(\alpha_2,\alpha_4) + \sum_{i=1}^{Z\ \text{protons}} \varepsilon_i^{\text{protons}}(\alpha_2,\alpha_4)$$

*The angle arguments in the Y are defined in stretched coordinates.

One should note that Nilsson uses the parameters

$$\varepsilon = \varepsilon_2 = \frac{2}{3} \alpha_2 \text{ and } \varepsilon_4 = - \alpha_4$$

The result of such a computation of the total energy as a function of ε and minimizing the energy with respect to ε_4 is shown in Fig. 10 (upper full curve). It is clear that the result is totally unrealistic since no fission barrier appears on the plot. The reason for this failure is now understood. It is related to the difficulty to insure the self consistency of the model for large deformations, or more precisely for a large range of deformations. It is only locally approximated by the condition that the volume inside an equipotential should be kept constant. Fig. 10 displays interesting wiggles especially for $\varepsilon = 0.2$ which corresponds to the ground state deformation of the actinide nuclei and around $\varepsilon \simeq 0.5$. These wiggles reflect the behavior of the last filled single particle levels. On the other hand the lack of self consistency is probably related to the bulk proton and neutron densities where all levels contribute.

The idea of Strutinsky was to use the liquid drop to take care of the self consistency requirement and the deformed shell model to keep the wiggles. The problem is therefore to extract the significant wiggles from the deformed shell model total energies.

As said above the total energy of the nucleus may be written:

$$E(\beta) = \sum_i^{(n,p)} \varepsilon_i(\beta) = \sum_i^{(n,p)} \varepsilon \delta(\varepsilon - \varepsilon_i(\beta)) d\varepsilon$$

where β is a generalized deformation parameter.

In essence the idea of Strutinsky was to spread the strength of each particle level over a finite interval, the width of which must exceed the inter-shell spacing. One then defines an average shell model energy

$$\tilde{E}(\beta) = \sum_i \varepsilon R_\Gamma(\varepsilon - \varepsilon_i) d\varepsilon$$

where R_Γ is the spreading function.

The shell correction term is then defined as

$$\delta E_{sh}(\beta) = E(\beta) - \tilde{E}(\beta)$$

and the total energy of the nucleus is taken as

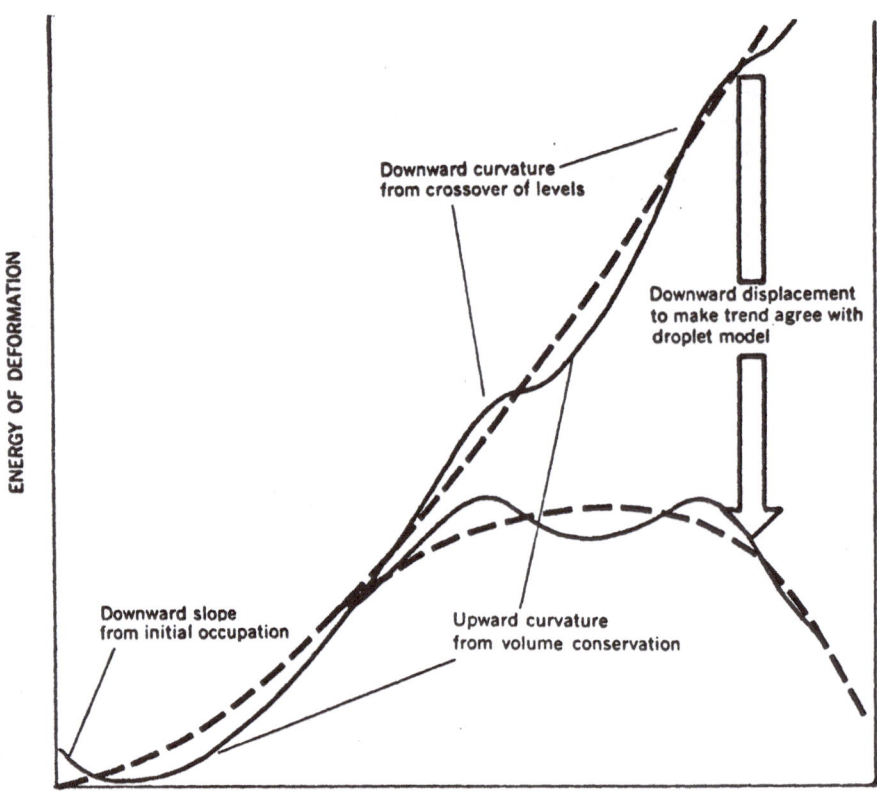

ELONGATION OF DEFORMED NUCLEUS, δ

Fig. 10.   A schematic extension of the Nilsson diagram shows features
            that help to explain the energy curves. The upper heavy
            curve is based on volume conservation within equipotential
            surfaces of the potential from which the single particle
            energies are calculated. The lower heavy curve is that ob-
            tained by requiring the general trend to be consistent with
            the liquid drop model as suggested by Strutinsky. Initial
            downward curvature corresponds to the first hump of the
            fission barrier; thereafter, the curve wavers upward and
            downward about an average curve shown in dashed form.
            From Ref. 19.

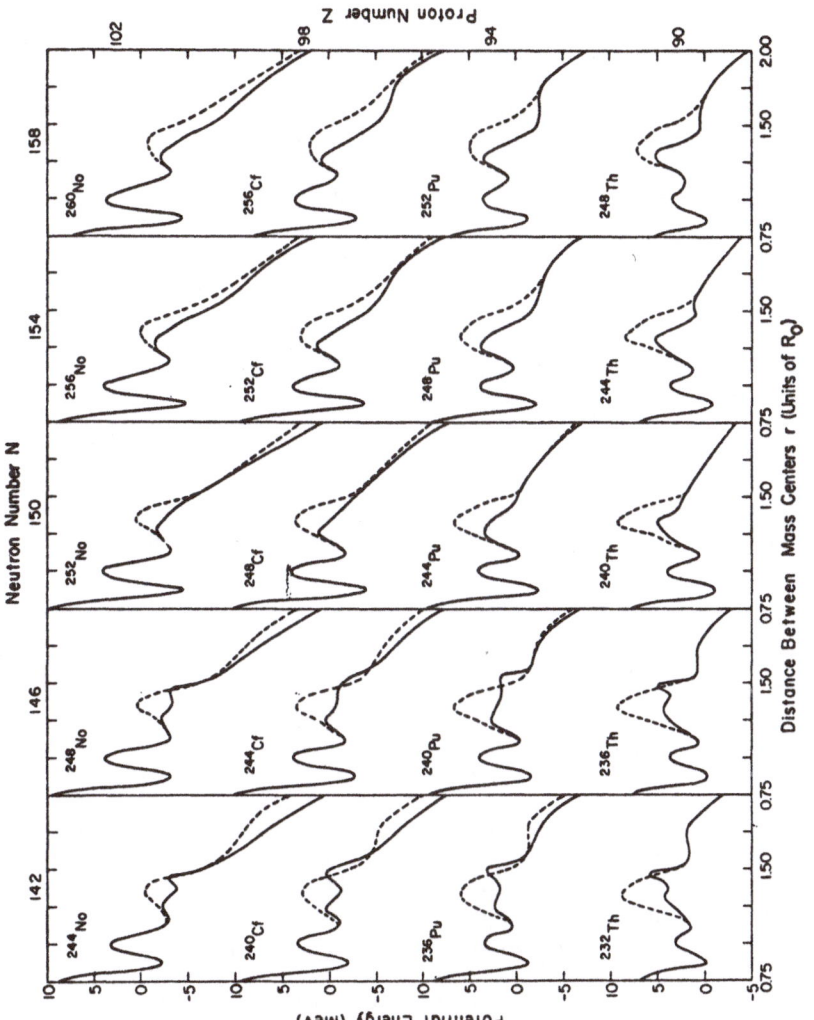

Fig. 11. Fission barriers for actinide nuclei, calculated with the folded Yukawa potential and the droplet model. The dashed curves give the potential energy for symmetric deformations as a function of the distance r between the centers of mass of the two nascent fragments. The solid curves give the potential energy along a path that leads over the mass–asymmetric second saddle point. This path is determined by minimizing the potential energy with respect to mass asymmetry for fixed values of τ or by the method of steepest descent. From Ref. 30.

$$E^{Strut}(\beta) = E_{L.D.}(\beta) + \delta E_{sh}(\beta)$$

the liquid drop energy being computed for a shape defined by an equipotential of the shell potential.

Fig. 10 illustrates the method. Calculated barriers are shown on Fig. 11.

The very important result obtained by Strutinsky and appearing on Fig. 11 is that the fission barrier splits in two in the actinide region. This is the so called double humped barrier. The second prolate minimum occurs for deformations where the ratio of the long axis to the short axis of the nucleus is close to 2.

4. Experimental consequences of the splitting of the fission barrier.

The conditions for the appearence of a secondary minimum in the fission barrier are :

1)- The occurrence of a minimum in the shell correction which occurs around certain values of the deformation (2 : 1 as said above) and around certain neutron or (and) proton magic numbers (deformed shells).

2)- This minimum has to coincide with the deformation of the liquid drop saddle point. This is due to the fact that the liquid drop barrier variations outside stationary points such as ground state or saddle point are so steep that they overcome the shell correction effects.

The competition between the shell correction and the liquid drop explains the features observed in Fig. 11. For small fissility parameters the second barrier coincides with the liquid drop saddle point and is, therefore, higher than the first barrier while the secondary minimum is rather shallow.

For large fissility parameter the liquid drop saddle point shape coincides with the first shell correction maximum giving a first barrier much higher than the second and a very shallow secondary minimum. It is in the U, Pu, Cm region that both barriers and secondary minimum are best developed.

4.1. Fission isomers -

The secondary minimum corresponds to the fission isomer configuration. The isomeric lifetimes reflect rather well the systematics of the computed fission barriers. For light systems like Th no isomer is seen and this is explained because the lifetime for back decay to the first minimum must be much smaller than for fissionning. For heavy systems like Cf the isomeric lifetime becomes too short to be measured. In between these regions the phenomenon of fission isomerism is well developed as seen in Fig. 7.

Direct confirmation of this explanation of fission isomers have been obtained by Specht,[28] and Metag,[29]. Specht observed electron conversion lines in the electron spectrum coincident with delayed fission fragments which were coming from isomeric fission of Pu^{240}. These lines agreed with a rotational sequence and their spacing allowed the determination of a rotational constant

$$\frac{\hbar^2}{2I} = 3.35 \text{ keV}$$

Twice larger than in the first well. Metag has made use of the fact that electron conversion produces a high degree of ionisation of the atomic shells to determine the lifetime of the rotational states by a charge plunger technique. Using this technique a quadrupole moment of 37 barns was obtained corresponding to the value 2/1 for the ratio of major to minor axis of the isomer.

Very recently K. Bemis[*] directly measured the quadrupole moment of the fissionning isomer Am^{240m}. He made use of a laser optical pumping technique and measured the hyperfine splitting of the optical transitions. Here again a value of the quadrupole moment of 33 barns was found.

4.2. Fission barrier systematics.

A rather good account of experimental fission barriers has been obtained using the Strutinsky approach. In fact experiment is able to provide the two barrier heigths as well as the second well depth. This is made possible by the measurement of isomer excitation functions. Fig.12 summarizes the state of the art as far as the two barriers E_A and E_B are concerned. The agreement with calculations is within 1 to 2 MeV. More important, the trends of experiment and theory are the same except for the first barrier in the Thorium and light Uranium isotopes. This is the region of the Thorium anomaly which we shall come back to.

[*]Private communication.

4.3. Structures in the fission probabilities.

The double humped barrier provides a natural frame for explaining both the strong resonances in the fission excitation
function and the intermediate structures. One considers that the
two configurations corresponding to the first and second well
respectively are very different. One can therefore distinguish
two classes of states according to the fact that most of the amplitude of the wave function lies in the first well or in the
second well. Compound nucleus formation deals with the first well
states. These states are only weakly coupled to the second well
states. It is then well known from perturbation theory that the
transition from a class I (first well) state to a class II state
will be very much enhanced if the two states happen to have the
same energies. The effective excitation energy in the second well
is smaller than that in the first well by the isomeric state excitation energy. Therefore the density of class II states is expected to be less than that of class I states. Further, class II states have much higher fission probabilities than class I states and
act, therefore, as doorway states towards fission. This is examplified in Fig. 9b where the different groups of class I states which
do fission correspond to specific class II states.

If the class II states have a strong vibrational component
in the fission direction their fission width is increased. This
may be responsible for some of the broad structures observed in
the low resolution excitation function. These structures are
usually assumed to be vibrational resonances which correspond
to β-vibrational states in the second well. However in the Th^{232}
case this interpretation is probably not correct. Here again we
meet the Thorium anomaly.

The Thorium anomaly and further barrier splittings.

Including mass or reflection asymmetric shapes in the barrier
calculations it was found that the second barrier height was significantly reduced for mass asymmetric nuclear shapes (for a
review of the calculations see Ref.30) especially in the actinides region. Furthermore it was also found that this barrier became
very flat and that even a shallow third minimum might develop on
top of it. It was suggested that in Thorium isotopes experiment
provided, in fact, the height of the second and third barriers.
Recently Blons et al.[31] have shown that the most prominent of the
so called vibrational resonances observed in Thorium^{232}(n,f) and
Thorium^{230}(n,f) could be resolved into a series of states. These
states could be arranged in two separate rotational sequences
displaced by a constant energy. Bohr and Mottelsson,[32] have
predicted that, for reflection asymmetric nuclear shape a splitting of the rotational band into two bands with different pari-
ties should occur. It seems more and more likely that this is

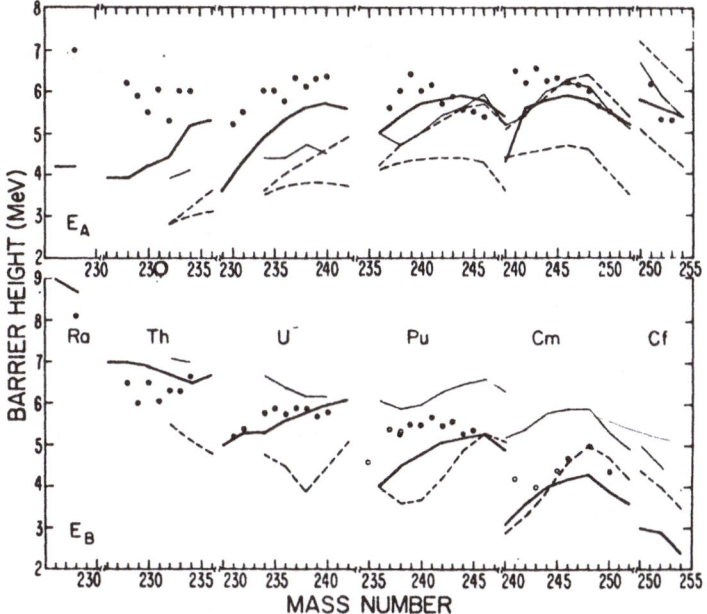

Fig. 12.   Comparison of experimental fission barriers [37] to various
theoretical calculations from Möller (solid line), Pauli
and Ledergerber (thin solid line) and Möller and Nix
(dashed lines). E_A has been corrected for effect of axial
asymmetry. From Ref. 33.

what has been observed in the Thorium isotopes. Furthermore the
rotational constant obtained from these bands amounts to about
2 keV, in agreement with the expected deformation at the second
saddle point. The rotational bands would be built on low lying single particle states in the third well. The fact that the structures
are more apparent in odd nuclei than in even-even ones would also
fit with this picture.

The Thorium anomaly, therefore, seems to be related to an increased complexity of the fission barriers. In this case the second barrier was involved. It seems that for the Californium isotopes the first barrier as well is more complex than anticipated.
This is apparent on Fig. 8   where the fission probability function of Cf^{252} shows   very distinct resonances. It was in fact
expected that no structures at all would appear in this case since
only the first barrier would be of importance in determining fission probabilities.

Relaxing the constraint on axial symmetry it was found that
the first barrier was probably axially asymmetric. In the Californium 252 case the liquid drop barrier deformation coincides with an
antishell structure and therefore a very flat first barrier is expected in analogy with the Thorium case (for the second barrier).
The structures observed in the Cf^{252} fission probability curve
might be due to single particle states built on an additional
shallow minimum sitting on top of the first barrier. For even more
complexities in the potential energy surface see Ref.33.

5. The origin of shell effects in deformed nuclei.

At first sight it might seen that the use of different deformed independent particle models in the Strutinsky shell averaging
procedure would lead to different deformations for the isomeric
well and different predictions of the deformed magic numbers.
Bloch and Balian,[14] were the first to point out that the occurrence of shell effects, i.e. a bunching of the single particle energy
levels, was intimately related to symmetries in the nuclear potential. More precisely shell effects are expected when there
exist closed stationnary semi classical trajectories. This question has been further investigated by several authors, among them
Bohr and Mottelson,[32] and Strutinsky,[34]. Here after we follow the
approach of Bohr and Mottelson,[32]. Let us first examine

5.1. The anisotropic harmonic oscillator case.
For an axially symmetric oscillator characterized by its two
frequencies $\omega_\perp$ and ω_z the energy writes

$$\frac{E}{\hbar} = (n_\perp + 1)\omega_\perp + (n_z + \tfrac{1}{2})\omega_z$$

with $\qquad n_\perp + n_z = N$

n_z and $n_\perp$ being the number of quanta in the direction parallel and perpendicular to the symmetry axis respectively. There always exist a degeneracy of order $2(n_\perp + 1)$ of the eigenstates labeled by n_z and $n_\perp$. Additional degeneracy will occur if two states with different values of n_z and $n_\perp$ coincide in energy

$$\frac{E}{\hbar} = (n_\perp^{(o)} + 1)\omega_\perp + (n_z^{(o)} + \tfrac{1}{2})\omega_z = (n_\perp^{(1)} + 1)\omega_\perp$$
$$+ (n_z^{(1)} + 1)\omega_z$$

which leads to $\omega_\perp \Delta n_\perp + \omega_z \Delta n_z = 0$

where clearly $\Delta n_\perp$ and Δn_z are integers which we denote i and $-j$ so that

$$\omega_c = \frac{\omega_z}{i} = \frac{\omega_\perp}{j}$$

it follows that all states with n_z and $n_\perp$ such that

$$N_c = n_z\, i + n_\perp\, j$$

will have the same energy

$$\frac{E}{\hbar} = (n_z + \tfrac{1}{2})\omega_z + (n_\perp + 1)\omega_\perp$$

$$= (n_z + \tfrac{1}{2})i\,\omega_c + (n_\perp + 1)j\omega_c$$

$$= \omega_c(N_c + j + \tfrac{1}{2}\,i)$$

We also note that all levels belong to one of these shells which can be labeled by the quantum number N_c.

On the other hand if $\frac{\omega_z}{\omega_\perp}$ is not rational there exists no other degeneracy in the spectrum than that related to $n_\perp$. The magnitude of the shell effects depends on ω_c.

Since $\omega_\perp$ and ω are related by

$$\omega_\perp^2\, \omega = \omega_o^3$$

one obtains

$$\omega_c = \omega_o (j^2 i)^{-\frac{1}{3}}$$

Prolate shapes are obtained for $j > i$. One sees that, for the same set of integers oblate shapes develop stronger shell effects than prolate ones. However these shell effects have not been observed due to the behavior of the liquid drop energy. For prolate shapes the stronger shell effect is obtained for $j = 2$ and $i = 1$. The distance between shells is equal to approximately two third of that of the spherical configuration.

We note that the classical trajectories are given by

$$Z = a \sin(\omega_Z t + q_1)$$

$$X = b \sin(\omega_\perp t + q_2)$$

$$Y = c \sin(\omega_\perp t + q_3)$$

The condition for occurrence of shells therefore means that the classical trajectories close upon themselves after a period

$$T_C = \frac{2\pi}{\omega_c} = i\, T_Z = j\, T_\perp$$

This is an example of the general result obtained by Bloch and Balian,[14].

We have shown that shell effects occur in the anisotropic oscillator when the ratios of frequencies $\omega_Z/\omega_\perp$ is rational.

This ratio is, of course, the ratio of the axis of the isopotential ellipsoids. Due to the self consistency required from nuclear independent particle models it is also the ratio of the axis of the nuclear matter density profiles. However it is not the ratio of the axis of the isoprobability surfaces of individual wave functions. From the properties of the harmonic oscillator one has

$$<z^2>_{n_Z} = \frac{\hbar(n_Z + \frac{1}{2})}{M\omega_Z}$$

$$<x^2 + Y^2>_{n_\perp} = \frac{\hbar(n_\perp + 1)}{M\omega_\perp}$$

so that for fixed n_Z, $n_\perp$ the isoprobability surfaces are ellipsoids with axis ratio varying as

$$\sqrt{\frac{\omega_\perp}{\omega_Z}}$$

The self consistency is only obtained through a

rearrangement in the oscillator occupation numbers. Let $n_Z^{(k)}$, $n_\perp^{(k)}$ the quantum numbers of the states occupied by nucleon k. The self consistency requirement reads

$$\sum_{k=1}^{A} \langle Z_k^2 \rangle = \frac{\hbar}{M\omega_z} \sum_{k=1}^{A} (n_Z^{(k)} + \frac{1}{2}) = \frac{\alpha}{M\omega_Z^2}$$

and

$$\sum_{k=1}^{A} \langle X_k^2 \rangle = \frac{\hbar}{M\omega_\perp} \sum_{k=1}^{A} (n_X^{(k)} + \frac{1}{2}) = \frac{\alpha}{M\omega_\perp^2}$$

$$\sum_{k=1}^{A} \langle Y_k^2 \rangle = \frac{\hbar}{M\omega_\perp} \sum_{k=1}^{A} (n_Y^{(k)} + \frac{1}{2}) = \frac{\alpha}{M\omega_\perp^2}$$

α being a normalization constant.

We have expressed, here, the proportionality between the rms length of the nuclear matter density and that of the axis of the isopotential surfaces. Denoting by E_z and E_y the energy stored in the relevant degrees of freedom we obtained, therefore, the condition

$$E_Z = \hbar \Sigma \omega_Z (n_Z^{(k)} + \frac{1}{2}) = \hbar \Sigma \omega_\perp (n_X^{(k)} + \frac{1}{2})$$

$$= \hbar \Sigma \omega_\perp (n_Y^{(k)} + \frac{1}{2}) = E_X = E_Y$$

expressing the equipartition of energy. This condition is fulfil-led by the rearrangement of nucleons at the level crossings. The equipartition of energy between the three degrees of freedom of the oscillator implies that the nucleons have the same root mean square average velocities with respect to the three axis and, there-fore, that the amplitude of their motion along these are just proportional to the period of the relevant motion that is

$$\frac{\ell_Z}{T_Z} = \frac{\ell_\perp}{T_\perp} \tag{5.2}$$

This is nothing more than the self consistency condition, derived from the equipartition of energy rule. We now try to gene-ralize the results just obtained for the harmonic oscillator.

5.2. More general potentials.

We restrict ourselves to the consideration of separable potentials. Let q_i be one of the separable variables for the motion of the studied system. If the system motion is finite, it is, periodical with a period T_i and takes place between two extrema $q_i^{(1)}$ and $q_i^{(2)}$. Then clearly

$$T_i = \int_{q_i^{(1)}}^{q_i^{(2)}} \frac{dq_i}{|v_i|} + \int_{q_i^{(2)}}^{q_i^{(1)}} \frac{dq_i}{|v_i|} = \oint \frac{dq_i}{v_i} ; (q_i^{(2)} > q_i^{(1)})$$

where $v_i = q_i = \dfrac{dq_i}{dt}$

Assuming a constant mass parameter m_i for the motion clearly

$$T_i = \oint \frac{dq_i}{\frac{p_i}{m}} = \oint \frac{dq_i}{\frac{\partial E}{\partial p_i}} = \frac{\partial}{\partial E} \oint p_i \, dq_i$$

Thus

$$T_i = 2\pi \frac{\partial I_i}{\partial E}$$

where

$$I_i = \frac{1}{2\pi} \oint p_i \, dq_i$$

is known as the action variable and is routinely used in the Bohr Sommerfeld quantization rule. One also gets

$$\omega_i = \frac{2\pi}{T_i} = \frac{\partial E}{\partial I_i}$$

Developing tne energy as a function of the action variable gives

$$E(I_1..I_2...) = E(I_1^{(o)} ... I_k^{(o)})$$
$$+ \sum_i (I_i - I_i^o) \, \omega_i(\{I_k^{(o)}\}) + \mathcal{O}^2(I_i - I_i^{(o)})$$

Bunchings of energy levels, or shells, occur for values of $I_1^{(o)} ... I_i^{(o)}$ such that the linear term cancels

$$\sum_i (I_i - I_i^{(o)}) \, \omega_i(I_1^{(o)} .. I_k^{(o)}..) = 0$$

One recognizes a condition similar to condition 5.1 obtained for the harmonic oscillator if one sets

$$I_1 = \hbar\, n_\perp$$
$$I_2 = \hbar\, n_z$$

Therefore, with general separable potentials shells occur when the ratios of the periods of the closed classical orbits are rational, with especially strong effects for small integer fractions.

In the deformed independent particle model the oscillator quantum numbers $n_\perp$ and n_z may still be used to label the nucleonic states together with Λ and $\Sigma,$ [32].

To first order the harmonic oscillator frequencies $\omega_\perp$ and ω_z are still equal to the classical paths frequencies. The condition for shell occurrence is still:

$$\frac{\omega_\perp}{\omega_z} = \frac{i}{j} \ .$$

If one assumes that the self consistency requirement implies the equipartition of energy between the three approximately separated variables X Y and Z relation 5.2 should be approximately valid for the realistic Independent Particle Models such as Nilsson model. Realistic calculations show indeed the occurrence of deformed shells with 2/1 ratio for prolate shapes corresponding to the fission shape isomers. The magic proton and neutron numbers relevant to these shapes are 86-88 and 148 respectively.

6. Origin of the asymmetry of fragments mass distributions.

A full dynamical theory of nuclear fission has still to be developed and only such a theory could provide an understanding of the features of fission fragments' mass distributions. However different static models have met some success in reproducing the general features of these distributions.

These models share a common approach in that they choose a quasi-static configuration and assume that mass yields are determined by the corresponding available phase space. The reference configuration has been chosen either as the saddle-point or as the so called scission point considered as two touching fragments. Two different assumptions have been made concerning the available phase space depending upon the amount of coupling expected between the collective and intrinsic nuclear excitations. We shall now consider three approaches which encompass these features.

6.1. The saddle point model.

C.F. Tsang and J.B. Wilhelmy,[35] pointed out that the reflection assymetric shape of the second saddle-point might be responsible for the asymmetric mass distributions of fission fragments.

The minimum potential energy at the second saddle point is found to correspond to reflexion asymmetric shapes. Tsang and Wilhelmy define an asymmetric fission barrier B_a which corresponds to this minimum while a symmetric fission barrier B_S corresponds to the height of the second saddle point obtained with the constraint of reflection symmetry put upon the nuclear shapes. For excitation energies well above both barriers the ratios of symmetric mass splits to most probable asymmetric mass splits are determined by

$$\frac{P}{v} = \frac{e^{\sqrt{a_A(E^{x}-B_A)}}}{e^{\sqrt{a_S(E^{x}-B_S)}}}$$

where a_A and a_S are the level density parameters for asymmetric and symmetric saddle point configurations. $\frac{P}{v}$ is the peak to valley ratio of the mass distributions. Here full equilibration between the collective and intrinsic degrees of freedom is assumed. Below the barrier and more especially for spontaneous fission Tsang and Wilhelmy write

$$\frac{P}{v} = \frac{e^{-\frac{B_A}{\hbar\omega_A}}}{e^{-\frac{B_S}{\hbar\omega_S}}}$$

where $\hbar\omega_A$ and $\hbar\omega_S$ are characteristic of the barrier penetrabilities. Tsang and Wilhelmy achieved a rather good correlation between the experimental and computed values of p/v with reasonable values for the level density parameters and barrier penetrability. They predict a switch to symmetric fission in the Fermium region and observe two saddle points, one symmetric, one asymmetric in the Radium region which might explain the triple humped mass distributions in this mass region. The shell effects are predicted to wash out when the excitation energy increases.

This washing out stems from the nuclear level density behavior. At low excitation energy the level density increases less steeply for closed shell or near closed shell configurations so that $a_A < a_S$

However $B_A = B_{LD} - \varepsilon_{sh} < B_S$

so that, for given excitation energy $E^{\ast}$

$$e^{\sqrt{a_A [E^{\ast} - (B_{L_D}^{(A)} - \varepsilon_{sh})]}} > e^{\sqrt{a_S (E^{\ast} - B_S)}} = e^{\sqrt{a_F (E^{\ast} - B_{L_D}^{(S)})}}$$

Here ε_{sh} is the shell correction to the liquid drop barrier and we assume that for symmetric configuration the level density is the Fermi gas level density while the shell correction may be neglected. For high excitation energies the level density for asymmetric saddle point shapes approaches,[32]

$$\rho_A = e^{\sqrt{a_F [E^{\ast} - (B_{L_D}^{(a)} - \varepsilon_{sh}) - \varepsilon_{sh}]}} = e^{\sqrt{a_F (E^{\ast} - B_{L_D}^{(A)})}}$$

It should be noted that for negative shell effects ($\varepsilon_{sh} > 0$) the level density always remains larger or, at high excitation energies, at least equal to the Fermi gas density obtained without shell effects.

In the frame work of the original saddle point model of Tsang and Wilhelmy the behavior of the mass distributions as a function of excitation energy in the Radium region remained puzzling because the raise of the symmetric component was much too rapid to be accounted for by the washing out of the shells. A possible explanation of this behavior has been found when it was realized that the symmetric saddle point responsible for the symmetric bump in the mass distribution has lost the axial symmetry. This leads to a steeper rise of the level density,[36] than for the mass asymmetric axially symmetric saddle point.

6.2. The scission point models.

The first scission point model, and in fact the first partially successful quantitative model of fission was introduced by P. Fong,[37]. The basic idea of P. Fong was that the fission fragments' properties were determined at a very late stage in the fission process due to the slowness of the process itself. This late stage could be approximated by two nascent fragments in contact. The fragments could be deformed and were submitted to their mutual polarizing influence.

In Fong's model this influence was purely Coulombic. Wilkins et al,[38] add a nuclear interaction term and, relaxing the condition that the two fragments should be touching, introduce a distance d between the two tips of the fragments. The potential energy of the "scission configuration" can then be written

$$V(A_1, Z_1, A_2, Z_2, \beta_1, \beta_2, d) = M_o(A_1 Z_1) + M_o(A_2, Z_2)$$
$$+ D(A_1, Z_1, \beta_1) + D(A_2, Z_2, \beta_2) + V_c(A_1, Z_1, A_2, Z_2, \beta_1, \beta_2, d)$$
$$+ V_N(A_1, Z_1, A_2, Z_2, \beta_1, \beta_2, d)$$

where β_1 and β_2 are the set of deformation parameters of fragments 1 and 2, $M_0(A,Z)$ the ground state mass of fragment (A,Z), $\mathcal{D}(A,Z,\beta)$ the deformation energy with respect to ground state of fragment (A,Z) with deformation β, V_c and V_N the Coulomb and Nuclear interactions between the fragments.

The available or free energy in the system amounts to

$$E^{\ddot{x}}(A_1, Z_1, A_2, Z_2, \beta_1, \beta_2, d) = M_0(A_1 + A_2, Z_1 + Z_2) + E^x$$

$$- V(A_1, Z_1, A_2, Z_2, \beta_1, \beta_2, d) = E^x_1 + E^x_2$$

where E^x is the initial excitation energy of the fissioning nucleus $(A_1 + A_2, Z_1 + Z_2)$.

E^x_1 and E^x_2 are the respective part of the free energy shared between the fragments.

In both Fong's and Wilkins' models the probability that a given scission configuration is realized is given by

$$Y(A_1, Z_1, A_2, Z_2, \beta_1, \beta_2) \simeq e^{\frac{E^{\ddot{x}}}{T}} \simeq e^{-\frac{V}{T}}$$

It can be seen here that d is considered as a free parameter.

According to Fong the temperature T is calculated according to the standard statistical theory namely

$$T^2 = \frac{E^x_1}{a_1} = \frac{E^x_2}{a_2} = \frac{E^x_1 + E^x_2}{a_1 + a_2}$$

where a_1 and a_2 are the level density parameters of nuclei 1 and 2. To some extent the dependance of a_1 and a_2 upon the shell structure of the fragments will counteract the influence of shells in the potential energy surface. We shall show later that the full statistical assumption of Fong cannot explain the even-odd effects presented by the fragments' charge distribution. These effects find a natural explanation in the frame of the thermodynamical model of W. Nörenberg,[39] which has been applied by B. Wilkins et al.[38]. This model makes a distinction between intrinsic such as two quasi particles, and collectives excitations. It assumes that during the descent from the saddle point to scission the collective excitations are strongly coupled between themselves and to the fission motion itself while the coupling to intrinsic excitation is weak. The amount of intrinsic excitation is, therefore, essentially determined at the saddle point and can be characterized by an intrinsic temperature τ_{int} which is essentially kept down to scission. The intrinsic temperature has a direct influence on the amount of shell correction to be included in the deformation energies $\mathcal{D}(A,Z,\beta)$ which become, therefore, dependant on τ_{int}. The strong coupling between the collective excitations leads to a thermal equilibrium between them characterized by a collective temperature T_{coll}. Therefore the probability for observing a given scission configuration becomes

$$Y(A_1,Z_1,A_2,Z_2,\beta_1,\beta_2) \simeq e^{-\frac{V(\tau int)}{T_{coll}}}$$

Wilkins et al. were able to obtain a satisfactory account of the trends of the mass distributions from the triple humped Radium to the Fermium ones. The tendancy towards symmetry with increased excitation energy is obtained through the induced increase of the intrinsic temperature τ_{int}.

The liquid drop potential energy minimum is obtained for quadrupolar deformations β of the fragments around 0.6. Therefore the 2:1 shell effects in the fragments, corresponding to the same value of β are very effective. The shells responsible for the mass assymetry of the fragments' distribution were found to be those with N = 86 and N = 64 which give rise to pronounced minima in the potential energy surfaces. Spherical shells at N = 82 Z = 50 and N = 50 play an important role for determining the kinetic and excitation energies of the fragments. The maximum kinetic energy, for example is obtained when the heavy fragment has Z = 50 and a very low excitation energy. These features are, indeed, observed in experiment.

A possible explanation of the successes of both the saddle point and scission point approaches may be found in the work of Mustafa et al.[40] These authors used a two center shell model to follow the potential energy surface from the saddle point to the scission configuration. Their calculations seem to indicate that the shell structures in the final fragments influence the potential energy surfaces already at the second saddle point as shown in Fig.13. Mustafa,[41] points out that the dynamics of the descent from saddle point to scission might play an important role in determining the asymmetric or symmetric character of the mass distributions in the Fermium region. The experimental evidence in this mass region seems to favor a slow motion towards fission. However except for that case, and possibly in the Polonium region, it appears that mass distributions are not very sensitive to the fission dynamics. We now show that charge distributions and more precisely the even-odd straggling which some of them present may be the most spectracular manifestation of these dynamics.

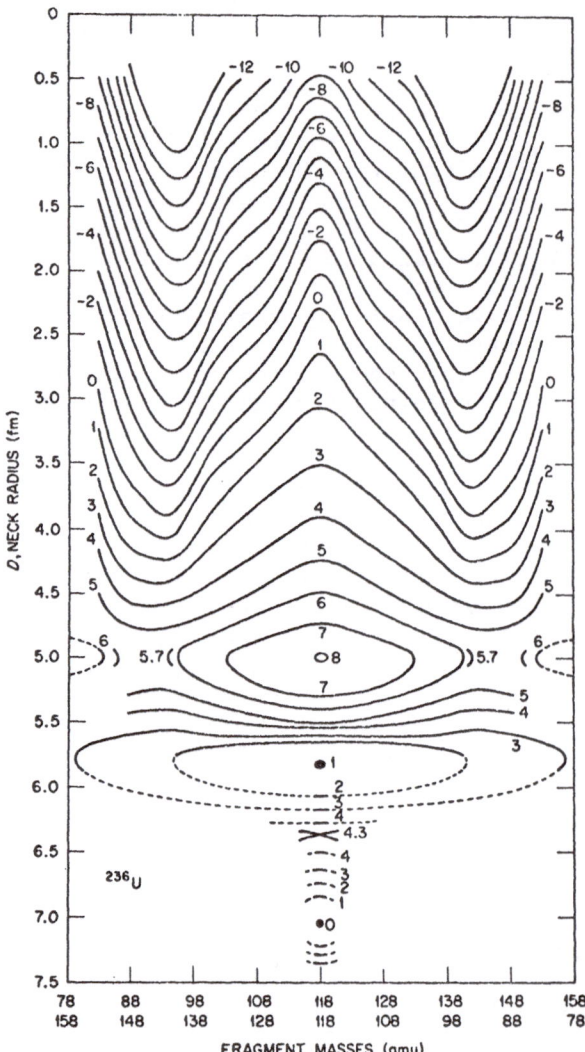

Fig. 13. Potential energy of ^{236}U, calculated by Mustafa, Mosel and
Schmitt with a modified two-center oscillator potential
[36]. Contours of constant potential energy are plotted as
functions of the neck radius D and the masses of the two
nascent fragments. The contours are spaced at intervals of
1 MeV and are labelled by the energy (in MeV) relative to
the ground-state minimum potential energy; an additional
contour is included near each saddle point. The dashed lines
represent interpolated or extrapolated values.

7. <u>The signifiance of the even-odd effects in charge distributions.</u>
   Let us first remark that charge distributions are not affected
by evaporation processes after scission so that they allow to draw
directly conclusions on the fission mechanism.
Fragments with even charges are produced in larger quantities than
with odd charges in the thermal neutron induced fission of U^{235}.
The magnitude of this enhancement can be measured by the even-odd
effect parameter

$$\delta = < \frac{Y_e - Y_o}{Y_e + Y_o} >$$

Y_e and Y_o being the yields of
fragments with even and odd
charges respectively.

where the average is taken over the full charge distribution.

   Table 7.1 summarizes our knowledge of the even-odd effects which
have been measured to date

Table 7.1

Fissioning System	Even-odd Effect δ in %
$Th^{229} + n_{th}^{(42)}$	35 ± 5
$Th^{232} + n_{3MeV}^{(46)}$	30 ± 12
$U^{235} + n_{th}^{43,44)}$	22 ± 7
$U^{233} + n_{th}^{(44)}$	23 ± 5
$Pu^{239} + n_{th}^{(44)}$	3 ± 5
$Pu^{241} + n_{th}^{(44)}$	4 ± 5
$Cf^{252}(s,f)^{(42)}$	12 ± 2
$U^{235} + n_{3MeV}^{(42)}$	5 ± 3
$U^{235} + n_{2MeV}^{(45)}$	8 ± 4

   Note that our discussion deals only with even Z fissioning
species.
   Fig.14 shows the charge distribution of $Th^{229}(n_{th},f)$ which displays the most spectacular even-odd effects observed to-date.
   The Table shows two very interesting trends.
   The even-odd effects seems to depend very much on the nature
of the fissioning nucleus. They decrease dramatically from Uranium
to Plutonium.
   They also vary strongly with the excitation energy of the fissioning nucleus to the extent that they almost disappear with only 3 MeV extra excitation in U^{235} fission.
   This behavior is not compatible with such approaches as
Fong's statistical model. Qualitatively this stems from the fact

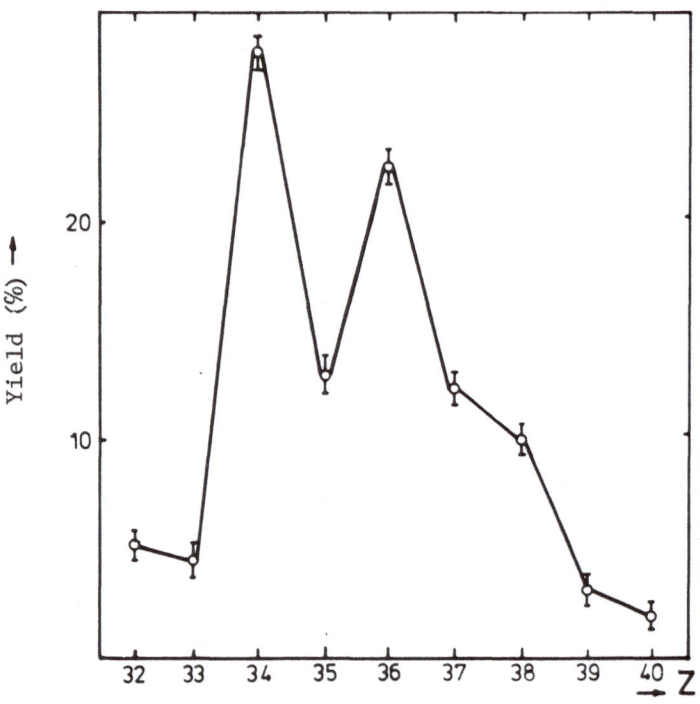

Fig. 14.   Light fragments' charge yields in $Th^{229}(n_{th},f)$.

that the statistical model emphasizes the properties of the final
fragments. It is hard to understand what could make the difference
between Plutonium and Uranium fission fragments as far as the parity of their charges is concerned. Similarly an increase of 3 MeV
in the total excitation energy is no reason to change dramatically the fragments properties. A full quantitative analysis of the
statistical model predictions with respect to even-odd effects is
given in Ref.42. We shall content ourselves, here, with a simplified approach. Within the statistical scheme the fragments yields
are given by

$$Y \simeq e^{(Q-V)/T}$$

where Q is the final energy balance of the reaction

$$Q = M_o(A_1+A_2, Z_1+Z_2) - M_o(A_1,Z_1) - M_o(A_2,Z_2)$$

and V is the potential energy at scission, T being a temperature.
The energy balance when the two fragments have even charges is higher than when they have odd charges

$$Q_e - Q_o = 2 \Delta$$

Δ being the average pairing gap in the fragments.

Thus

$$\frac{Y_e}{Y_o} = e^{(Q_e - Q_o)/T} = e^{2\Delta/T}$$

assuming no even-odd effect on V.

Practically, because the level density (or temperature) has
a dependance upon the even-odd character of the fragments (this
dependance tends to wash out the even-odd effects) one may define
an effective pairing correction

$$\frac{Y_e}{Y_o} = e^{D/T}$$

Where D should be at most of the order of 2Δ (2.5 MeV).

If one considers the even-odd effects observed for thermal
and 3 MeV neutrons induced fission of U^{235} one gets

$$(\frac{Y_e}{Y_o})_{th} = 1.5 = e^{\frac{D}{T}th}$$

thus $$T_{th} = \frac{D}{0.4}$$

and

$$(\frac{Y_e}{Y_o})_{3MeV} = 1.11 \qquad\qquad T_{3MeV} = \frac{D}{0.1}$$

giving $$\frac{T_{3MeV}}{T_{th}} = 4$$

a very unreasonable value. In other words the strong variations
of the even-odd effects with excitation energy would imply a much
too fast variation of temperature with energy.

A simple explanation of both the observed even-odd effects
and of their variations with excitation energy and nuclear species
is obtained if one assumes that the total number of proton pairs
breaking during the fission process is small. To keep the argu-
ment as simple as possible we first consider that no additional
pair breaking occurs after saddle-point. If the nucleus is completely paired at the saddle-point, like in below-the barrier fission,
only fragments with even-Z will be produced since an odd-Z oddZ
split requires that at least one pair has been broken. If at least
one pair is broken at saddle point the number of oddZ-oddZ and
evenZ-evenZ splits will be equal since the now uncorrelated nucleons may end up indifferently in any of the two fragments. In
this picture the observed excess of even-Z fragments is therefore
a measure of the probability that the saddle point configuration
has no broken proton pair.

$$P_o^{(p)} = \delta = \frac{Y_e - Y_o}{Y_e + Y_o}$$

This relationship holds if one relaxes the assumption that
pairs are not broken after the saddle-point provided one keeps
the requirement that the unpaired nucleons end in the same fragment with the same probability as in different ones.

Along this line the variation of the even-odd effects observed in U^{235} neutron induced fission is readily understood. Increasing the excitation energy at saddle-point by 3 MeV decreases
strongly the probability that no proton pair is broken. More quantitatively 3 MeV corresponds to a little more than two additional
broken pairs at saddle-point. Those can be either proton or neu-
tron pairs with respective approximate probabilities $\frac{Z}{A}$ and $\frac{N}{A}$.

The probability that the two broken pairs be neutron pairs
is, therefore, $(\frac{N}{A})^2$ also equal to the probability that no additional proton pair is broken. For U^{236} this leads to 0.37 reduction of the probability of no proton pair breaking and correspondingly to an expected decrease of the even-odd amplitude from 22%

for thermal neutrons to about 8%. A smaller value is obtained if,
the relative probabilities of neutron to proton pair breaking is
taken to be proportional to the densities near the Fermi surface
rather than to the bulk densities. Then the reduction amounts to

$(\frac{N^{2/3}}{Z^{2/3}+N^{2/3}})^2$,   leading to an even-odd amplitude of 7%

       These values are close to the experimentally observed ones.

       The important point in this simple picture is that whenever
pairs are broken at saddle they remain so down to the scission,
in accordance with the thermodynamical model of Nörenberg.
Additional pair breaking can occur after saddle point formation
and it is assumed that their number does not depend strongly upon
the excitation energy at saddle point.

       The fact that pair breaking does occur past the saddle-point
configuration is clearly evident from the production of odd-Z elements in spontaneous fission. It would be of importance, as we
shall see later, in order to understand the dynamics of fission,
to know when, in the process, those pairs are broken. We say that
a pair is broken early *if* the two unpaired nucleons behave independently as far as their final location in fragments is concerned;
we say that a pair is broken late when the two unpaired nucleons
end in different fragments. This is the situation which we expect
to occur when pairs are broken due to a sudden snap off of the neck
joining the two nascent fragments. The study of the fragments excitation energies (or kinetic energies) may provide a clue to the
question of when pairs are broken. Considering the case of early
pair breaking it is clear that evenZ-evenZ splits will lead to the
condensation, during the deexcitation of the fragments, of one
more proton pair than oddZ-oddZ ones. Therefore the total excitation energy for even-even splits should be about 2Δ higher than
for the odd-odd case. Due to energy conservation one does not expect any even-odd effect on the fragments kinetic energies. Let
$E_k^{(1)}$ and $E_k^{(2)}$ the kinetic energies corresponding to the cases
without and with pair breaking respectively. We may express the
kinetic energies expected, for early pair breaking, for odd-odd
splits and even-even splits respectively

$$E_k^{odd}(Z) = E_k^{(2)}$$

$$E_k^{even}(Z) = \left. P_o^{(p)} E_k^{(1)} + \frac{1}{2}(1-P_o^{(p)}) E_k^{(2)} \right/ \frac{1}{2}(1+P_o^{(p)})$$

putting $E_k^{(1)} - E_k^{(2)} = \alpha$

and recalling that $P_o^{(p)} = \delta$

$$E_k^{even}(Z) - E_k^{odd}(Z) = \frac{2\delta}{1+\delta}\,\alpha \quad (7.1)$$

This formula only holds if it is assumed that the kinetic energy for pair breaking $E^{(1)}$ does not depend on the number of broken pairs.

Some quantitative predictions can be made under the assumption of late pair breaking. We make the hypothesis that n pairs may be broken with equal probability. Whenever an odd number of pairs is broken an odd-odd Z split is obtained while the reverse is true when no or an even number of pairs is broken. For one pair to be broken an amount α of fragments' total kinetic energy is required. Then one can show that

$$E_k^{even} - E_k^{odd} = \frac{2\,\delta\,\log\delta}{\delta^2 - 1}\,\alpha \quad (7.2)$$

The difference between the predictions of expression 7.1 and 7.2 is shown in the table 7.2.

Table 7.2 computed relative even-odd effects in kinetic energies according to two different hypothesis or the pair breaking mechanism.

δ	0.1	0.2	0.4	0.8
$(E_k^e - E_k^{odd})/\alpha$ Early pair breaking	0.18	0,33	0.57	0.95
$(E_k^e - E_k^o)/\dot\alpha$ Late pair breaking	0.47	0.67	0.87	0.99

From the table it is clear that the late pair breaking assumption gives a better account of the small variations of the differences

$$E_k^e - E_k^o$$

It is also possible to estimate the average number of broken pairs $\bar{\mu}_p$. In the early pair-breaking assumption

$$\bar{\mu}_p = -\log\delta$$

while in the late pair breaking assumption

$$\bar{\mu}_p = -\frac{1}{2}\log\delta$$

An estimate of the excitation energy in the fragments due to broken proton pairs is obtained by

$$X^{(1)} = -2 \, \Delta \, (\log \delta + (1-\delta)/2)$$

in the early pair breaking case and

$$X^{(2)} = -\Delta \, (\log \delta + (1-\delta)) \quad \text{in the late pair breaking one}$$

Here Δ is the pairing gap.

To this contribution one must add a similar one due to neutron pair breaking. The computed excitation energies for $Cf^{252}(sf)$, $U^{235}(n_{th},f)$ and $Th^{229}(n_{th},f)$ are shown in Table 7.3

Table 7.3

	$Cf^{252}(s,f)$	$U^{235}(n_{th},f)$	$Th^{229}(n_{th},f)$
$X_T^{(1)}$ (MeV)	8	5.5	3.5
$X_T^{(2)}$ (MeV)	3.1	1.8	1

Computed intrinsic excitation energies according to two hypotheses

It is seen that the assumption made on the pair breading mechanism has strong consequences on the amount of damping of the collective modes in the fission process. The late pair breaking hypothesis assumes, of course, a very small damping in the first stage of the descent from the saddle point.

Obviously more experimental results are required to allow definite conclusions concerning the best assumption to make.

Especially, studies of the even-odd effects as function of excitation energy above the saddle point should give fruitful information.

CONCLUSION -

During the course of these lectures we have found that our basic understanding of the statics of the fission process is rather satisfactory although very subtle and interesting effects are still being found related to the potential energy surfaces. Owing to fission a spectroscopy of nuclei in extreme state of deformation has been made possible.

The long standing problem of the asymmetry of mass distributions appears to find its solution in the influence of deformed or spherical shells in the fragments all the way from the second saddle-point.

On the other hand our understanding of the fission dynamics is still rather poor. We may hope that the careful study of the charge distributions of fission fragments will shed light on this problem.

Bibliography

1. O. Hahn and F. Strassmann, Naturwisess.27, 11 (1939).
2. E. Fermi, Nature 133, 898 (1934).
3. L. Meitner and O.R. Frisch, Nature 143, 239, 47 (1939).
4. J. Frenkel, Phys. Rev. 55, 987 (1939)
                   J. Phys USSR 1, 125 (1939).
5. N. Bohr and J. Wheeler, Phys. Rev. 56, 426 (1939).
6. H. von Halban, F. Joliot and L. Kowarski, Nature 143, 470 (1939).
7. O. Haxcel, J.M.D. Jensen and H.E. Suess, Phys. Rev. 75, 1766
   (1949).
   M.G. Mayer, Phys. Rev. 75, 1969 (1949).
8. S. Polikanov, V. Druin, V. Karnankov, V. Mikleev, A. Pleve,
   N. Skobelev, V. Subbotin, G. Ter Akopyan and F. Fornichev,
   Sov. Phys. JETP 15, 1016 (1962).
9. A. Bohr and B. Mottelson, Kgl. Dan. Vid. Selsk. Mat. Fys. Medd.
   30, No 1 (1955).
10. S.G. Nilsson, Kgl. Dan. Vid. Selsk. Mat. Fys. Medd. 29, No 16
    (1955).
11. See for example
    W.D. Myers in "Dynamic Structure of Nuclear States", Proc. of
    the 1971 Mont Tranblant, Int. Summer School (Univ. of Toronto
    Press 1977).
12. W.D. Myers and W.J. Swiatecki, Nucl. Phys. 81, 1 (1966).
13. V.M. Strutinsky, Yad. Fis. 3 , 614 (1966), Nucl. Phys. A 95,
    420 (1967), A 122, 1 (1968).
14. R. Balian and C. Bloch, Ann. Phys. (N.Y.) 60, 401 (1970).
15. M. Brack and P. Quentin, International Conference on Nuclear
    Self Consistent kields, ICTP, Triest 1975, p. 353.
16. L. Wilets, Theories of Nuclear Fission, Clarendon Press Oxford
    (1964).
17. J.R. Nix, Nucl. Phys. A 130, 241 (1969).
18. See for example : L. Moretto, Lectures at the Erice School
    On Heavy Ion Interactions at High Energies (1979).
19. R. Vandenbosch, J. Huizenga, Nuclear Fission (1973) Academic
    Press New York.
20. H.C. Pauli and T. Ledergerber, 3rd Symposium on Physics and
    Chemistry of Fission, IAEA, Rochester, Vol.I, p.463 (1973).
21. D. Paya, J. Blons, H. Derrien, A. Fubini, A. Michaudon and
    P. Ribon, J. Phys. and Radium, Suppl. C1.159 (1967).
22. E. Migneco and J. Theobald, Nucl. Phys. A.112, 603 (1968).
23. J. Blachot, J. Crançon, C. Hamelin, A. Moussa, 4th Int.
    Symposium on Physics and Chemistry of Fission, IAEA, Jülich
    (1979).
24. R. Brissot, J. Crançon, C. Ristori, J.P. Bocquet and A. Moussa,
    Nucl. Phys. A 282, 109 (1977).
25. H. Nifenecker, J. Blachot, J.P. Bocquet, R. Brissot, J.Crançon,
    C. Hamelin, G. Mariolopoulos, C. Ristori, 4th Int. Symposium
    on Physics and Chemistry of Fission, IAEA, Jülich (1979).
    IAEA Vienna (1980) Vol.II p.35.

26. S.E. Larsson, I. Ragnarsson and S.G. Nilsson, Phys. Lett. 38B 269 (1972).
27. M. Brack, Nuclear Theory for Applications, ICTP, Triest 1980, IAEA SMR 43, p.327.
28. H.J. Specht, J. Weber, E. Konecny and D. Heunemann, Phys. Lett. B.41, 43 (1972).
29. D. Habs, V. Metag, H.J. Specht and G. Ulfert, Phys. Lett. 38, 387 (1977).
30. P. Moller and R. Nix, 3rd Int. Symposium on Physics and Chemistry of Fission, IAEA, Vol. I 103, Rochester (1973).
31. J. Blons, C. Mazur, D. Paya, M.Ribrag and H. Weigman, Phys. Rev. Lett. 41, 1282 (1978).
32. A. Bohr and B. Mottelson, Nuclear Structure, Vol.II, Benjamin 1975.
33. H.C. Britt, 4th International Symposium on Physics and Chemistry of Fission, IAEA, Jülich (1979). IAEA Vienna(1980) Vol.I.p.3.
34. V.M. Strutinsky, 4th International Symposium on Physics and Chemistry of Fission, IAEA, Jülich (1979).
35. C.F. Tsang and J.B. Wilhelmy, Nucl. Phys. A184, 417 (1972).
36. S. Bjornholn, A. Bohr, B. Mottelson, 3rd Int. Symp. on Phys. and Chemistry of Fission, IAEA, Vol.I p.367, Rochester (1973).
37. P. Fong, Phys. Rev. 102 (1957) 434.
38. B.D. Wilkins, E.P. Steinberg and R.P. Chasman, Phys. Rev. C14, 1832 (1976).
39. W. Nörenberg, Proc. 2è IAEA Symp. on Chemistry and Phys. of Fission, p.51, Vienna 1969.
40. M.G. Mustafa, U. Mosel and H.W. Schmitt, Phys. Rev. 1519 (1973).
41. M.G. Mustafa, 4th International Symposium on Physics and Chemistry of Fission, IAEA, Jülich (1979).
42. G. Mariolopoulos, Ch. Hamelin, J. Blachot, J.P. Bocquet, P. R. Brissot, J. Crançon, H. Nifenecker and Ch. Ristori, To be published in Nuclear Physics.
43. H.G. Clerc, W. Lang, H. Wohlfarth, H. Schräder, K.H. Schmidt, Proc. 4th IAEA Symp. on Physics and Chemistry of Fission Jülich (1979), IAEA Vienna (1980) Vol.II p.65.
44. R. Brissot, J. Crançon, Ch. Ristori, J.P. Bocquet and A. Moussa Nucl. Phys. A282 (1977) 109.
45. S. Amiel, H. Feldstein, Phys. Rev. 11 (1975) p.845.
46. F. Izak Biran, S. Amiel, Phys. Rev.C, Vol.16 n°1 p.266.

MUON INDUCED FISSION

S. Polikanov

CERN
Geneva, Switzerland

INTRODUCTION

The title of my lecture implies some general remarks related
both to fission and muonic atoms. It is relevant to start with
fission because the main goal of the experiments we are going to
discuss was to learn more about fission itself. But one has to keep
in mind also that the fission processs is well understood now and
it is tempting from time to time to consider it as a rather good
tool for the study of other processes.

FISSION-BULK PROPERTIES OF NUCLEAR MATTER + SHELL EFFECTS

Soon after the discovery of fission by Hahn and Strassmann,
this process was explained by N. Bohr and Wheeler on the basis of
a very simple model. According to this model a nucleus was considered as an electrically charged incompressible liquid drop (LDM).
Coulomb repulsion of protons tends to elongate the nucleus. Nuclear
forces resist that and this effect can be described as surface
tension. In this model fission is nothing else but a partition of
a liquid drop into two drops of smaller size. Though at first sight
quite unrealistic, LDM in fact explains reasonably well many features of fission. That is not by chance. LDM includes some very
important bulk properties of nuclear matter - its incompressibility
as well as Coulomb repulsion of protons.

It appeared, however, that LDM was not sufficient to explain all known facts about fission and with the further development of the theory more attention was paid to the role of the single particle effects.

A very important step was made in 1956 by A. Bohr who introduced transitional states at the saddle point. This concept was accepted soon as at an excitation energy equal to the height of the fission barrier, the nucleus is practically cold at the saddle point. One can expect the appearance of separated collective states and single particle excitations. A. Bohr's suggestion has found good support in measurements of angular distribution of fission fragments.

The next step which provided the possibility to explain in a rather natural way some equilibrium states at large deformation was done by Strutinsky[1]. Strutinsky has taken into account the fluctuations of the level density near the Fermi surface. The nature of such fluctuations is seen especially clearly for an anisotropic harmonic oscillator (Fig. 1). One can see how the level density depends on the shape of a nucleus. It follows from Fig. 1 that for nuclear axis ratios $\sim$ 2 new shells appear.

According to Strutinsky, nuclear masses can be evaluated as a sum of two terms. The first one is the nuclear mass estimated on the basis of LDM. It takes into account the bulk properties of nuclear matter. The second represents the shell corrections, that is, the behaviour of levels near the Fermi surface is taken into consideration. Figure 2 shows the fission barrier as it looks after shell corrections are introduced. The existence of the second minimum makes it possible to explain the following phenomena:

1. fission isomers;
2. sub-barrier fission resonances.

Shell corrections are very large for superheavy elements. In fact, the fission barrier of superheavies is to be determined only by the shell effect.

At present there are many experimental data which might be taken as supporting the idea of the two-humped fission barrier.

$$V = \frac{1}{2} M \left(\omega_\perp^2 \left(x_1^2 + x_2^2 \right) + \omega_3 \, x_3^2 \right)$$

$$E = \hbar \, \omega_\perp \left(n_\perp + 1 \right) + \hbar \, \omega_3 \left(n_3 + \frac{1}{2} \right)$$

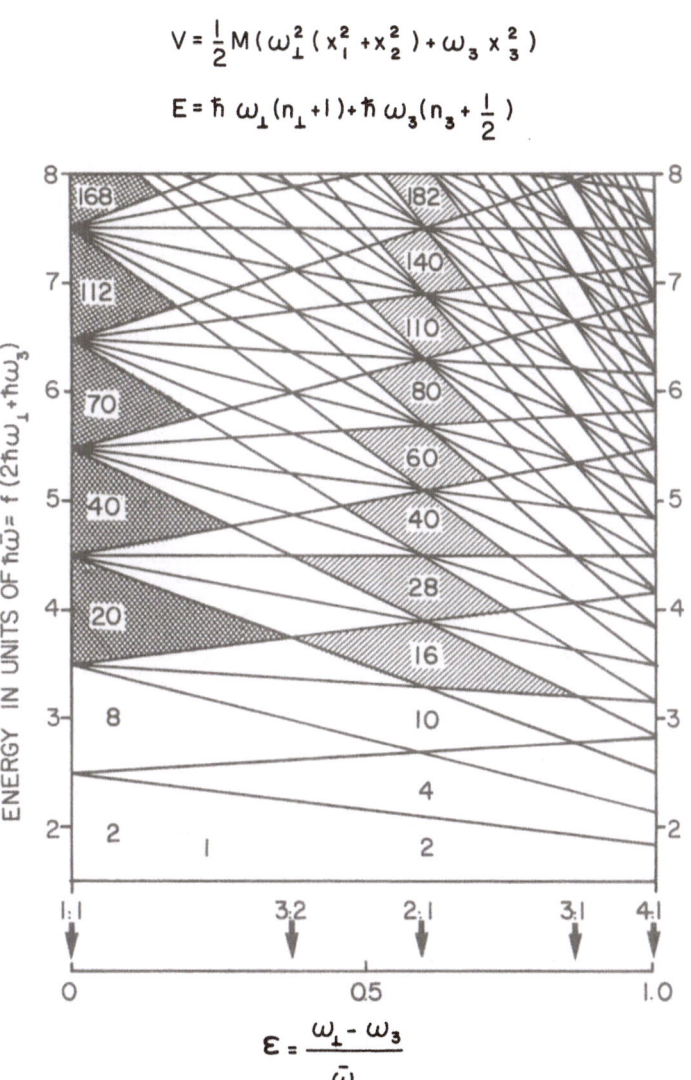

Fig. 1   The levels for the anisotropic harmonic oscillator.

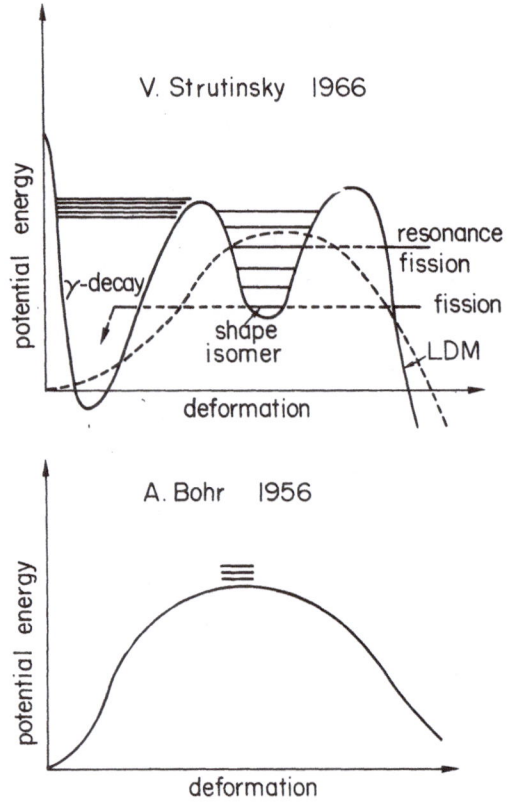

Fig. 2   Two-humped fission barrier.

MUONIC ATOMS OF HEAVY ELEMENTS

Some of the known elementary particles (μ^-, π^-, K^-, $\bar{p}$, Σ^-, Ξ^-, Ω^-) are stable enough to be slowed down in the matter and from the continuum enter into a discrete spectrum replacing one of the electrons.   However, only in the case of a muon a relatively stable atom is formed. Due to the strong interaction all other elementary particles are absorbed in a very short time.   For heavy elements they cannot even enter the 1s orbit being captured from orbits with higher n.

For heavy elements the radius of the 1s muonic atom orbit is of the same order of magnitude as the nuclear radius. Thus the muon spends a lot of time inside the nucleus. Finally, the muon disappears in the following process:

$$\mu^- + p \rightarrow n + \nu$$

Most of the energy released is taken by the neutrino. Nevertheless, the residual nucleus is excited usually up to energies of about 20 MeV. The compound-nucleus either evaporates neutrons or decays through the fission channel. As soon as the muon absorption is governed by weak interaction the typical lifetimes for the muonic atom of fissile isotopes are close to 80 nsec.

It was pointed out first by Wheeler[2] that during the atomic de-excitation the energy of a transition can be transferred to the nucleus without X-ray emission. For uranium, the energy of 2p - 1s is about 6.2 MeV. At that energy the nuclear level density is rather high. The estimates done by Zaretsky and Novikov[3] show that the probability of 2p - 1s radiationless transition is about 0.24. One cannot, in principle, eliminate 3d - 1s, 4p - 1s radiationless transitions either.

Until now the radiationless transitions were not studied with good accuracy. In the early experiments of Diaz et al.[4] fission induced by radiationless muonic transitions was observed for ^{238}U. Later, some other isotopes were studied. Hargrove et al.[5] have observed prompt neutron emission for negative muons, interacting with lead and bismuth. The analysis showed that the 3d - 1s transition is of importance.

As for fissile targets, until recently only the probability of prompt fission was measured. In fact, a more detailed study of prompt fission induced by radiationless transitions is of certain interest as soon as one can treat it as photofission in the presence of a muon.

MUON INDUCED FISSION

Before discussing some results obtained in recent years let us consider what is quite specific for muon induced fission. It seems useful to refer here to the following problems:

1. fission barrier augmentation in the presence of a muon;

2. muon attachment to fission fragments.

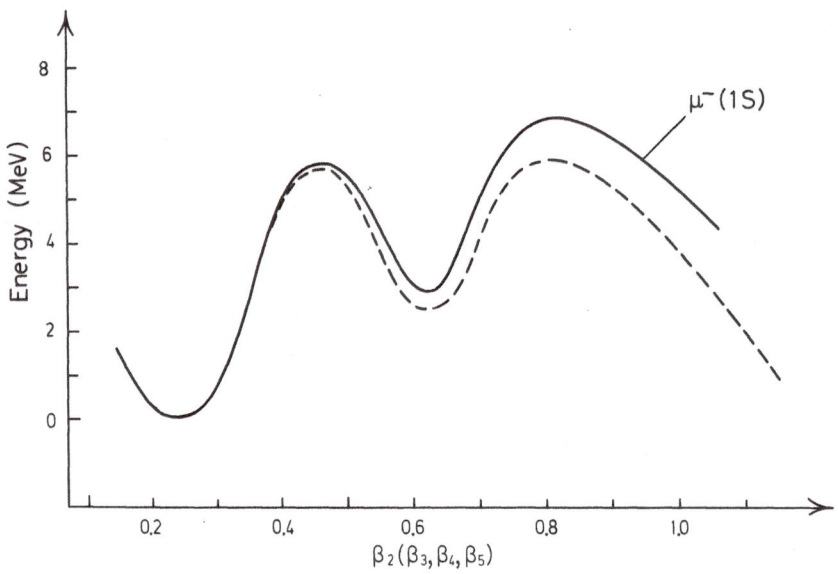

Fig. 3   Fission barrier augmentation in the presence of a muon

Fission barrier augmentation

As was explained earlier, negative muons in the 1s orbit move
close to the nucleus, penetrating deep inside of it. That changes
the Coulomb energy of the nucleus and consequently that of the
fission barrier too. One can easily understand this, keeping in
mind that all the effect is due to the decrease of the muon binding
energy, when the nucleus is elongated.

Leander and Möller[6] estimated fission barrier augmentation in
the presence of a muon in the 1s orbit. Figure 3 presents results
of their calculation for ^{236}U. One can see that the inner fission
barrier practically is not influenced. On the other hand, the
outer fission barrier increases in height. An increase is expected
of 0.7 - 1.2 MeV being very sensitive to the position of the saddle
point. An isomeric shift of $\sim$ 600 keV should take place. That
will drastically change the properties of the shape isomer.

Muon attachment to fission fragments

Fission is a slow process proceeding in time not shorter than
10^{-19} s. When fission takes place in the presence of a muon its
atomic orbit will be adjusted adiabatically with the change of the
nuclear shape. The question immediately arises to which fission

fragment will the muon be attached after scission. In fact, muon
attachment was first observed in experiments done at CERN[7] and
only later the theory of this process was developed.

The most simple approach was used by Olanders et al.[8] who
represented a fissioning muonic atom by two point charges. The
time dependent Schrödinger equation for the muon in the field generated by two separating point charges was studied. Until the process is adiabatic one can expect that the muon will find the deeper
well created by electrical charge of the heavy fission fragment.
But still one cannot neglect the probability of muon attachment to
the light fission fragment. As the analysis shows, at the part
of the trajectory preceding scission, the velocity of the system
is high enough and there still is a chance for a muon to be attached
to the light fragment.

As soon as the velocity of the nuclear matter depends on its
viscosity, the muon attachment to the light fragment will depend
on the viscosity, too (Fig. 4 [9,10]).

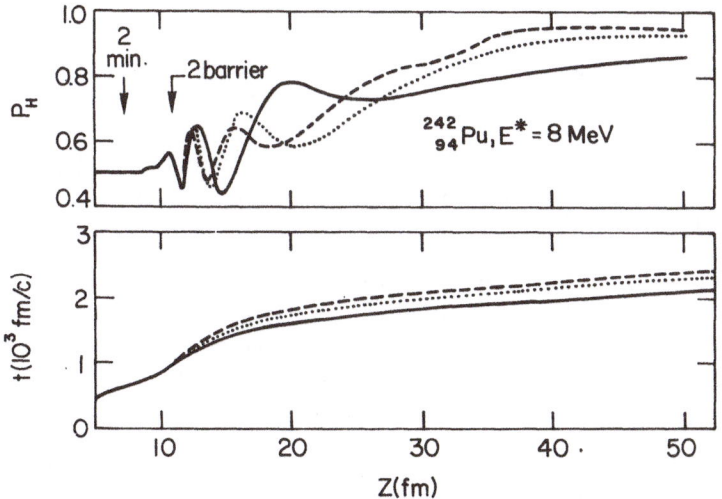

Fig. 4   Muon attachment P_H to fission fragments as a function of
         time and of their distance Z for low (—) and high (---)
         friction parameter.

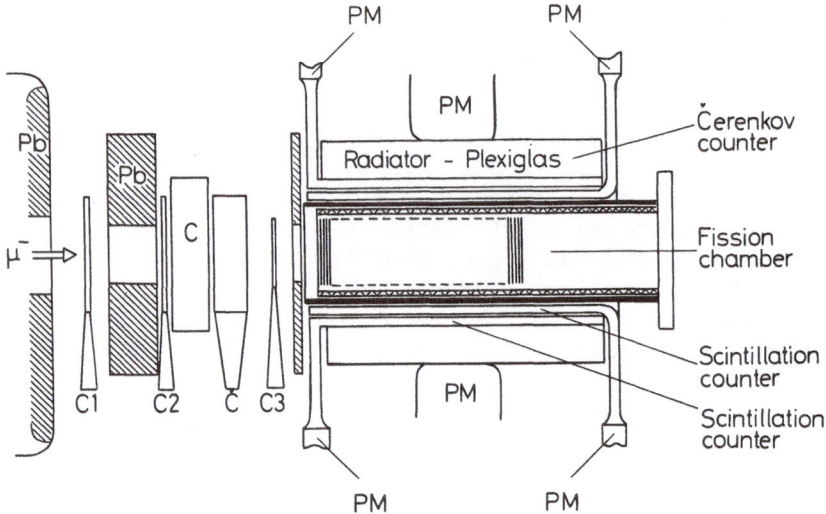

Fig. 5   Schematic view of the apparatus for the study of muon
          induced fission.

EXPERIMENTAL RESULTS

Muon attachment to fission fragments

     As was mentioned above, the muon attachments to fission frag-
ments were first observed in the experiments at CERN[7]. The experimental set-up is shown in Fig. 5. Fission fragments were detected
by a multiplate ionization chamber. Plastic and Cerenkov counters
surrounding the ionization chamber were used to detect electrons
from the muon beta-decay.

     The time distribution of fission events is presented in Fig. 6.
The prompt peak is due to radiationless muon transition; the delayed
fission is caused by muon capture. The muon survives in prompt
fission and its later fate is determined by the properties of the
muonic atoms of the fission fragments.

     Figure 7 shows the time distribution of electrons from muon
beta-decay as measured in coincidence with the prompt fission. A
lifetime of 134 ns is observed for the ^{238}U target. Figure 8 presents the so-called Primakoff plot. The muon capture probability
is given as a function of N/2A (N - neutron number, A - mass number).
Two shadowed squares correspond to the two extreme possibilities.
The muon is either attached to the heavy or to the light fission

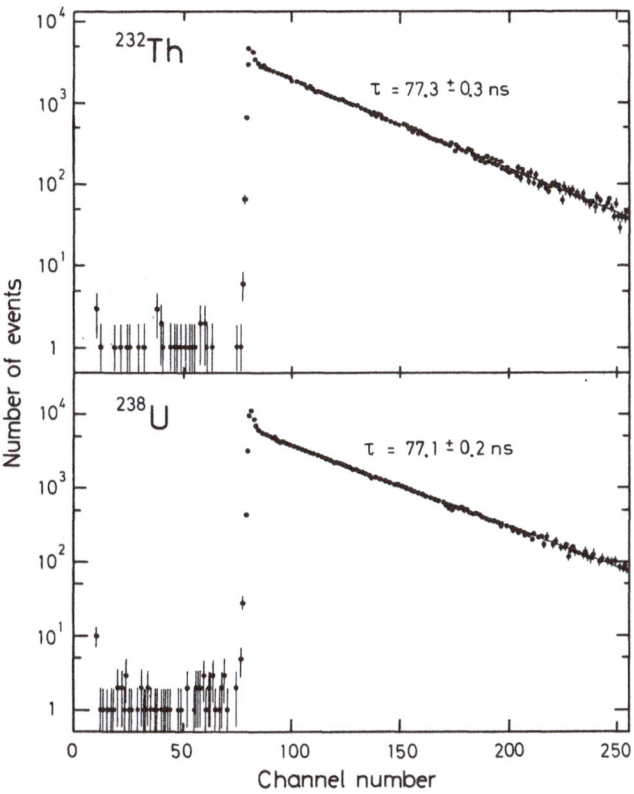

Fig. 6 Time distribution of fission events.

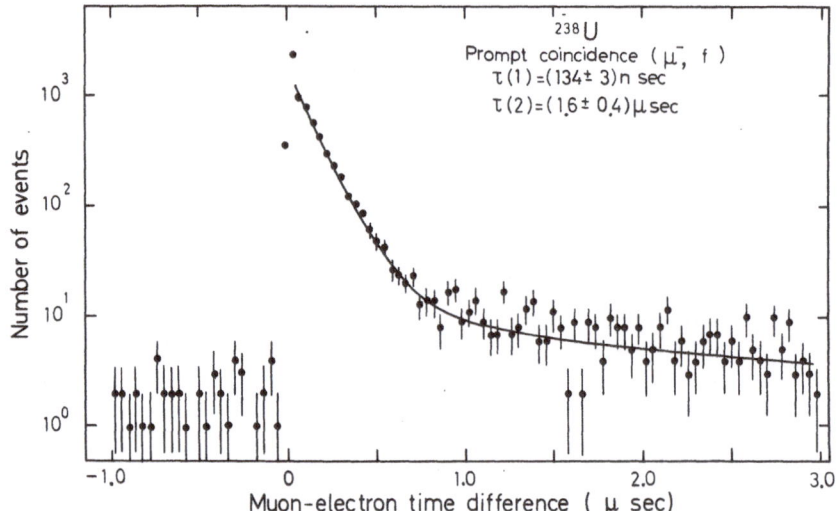

Fig. 7   Time distribution of electrons from muon beta-decay
         (prompt fission).

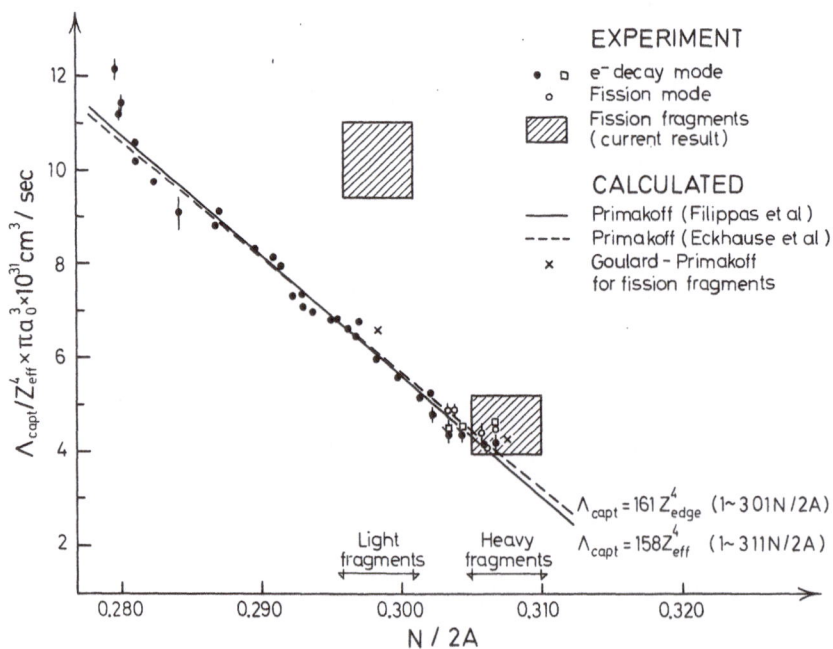

Fig. 8   Primakoff plot.

fragment. From the experimental value $\tau = 134 \pm 4$ ns the conclusion is drawn that muons are attached mainly to the heavy fragments. At the present level of accuracy one cannot eliminate a few percent of muons attached to the light fission fragments.

The Rochester group which have studied neutrons emitted when muons interact with uranium have measured the lifetime for the muonic atoms in fission fragments and observe a value consistent with that measured at CERN[11].

Muon conversion

As follows from the CERN results (Fig. 7) the component in the electron time distribution with the lifetime 1.9 ± 0.6 μs is present too. This lifetime might be explained by suggesting that a muon is converted from the muonic atom of a highly excited fission fragment and stopped in aluminum, forming a muonic aluminum atom. At first sight the muon conversion is hardly expected as the muon binding energy is much larger than that for the electron. However, the overlap of the muon wave function with the nuclear wave function is considerable which makes muon conversion possible.

Some experiments with nuclear emulsions irradiated by muons support the CERN results[12]. It is claimed here that most of the muons converted have been emitted from the muonic atoms of the light fission fragments. This conclusion is not sufficiently justified and further experiments are needed.

Muon induced quadrupole photofission

In fact, fission due to radiationless muon transition is nothing more than photofission in the presence of a muon in the 1s orbit. Until recently, the role of different radiationless transitions for actinides was not clear. Recent experiments carried out at SIN[13] have shown the dominating role of 3d → 1s radiationless transitions for ^{238}U. This conclusion was drawn from the comparison of the muonic X-ray spectra measured in coincidence both with prompt and delayed fission. The spectra of X-rays are presented in Fig. 9. One can see that 3d → 2p transition is suppressed for prompt fission. The analysis shows that not less than 50% of all prompt fissions are induced by 3d → 1s radiationless transition. The comparison with the known data for photofission[13] indicates that Γ_n/Γ_f is increased by a factor ~ 10 for muon induced fission. If not, taking into consideration the possible hindrance of the fission channel for quadrupole giant resonance[14] (the energy of

3d → 1s transition equals 9.5 MeV and is on the slope of this resonance) and attributing all the effect to fission barrier augmentation due to muon Γ_n/Γ_f increase, this can be easily explained. In
fact, the level density at the top of the outer barrier is lower in
the presence of muons and consequently Γ_f decreases. At the same
time, neutron evaporation is not affected at all in the presence of
a muon in the 1s orbit.

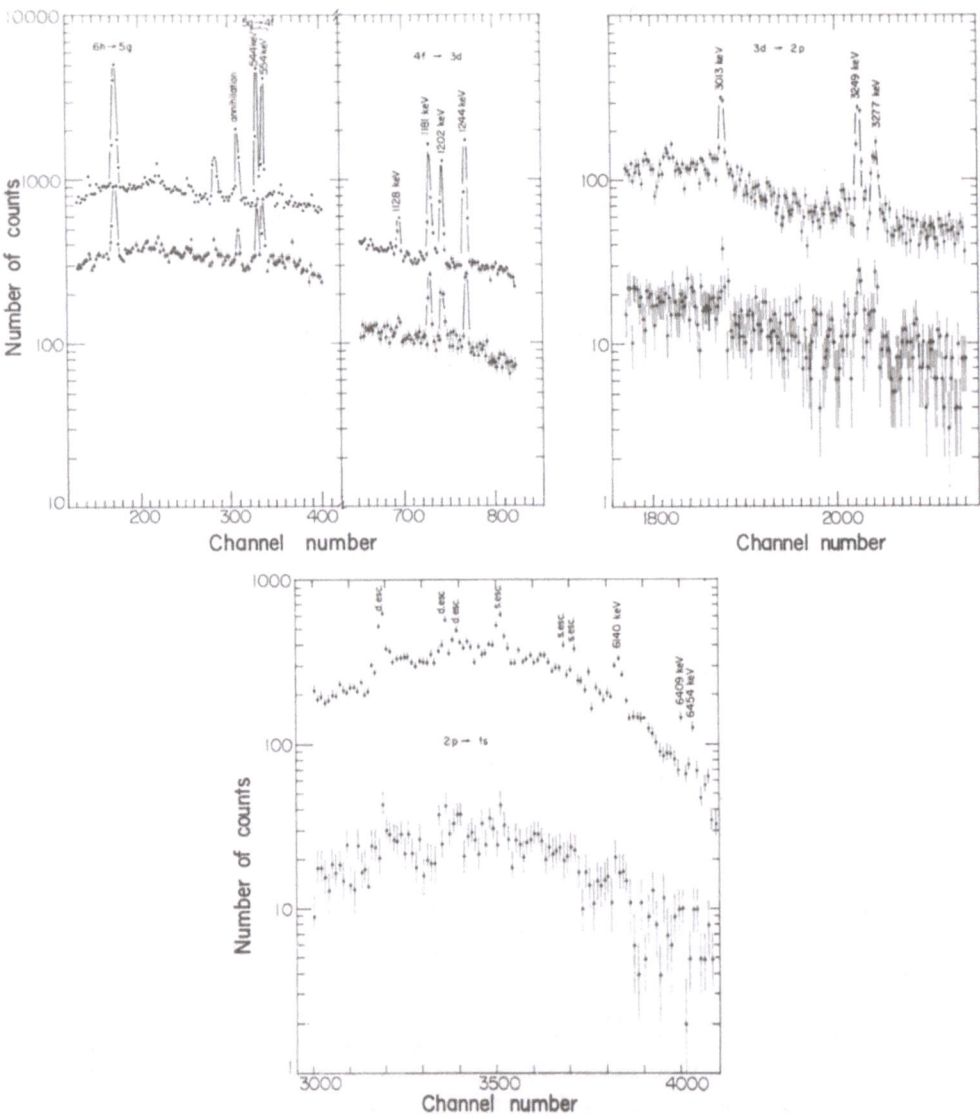

Fig. 9 Muonic X-ray spectra measured in coincidence with fission.

Muonic atoms of shape isomer

One can expect that with a certain probability the second well will be populated due to radiationless muon transitions. That will lead to the formation of muonic atoms of shape isomers. Until now there has been no good evidence of such a process. Some attempts to observe muonic atoms of ^{238}U shape isomer were undertaken at CERN[14]. There are three possible modes of the decay for muonic atoms of ^{238}U shape isomers:

1. spontaneous fission and subsequent beta-decay of the muon attached to one of the fission fragments;

2. beta-decay of the muon and subsequent spontaneous fission;

3. backtunneling (γ decay in the first well) and subsequent fission after muon capture.

In the experiments done at CERN triple coincidences were measured (μ-e-f).

Figure 10 presents the time distribution of fission events for the decay mode 1. A peak for prompt fission is clearly seen. There is no evidence for spontaneous fission isomers.

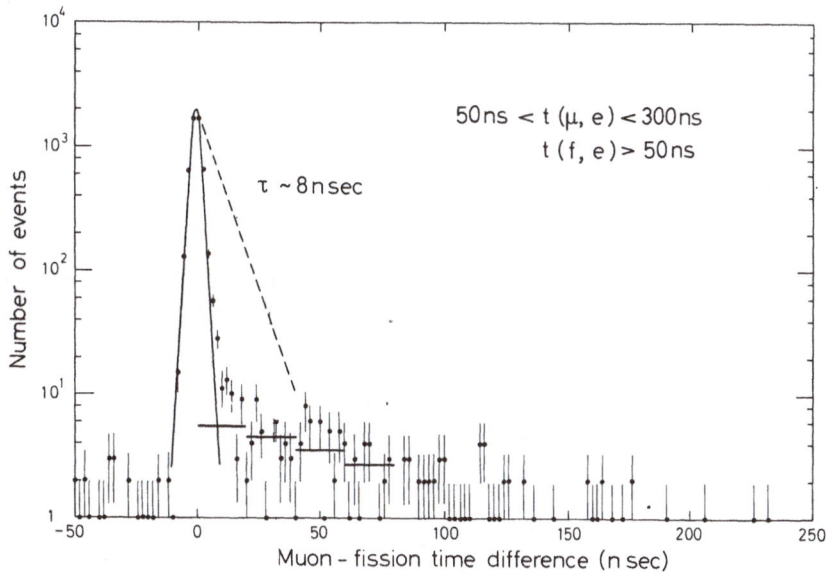

Fig. 10   Time distribution of fission events measured in coincidence with electrons from muon beta-decay.

SOME PROSPECTS

The better accuracy in experiments on the study of muon attachment to the fission fragments is desirable. One can hope that by
modifying the equipment in the proper way it will be possible to
find out to what extent the muon is attached to the light fission
fragment. Hopefully that is one of the ways to estimate the
viscosity of nuclear matter.

Better accuracy is needed also for missing X-ray studies.
Doubtless here ^{238}U is of interest to find out if $2p \rightarrow 1s$ radiationless transitions can result in fission.

It seems attractive to study fission induced by muon transfer
from muonic hydrogen atoms. In these experiments hydrogen should
be used as a target. The advantage of this method is the following.
Very thin targets of fissile isotopes can be deposited on the surface
of a fission fragment detector and this detector, placed in hydrogen, will be used to observe muon induced fission. That makes it
possible to study muon induced fission of many isotopes from very
unstable to alpha particle emission or spontaneous fission.

REFERENCES

1. V.M. Strutinsky, Nucl. Phys. A95 (1967) 420.
2. J.A. Wheeler, Rev. Mod. Phys. 21 (1949) 133.
3. D.F. Zaretsky et al., Nucl. Phys. 28 (1961) 177.
4. J.A. Diaz et al., Nucl. Phys. 40 (1963) 54.
5. C.K. Hargrove et al., Phys. Rev. Lett. 23 (1969) 215.
6. G. Leander and P. Möller, Phys. Lett. 57B (1975) 295.
7. Dz. Ganzorig et al., Phys. Lett. 77B (1978) 257.
8. P. Olanders et al., Phys. Lett. 90B (1980) 193.
9. Z.Y. Ma et al., Preprint, Nuclear Science Division Lawrence
    Berkeley Laboratory, University of California, Berkeley 1975.
10. J.A. Maruhn et al., Phys. Rev. Lett. 44 (1980) 1576.
11. W.W. Wilcke, Phys. Rev. C21 (1980) 2019.
12. J.T. Caldwell et al., Phys. Rev. C21 (1980) 1215.
13. T. Johansson et al., to be published.
14 J. van der Plicht et al., Phys. Rev. Lett. 42 (1979) 1121.

FEW PROBLEMS IN EXPERIMENTAL NUCLEAR ASTROPHYSICS *

Claus Rolfs

Universität Münster, Institut für Kernphysik, W. Germany
and
Ohio State University, Physics Department, USA **

LECTURE 1: " The Missing Solar Neutrinos "

Our planet earth is warmed by the prodigious energy output of
a nearby star, our sun. This typical and most important star has
been studied by scientists over the centuries with ever improving
techniques. From these conventional optical studies one knows its
mass, luminosity, radius, surface temperature, surface composition
and age more accurately than for any other star. An answer to the
question "what goes on in the interior of the sun" was not possible
until 1920, when Eddington suggested that the energy which powers
the sun comes from the fusion of the most abundant species hydrogen into helium

$$4 \text{ H} \rightarrow 1 \ ^4\text{He} + 2 \ e^+ + 2 \ \nu$$

with an energy release of about 27 MeV. The actual mechanism of
this fusion was elucidated not before the late thirthys, when
Bethe, Critchfield and Weizsäcker showed that two sets of reactions
can provide this fusion, namely the p-p chain and the CNO-cycles.
The mass of a given star dictates hereby which of the two sets represents the predominant stellar energy source.

---

* Two lectures delivered at The 1980 International Summer School
  on Nuclear Structure, Dronten, The Netherlands, 12-23 August 1980
  are scheduled for publication by Plenum Publishing Company.
**Present address

With this hypothesis on the solar energy source, the well-known external properties of the sun and a few standard physical and astrophysical principles, a solar model can be constructed. The more recent standard solar model calculations of Bahcall and others (1) lead to a temperature of $T = 15 \times 10^6$ ^{0}K and a pressure of $P = 2.2 \times 10^{11}$ bar at the center of the sun. These calculations suggest also that the sun runs mainly on the p-p chain as the solar energy source.

These interesting results required however a feature commonly applied in physical praxis: the experimental verification. Conventional astronomical instruments can only record the photons emitted by the outermost layers of the sun. The energy created by the thermonuclear furnace deep in the sun's interior is transformed into photons, which have there a mean free path of much less than a centimeter. The energy transport to the sun's surface involves furthermore a time scale of the order of several million years. Therefore, the sunlight observed on earth does not provide an unmodified picture of the sun's interior (hidden by an enormous mass of cooler material) nor of its properties today.

Of the particles released by the assumed thermonuclear reactions in the solar interior, only one type has the ability to penetrate directly and unmodified from the center of the sun to its surface and escape into space, where a fraction of them will arrive (about 8 minutes later) on earth: the neutrinos. Thus neutrinos are the only solar spies, which offer a unique possibility of "looking" into the solar interior today and testing the above results of the solar model calculations. Moreover, the sun is in the best understood stage of stellar evolution, the quiescent main sequence phase. If one has to have confidence in the many astronomical and cosmological applications of the theory of stellar evolution, an experiment designed to capture the solar neutrinos ought at least to give the right answers for the sun.

R.Davis,Jr. and his coworkers (2) have developed a radio-chemical method of detecting solar neutrinos via the basic reaction
$$\nu + {}^{37}\mathrm{Cl} \rightarrow e^- + {}^{37}\mathrm{Ar}$$

using a huge tank with 610 tons of perchlorethylene, placed in the Homestake gold mine (1.5 km underground) in Lead,South Dakota. This detection method is particularly sensitive (1) to the high energy neutrinos from the ^{8}B decay, where ^{8}B is produced in the p-p chain via a small reaction branch $^3\mathrm{He}(\alpha,\gamma)^7\mathrm{Be}(p,\gamma)^8\mathrm{B}$. In turn, these neutrinos are quite sensitive to the solar temperature (and solar model). Therefore, the detector has been named often a solar thermometer.

The observed neutrino rate (up to 1978) of N_{exp} = 2.2 ± 0.4 SNU*
is in conflict with the predicted rate (1) of 5.0 ± 1.4 SNU. This
discrepancy is known as the "Solar Neutrino Problem" and has led to
a crisis in the theory of stellar evolution; many authors have open-
ly questioned some of the basic principles and approximations in
this supposedly "solved" subject.

What went wrong? Many solutions have been suggested, which led
to a new field in physics, the "Fantastronomy". Is it (i) the radio-
chemistry (ν-detection), (ii) the nuclear physics (ν-production),
(iii) the atomic physics (opacity, energy transport), (iv) the ν-
properties (decay or oscillations), (v) the astrophysical principles (solar model), (vi) the Solar-OPEC (energy production reduced
today in the sun), (vii) the sun as a very very special and not-so-
typical star (theology), etc. ??? Are there some NEW physical laws
covered under the sun's surface?

With regard to possible nuclear physics aspects of the problem,
one has to realize an important drawback: the cross sections of the
nuclear reactions involved in the p-p chain cannot be measured in
the laboratory directly at the relevant solar energies (about 10-
20 keV), even with the best techniques available today. The usual
procedure involves therefore the measurement of a given reaction
over a wide range of beam energies and to extrapolate then the results down to the relevant stellar energies under the guidance of
theoretical and other arguments. Of course, if the direct measurements can be extended to still lower energies with sufficient precision, the extrapolated rates will stand on an improved ground.
Such measurements frequently require great ingenuity in the development of novel techniques and in the painful search and investigation of background and other problems as well as of systematic
errors in the analyses of the raw data.

The predicted capture rate in the Davis neutrino detector is
quite sensitive to the nuclear cross sections of the reactions
$^3He(\alpha,\gamma)^7Be$ and $^7Be(p,\gamma)^8B$ at solar energies. Both reactions have
been measured several times at the laboratory of Caltech. The
$^7Be(p,\gamma)^8B$ reaction measured as low as $E_{c.m.}$ = 160 keV (Kavanagh
et al.,1969,ref.3) has been rechecked by Wiezorek et al.(1977,ref.4)
at Münster, with results in fair agreement with the earlier (unpublished) Caltech work. The $^3He(\alpha,\gamma)^7Be$ reaction has also been investigated at Caltech as low as $E_{c.m.}$ = 180 keV (Parker and Kavanagh,
1963,ref.5) and $E_{c.m.}$ = 164 keV (Nagatani et al.,1969,ref.6). The
calculated neutrino rate depends approximately as $S^{0.8}$ from the
extrapolated astrophysical S-factor of this reaction. Although
the   measurements of the $^3He(\alpha,\gamma)^7Be$ reaction have been carried
out at the time with the best available techniques, the improved
* 1 SNU = 1 Solar Neutrino Unit = 10^{-36} captures per target atom
  per second

technologies available today made it desirable to recheck again
this reaction and to carry the measurements eventually to significantly lower beam energies.

Since there are no compound states in ^{7}Be near the ^{3}He+^{4}He threshold (Fig.1), the process will be dominated by the non-resonant,
direct capture mechanism into the groundstate and 429 keV first excited state of ^{7}Be, provided both states have a significant ^{3}He $\otimes$ ^{4}He
cluster configuration. One expects therefore to observe two primary

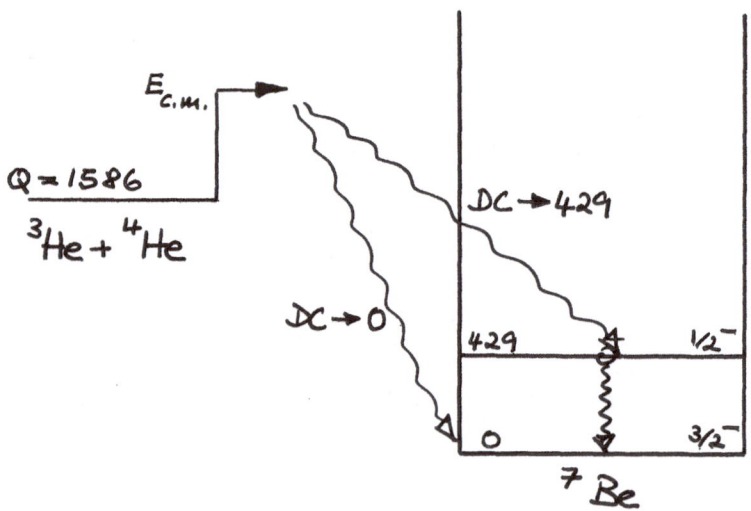

Fig. 1: Schematic diagram of the ^{3}He(α,γ)^{7}Be reaction.

γ-ray transitions, DC → 0 and DC → 429 keV (varying in energy with
the beam energy) and a 429 keV (isotropic) secondary transition.
Since the capture cross section above the Coulomb barrier is expected to be only a few μbarn (for C^2S= 1), measurements far below the
Coulomb barrier ($E_{c.m.} \leq$ 150 keV) will be in the nano-to-pico barn
range.

The experiments have been started in spring 1978 at the 350 kV
Münster accelerator (up to 150 μA particle current at the target)
and continued since then at Münster as well as at the 4 MV Dynamitron accelerator in Stuttgart. The experiments and data analyses
are still going on, but will get in a final stage by the end of
this year.

The experimental program can be grouped into three major activities:

(i)   the measurement of excitation functions over a wide range of
      beam energies and as low in energy as technically feasible
      (lowest measurement at $E_{c.m.}$ = 107 keV,June 1980);
(ii) the measurement of γ-ray angular distributions at several
      selected beam energies over the entire beam energy range in-
      vestigated (done yet at $E_{c.m.}$ = 148 and 197 keV,July 1980);
(iii) the measurement of absolute cross sections at low and high
      beam energies,involving two different types of target geo-
      metries (done yet at $E_{c.m.}$ = 145 and 197 keV,June/July 1980).

For reasons of length and time, not all aspects of these activities can be presented here. Rather, only a few examples out of
these activities will be presented here,illustrating the ways and
methods used in solving some of the many problems involved in this
"Münster-Monster" enterprise.

(i) Measurements of excitation functions

The measurements of excitation functions have been carried out
with an 80 cm^3 Ge(Li) detector at 90°, a disc-shaped target chamber and a three-stage differentially pumped gas target system
(Fig.2). Some details of this system are described in ref.7 and of
the experimental set-up and procedures in ref.8. In the more recent measurements (June-August 1980), 80 cm^3 Ge(Li) detectors have
been placed on each side of the target chamber and a fourth pumping stage has been added to the system.

Sample γ-ray spectra covering almost the entire beam energy
range investigated are shown in Fig.3. As the beam energy is varied,
the two primary transitions change their energy accordingly (e.g.,
E_γ(DC → 0) = Q + $E_{c.m.}$). All three γ-ray transitions are Doppler
broadened due to the extended target chamber used (Fig.2). At low
beam energies room and cosmic background reduces the signal-to-
noise ratio, while at high beam energies the neutron induced γ-ray
background from the $^{13}C(\alpha,n)^{16}O$ contaminant reaction (induced by
the α-beam on the Ta-canals and the beam stop of the system,Fig.2)
hampers the measurements.

Since at low beam energies the cross section drops very rapidly
with decreasing beam energy, an accurate knowledge of the effective
beam energy associated with the observed γ-ray yields is as important as the yield measurements themselves. Due to the thin targets
used (gas pressure in the target chamber = 2.0 Torr), the centroid e.g.
of the DC → 0 γ-ray transition provides this important information
to a high accuracy (better then 1.5 keV) "on the house".

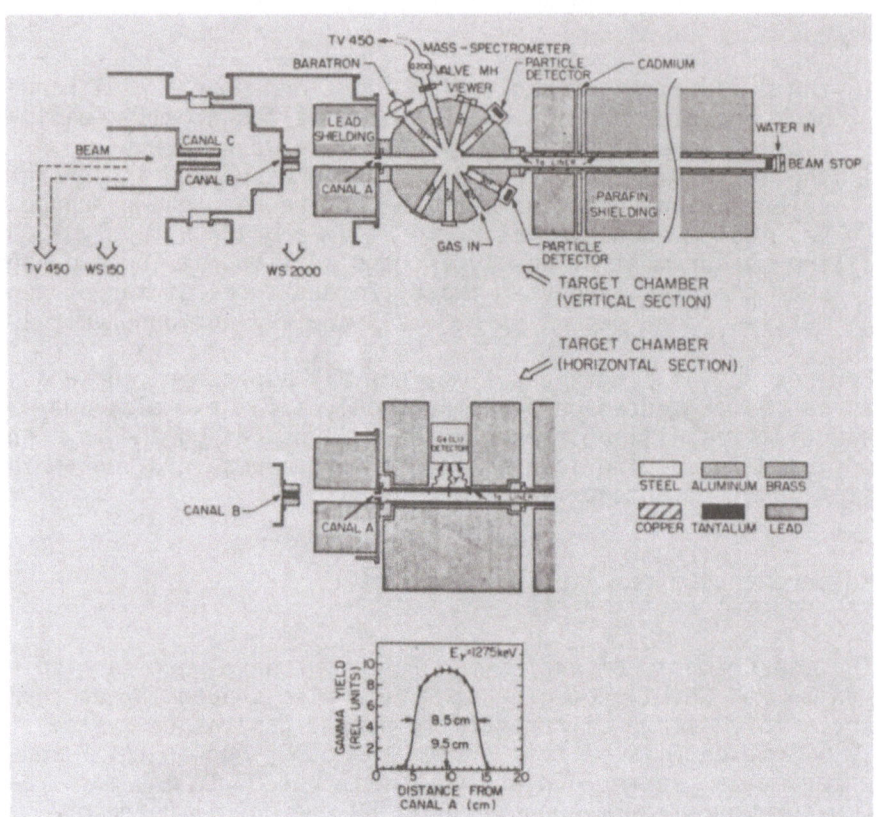

Fig. 2:Schematic diagram of the experimental set-up used in the
        measurements of excitation functions for the ^{3}He(α,γ)^{7}Be
        reaction. The 2.0 Torr target gas pressure is reduced
        to 5×10^{-2}, 1×10^{-4} and 5×10^{-7} Torr along the subsequent pum-
        ping stages. For details,see ref. 7 and 8.

    The conventional method requires a precise knowledge of the
absolute projectile energy, its energy loss in the gas target
system and target chamber as well as the influence of intense
beams on gas target densities along the beam axis. The latter feature has been investigated experimentally at Münster (9). The important quantity is the dissipated power of the beam in the gas and effects greater than 10% occur already for a dissipated power greater than
20 mWatt/mm (Fig.4). For the 2.0 Torr gas pressure used and a 150
μA beam current, only effects less then 10% are expected therefore
in the studies of the ^{3}He(α,γ)^{7}Be reaction.The absolute beam energy
of the Münster accelerator is known to an accuracy of ± 0.4 keV (10).

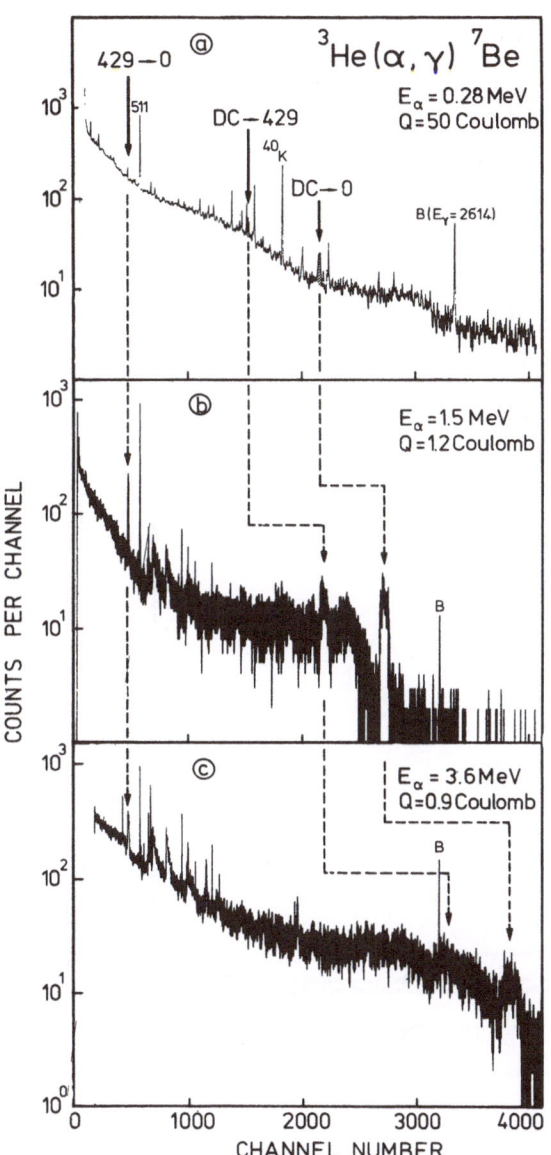

Fig. 3:Sample γ-ray spectra obtained in the set-up of Fig.2

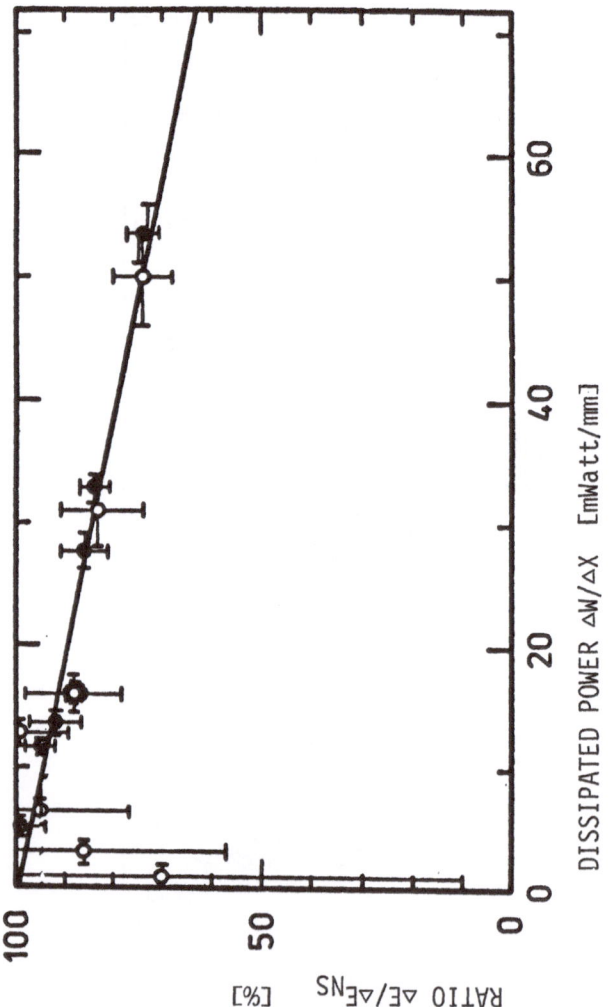

Fig. 4: The effects of intense beam currents on gas target densities are shown. The observed to expected energy loss of the beam in the gas (here nitrogen) is plotted as a function of dissipated power of the beam in the gas. The data have been obtained for different beam currents and target pressures , also in different target geometries. For details, see ref.9.

Experiments have also been carried out to obtain energy loss data at
low beam energies, with results deviating significantly from the
Northcliff-Schilling tables. All these investigations provided finally effective beam energies, which were in excellent agreement
with the results deduced from the above γ-ray method.

     For most of the measurements, the number of projectiles has been
determined via the elastic scattering yield at 30° (Fig.2), where
the deviation of the scattering yield from the Rutherford law had
been measured in additional experiments. Since June 1980, a beam
calorimeter has also been used in the experiments.

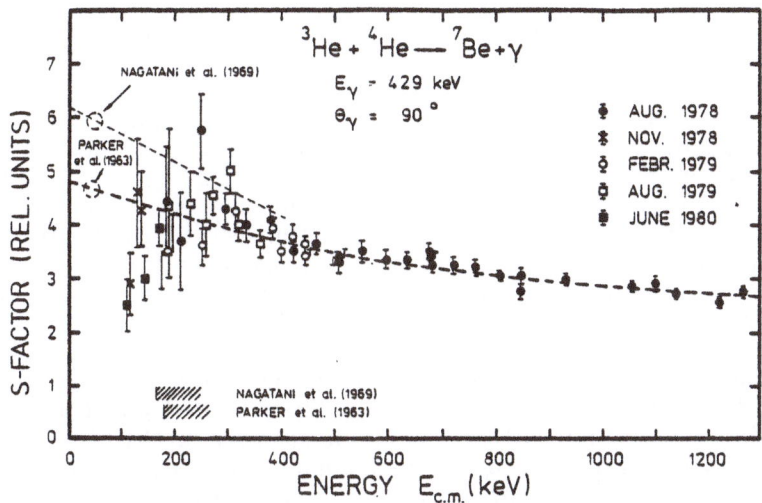

Fig. 5:Observed excitation function for the isotropic 429 keV secon-
       dary transition. The data are presented in form of the astro-
       physical S-factor (in relative units).

     Since the 429 keV secondary transition is isotropic (Fig.1), the
observed yields for this transition at 90° (Fig.2) represent already
angle integrated yields for the DC → 429 keV branch of the capture
reaction. The resulting excitation function for this branch only is
presented in Fig. 5 in form of the astrophysical S-factor. These

data suggest a flat if not decreasing S-factor at the extreme low
beam energies. The 90° data for the DC → 0 branch (Fig.1) have to
await the results of γ-ray angular distribution measurements (see
below).

(ii)   The supersonic JET gas target system for γ-spectroscopy
----

      For the measurements of γ-ray angular distributions in the
$^3He(\alpha,\gamma)^7Be$ reaction (Fig.1), a nearly point target of sufficient
density is required, which can be achieved by a supersonic JET system. The inlet gas at several bar pressure flows through a nozzle
(about 1 mm diameter), expands then freely over a distance of 6 mm
and is finally captured by a receiver, whose diameter has to be adjusted experimentally (about 6 mm diameter). The gas is then com-
pressed through a series of pumps up to the inlet pressure. The
gas outside the JET is pumped differentially and fed back into the
recirculation system.

      Gas recirculation is essential if isotopic enriched gases such
as ^3He are used, otherwise the required amount of ^3He gas would
exhaust a large fraction of the total research budget of the For-
schungsministerium (or DFG). Running an intense ^3He beam instead
over long experimental periods (several months) is also not cheap,
aside from the increased γ-ray background through $(^3He,p\gamma)$ or
$(^3He,n\gamma)$ contaminant reactions created by the ^3He beam on the canals
and the beam stop of the gas target system.

      For the gas recirculation, the following requirements on the
pumps had to be fullfilled: (i) high pumping speed and compression
ratio, (ii) small death volume, (iii) high leak tightness, (iv)oil-
freedom, (v)high stability over long running periods. For the JET-
target itself, one had to prove that a real point target was available for γ-ray measurements, i.e. that yield contributions from
the outside regions of the JET were negligible. It should be noted,
that these contributions cannot be eliminated by collimators or
shieldings as in particle spectroscopy measurements. This feature
had also to be verified especially for light gases.

      The technical development of such a JET system has been started
in spring 1979. The story of the many failures, new approaches and
various test phases is too long to be presented here in detail and
in fact it was not until this early summer that all the above requirements could be finally fulfilled. The JET gas target system
involves now 17 pumps (roots blowers,turbo pumps and compressors).
A schematic diagram of the system is presented in Fig.6. The pressure reduction in the system (for two different gases) is illus-

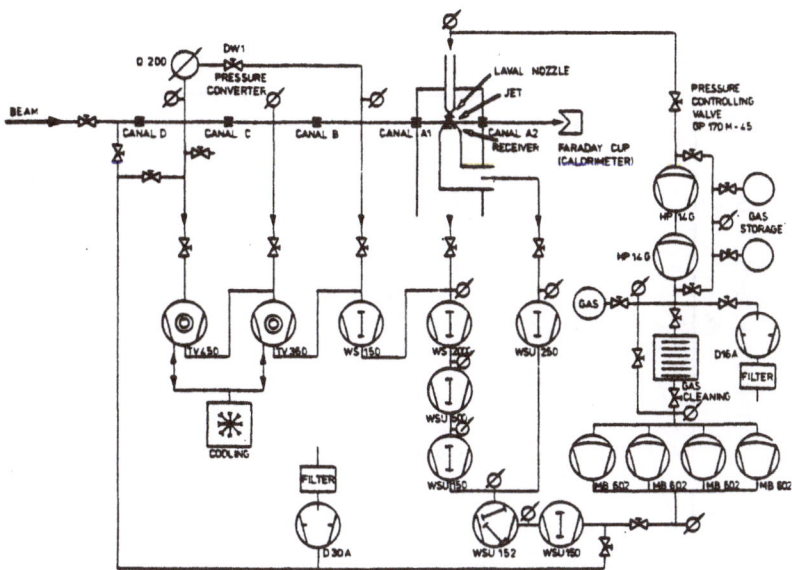

Fig. 6: Schematic diagram of the differentially pumped and recircu-
lating JET gas target system (5 stages).

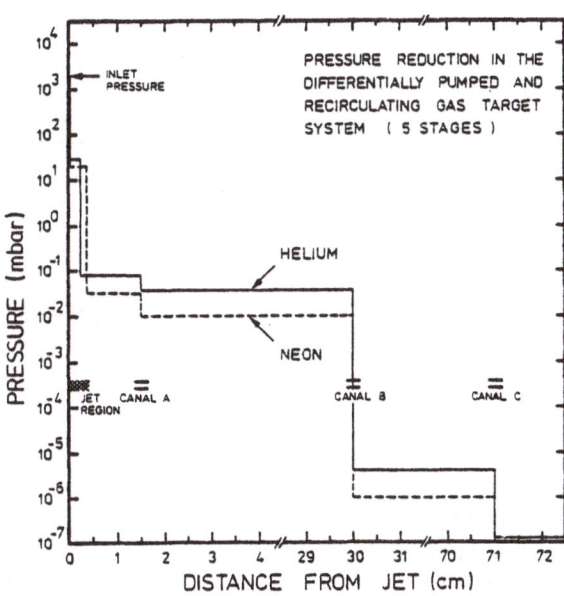

Fig. 7:Pressure reduction in the JET gas target system for an inlet pressure of 2 bar.

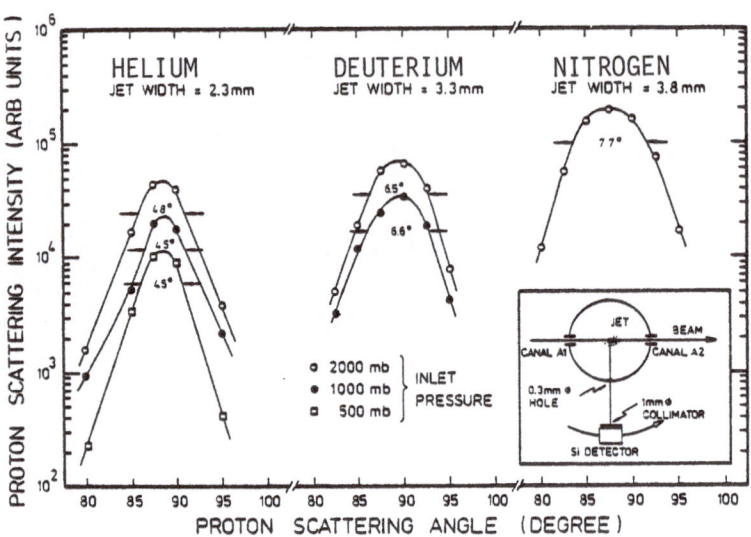

Fig. 8: Illustrating examples of the JET profile for a few gases.

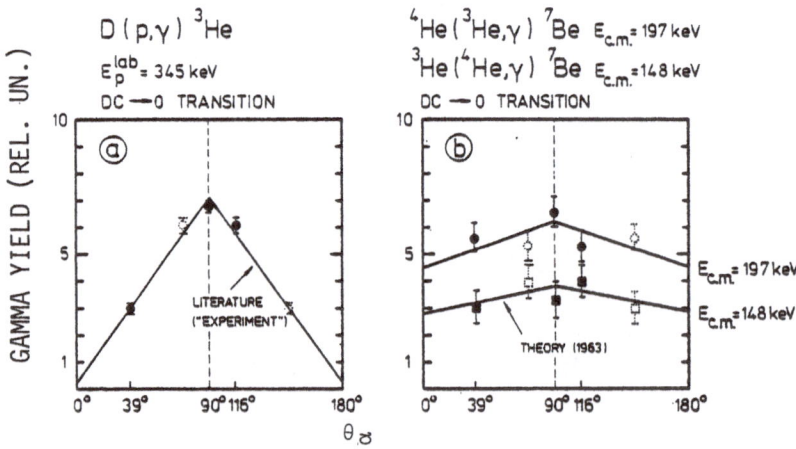

Fig. 9: The results of γ-ray angular distribution measurements using
the JET gas target system and three Ge(Li) detectors are
shown for the reactions d(p,γ)^{3}He and ^{3}He(α,γ)^{7}Be.

trated in Fig.7 and some of the tests on the JET profile in Fig.8. The resulting γ-ray angular distribution for the $d(p,\gamma)^3He$ reaction (a test of a light gas) is in good agreement with the literature (Fig.9a). Finally, the observed angular distributions of the DC $\rightarrow$ 0 transition in $^3He(\alpha,\gamma)^7Be$ at $E_{c.m.}$ = 148 and 197 keV are illustrated in Fig. 9b. The results are nearly isotropic, but also not in conflict with the predicted pattern (11). Such measurements will be continued at the higher beam energies (Stuttgart,Oct./Nov.1980).

As a summary of this first lecture, the many aspects in the experimental investigation of the $^3He(\alpha,\gamma)^7Be$ reaction appear now to be under control. However, only when all necessary data are obtained and analyzed, an answer will be possible to which extent the solar neutrino problem is localized in part in the $^3He(\alpha,\gamma)^7Be$ reaction. It is hoped, that a full description of all these investigations will be available in spring 1981 (to be submitted to Zeitsch. f.Phys.,ref.12).

LECTURE 2: " Heavy Ion Burning "

The theory of nucleosynthesis in the universe (13,14) has been able to explain many of the observed properties of stars and other astronomical objects, and in particular numerous details of the abundance distribution of the chemical elements and their isotopes. Under conditions of high temperatures and densities, such as in stellar interiors, nuclear reactions are believed to be predominantly responsible for the energy generation, the temporary stability and the dynamical evolution of these bodies. Nuclear burning proceeds from the conversion of hydrogen to helium (Lecture 1), then from helium to carbon and oxygen, and then from these to heavier elements through the intermediate operation of heavy ion burning: ^{12}C + ^{12}C, ^{12}C + ^{16}O and ^{16}O + ^{16}O. These heavy ion reactions play an important role in the behaviour of highly developed stars and they determine how stars evolve and produce many of the chemical elements, how massive stars explode as supernovae, and what sort of remnants these stars leave behind (black hole, neutron star or white dwarf).

(i) The ^{12}C + ^{12}C Fusion Reaction at Subcoulomb Energies

The ^{12}C + ^{12}C fusion reaction exhibits pronounced intermediate resonance structures near and below the Coulomb barrier. Independently of these resonance structures, the reaction yield - expressed in terms of the modified form of the S-factor, $\tilde{S}$ = $S(E)exp(-gE)$ with g = 0.46 MeV^{-1} - shows an anomalously steep rise with decrea-

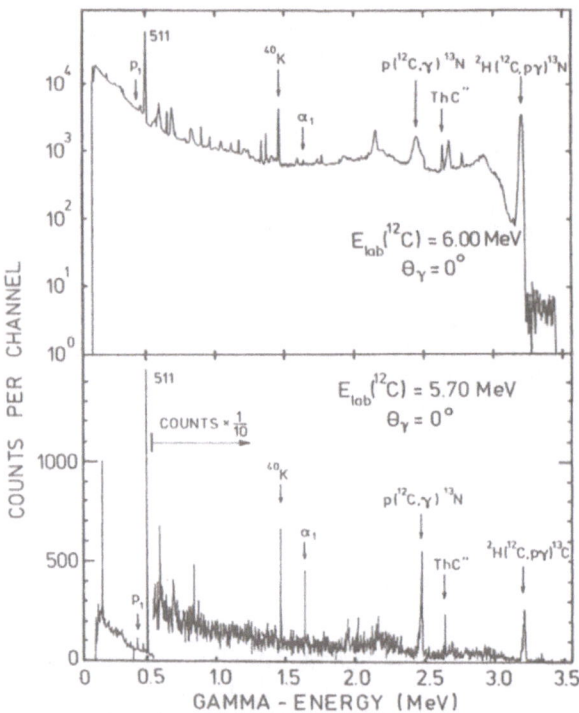

Fig.10: Sample γ-ray spectrum (top) obtained with a standard carbon
        target. The spectra are dominated by background radiation
        created by the interaction of the beam with the hydrogen
        contamination in the target (few atom %). The lower part
        shows a spectrum obtained with a carbon target of low hy-
        drogen contamination (0.02 atom %).

sing beam energy (ref.15 and Fig.11).This steep rise could not be
explained (14) in the standard Coulomb-barrier-penetration model
(including optical model parameters) and has been interpreted as
the high-energy wing of a broad resonance, so that the $\tilde{S}(E)$-factor
may be again a decreasing function of energy around $E_{c.m.}$ = 1 MeV.
However, Michaud (16) has cited this steep rise as evidence for
a new phenomenon known as "absorption under the barrier". This
phenomenon is related to the question whether the fusion of two
heavy nuclei can be initiated far out in the teneous tail of the
nuclear mass distribution, in what might be called the nuclear
stratosphere. In this case the rise in the $\tilde{S}(E)$-factor will continue, resulting in a quite different extrapolated rate at stellar
energies (different by more than two orders of magnitude) than
that rate obtained from the standard model.

The ^{12}C + ^{12}C fusion reaction was restudied (17,18) via γ-ray
spectroscopy with a 80 cm^3 Ge(Li) detector in close geometry (0°),

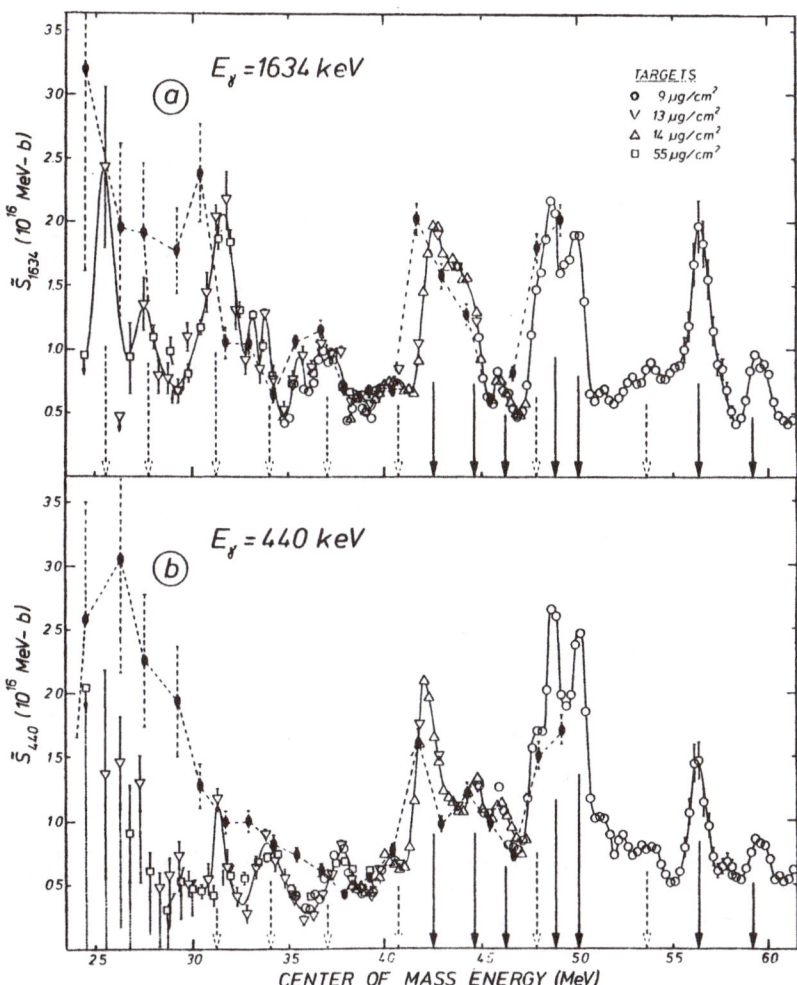

Fig.11:The results for the 1634 (a) and 440 (b) keV γ-ray lines are
compared with previous work (points connected by dashed lines,ref.15). The location of reported and new resonance
structures is indicated by solid and dashed vertical lines,
respectively.

using the intense ^{12}C beam (up to 15 particle μA) from the Bochum dynamitron tandem accelerator. Independent experiments have been carried out to determine with sufficient accuracy the absolute ^{12}C beam energies as well as the ^{12}C target thicknesses (evaporated onto Ta backings).

The ubiquitous hydrogen contamination in the targets (few atom %) seriously hampered the measurements below $E_{c.m.}$ = 3.5 MeV (Fig.10). Therefore, the production of targets with low hydrogen content (near 0.02 atom %) was necessary for the low energy measurements. This requirement was achieved (Fig.10) by (i) improved target production techniques, (ii) selection of graphit material with lowest intrinsic hydrogen contamination (...sorry,but the german graphit was the best: Ringsdorf Werke) and cooling the solid target by oil flow at an elevated temperature of 100° C. The results for the energy range investigated, $E_{c.m.}$ = 2.45 - 6.15 MeV, are shown in Fig.11. The resonance structures observed previously at higher beam energies persist down to low energies. These structures are superposed on a flat reaction yield, which does not show a strong increase at low energies, in contrast to previous work (15). There is some evidence that the disagreement is all or in part in the energy scales. The present results obviate the need of the hypothesis of absorption under the barrier at least down to $E_{c.m.}$ = 2.45 MeV and therefore support strongly the standard model (14).

The ^{12}C + ^{12}C fusion reaction has also been restudied recently (19) via particle spectroscopy at $E_{c.m.}$ = 2.75 - 6.30 MeV. The results confirmed the above conclusions.

If the observed resonances can be grouped into sets of rotational-vibrational bands, it will be essential to search in future experiments for intraband γ-ray transitions.

(ii) Comparison of the Fusion Reactions ^{12}C + ^{20}Ne and ^{16}O + ^{16}O
----

near the Coulomb Barrier

The existence of intermediate resonance structures in heavy ion induced reactions has been established for the fusion reactions ^{12}C + ^{12}C (see above) and ^{12}C + ^{16}O. However, experimental studies of other systems in the A = 9 - 28 mass range have given no clear indication for the presence of such structures. The physical explanation of the presence or absence of such structures is a long-standing problem in heavy ion physics and has been the subject of

continuing interest, speculations and many contrary opinions (20).
Michaud and Vogt (16,21) suggested a mechanism, in which the resonance structures correspond to intermediate alpha-cluster molecules.
On the basis of this model, it was suggested that the ^{12}C + ^{20}Ne
system should exhibit similar structures as the ^{12}C + ^{12}C system
due to the alpha-particle clustering of the low-lying states in ^{12}C
and ^{20}Ne.

In order to explore the influence of compound nucleus and entrance channel effects, which both can affect heavy ion induced
fusion cross sections, a comparison of different heavy ion reactions
forming the same compound nucleus should be available. The ^{12}C + ^{20}Ne
and ^{16}O + ^{16}O systems have similar reaction Q-values (Fig.12) and
represent therefore an attractive example in the search for such
effects, since nearly the same region of excitation energies in the
compound nucleus ^{32}S are reached. Since the colliding nuclei involve in one case deformed nuclei (^{12}C + ^{20}Ne) and in the other case
spherical nuclei (^{16}O + ^{16}O), these two reactions are also of parti-

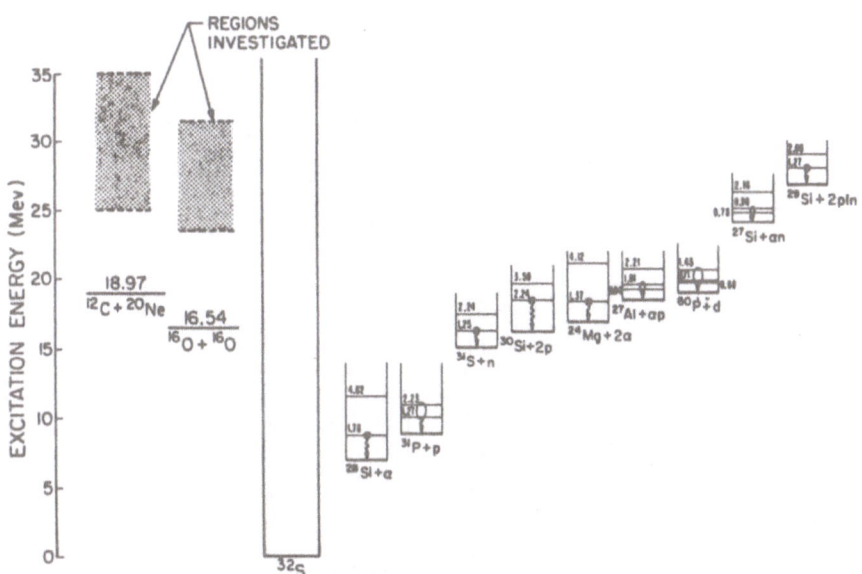

Fig.12: The energy regions investigated are shown for both the ^{12}C +
^{20}Ne and ^{16}O + ^{16}O fusion reactions. Indicated are also most
of the heavy residual product nuclei formed by particle evaporation of the compound nucleus ^{32}S. The key γ-ray transitions are shown, which characterize the heavy product nuclei.

cular interest in the search for geometrical effects (rigid or dyna-
mically induced deformation) in the entrance channel on fusion cross
sections. Such effects are expected to be visible in the energy dependence of both fusion cross sections particularly at subcoulomb
energies.

Due to the temporary presence of ^{12}C and ^{20}Ne nuclei in the core
of a star, ^{28}Si nuclei might also be synthesized in part by the
^{12}C + ^{20}Ne fusion reaction. In order to probe its astrophysical
importance, a knowledge of the ^{12}C + ^{20}Ne fusion cross sections at
subcoulomb energies was therefore desirable.

In view of the above nuclear and astrophysical aspects of the
^{12}C + ^{20}Ne fusion reaction, this process has been studied via γ-ray
spectroscopy with a Ge(Li) detector (Fig.2), at $E_{c.m.}$ = 6-15 MeV.
The ^{16}O + ^{16}O system has also been investigated at $E_{c.m.}$ = 7-14 MeV
in the same experimental set-up. The latter reaction was restudied
due to (i) its astrophysical importance, (ii) the existing discre-
pancies in the absolute fusion cross section and (iii) a reliable
comparison with the ^{12}C + ^{20}Ne data. The studies at subcoulomb energies were possible (8) with the use of intense heavy ion beams
(up to 10 μA particle current from the Bochum accelerator) in connection with pure and indestructable gas targets as well as of
high resolution Ge(Li) detectors.

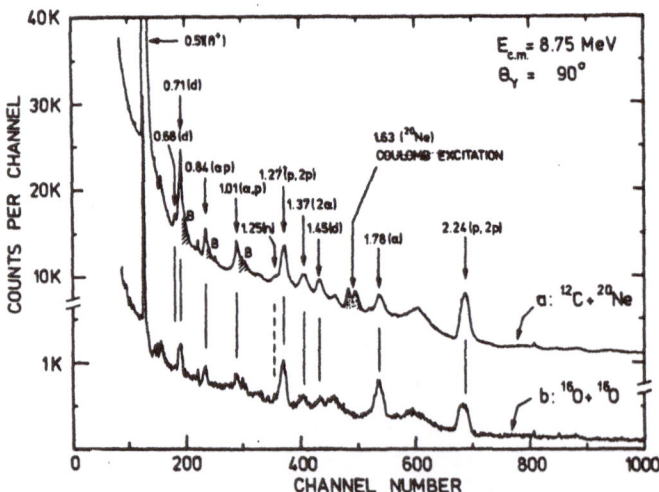

Fig.13: Shown are spectra obtained with the Ge(Li) detector for both
fusion reactions at the same energy.The lines are identified
by their energy and their association with the light particles
emitted in the exit channels (Fig.12). All lines exhibit substantial Doppler broadening.

Sample γ-ray spectra from both reactions at the same C.M. energy are shown in Fig.13. Due to similar reaction Q-values (Fig.12),the γ-ray spectra are expected to be nearly equal as is evident from the data. The exception is the 1.63 MeV line in the ^{12}C + ^{20}Ne system, which arises from Coulomb excitation of the first excited state in ^{20}Ne. This perniceous background line served however as an intrinsic monitor (at energies below the "safe energy", $E_{c.m.} \leq 8.1$ MeV) to obtain the absolute cross sections for the other γ-ray lines in both reactions. This internal monitor was quite welcome in view of the complex experimental set-up (Fig.2).

The resulting excitation functions for the ^{12}C + ^{20}Ne system are shown in Fig.14. They are rather smooth functions of energy and give no evidence for the existence of pronounced resonance structures. It is concluded that the alpha-cluster model of Michaud and Vogt alone does not provide a physical explanation for the presence or absence of intermediate resonance structures in heavy ion induced reactions. The observed smooth excitation functions for the ^{16}O +^{16}O system are in good agreement with the earlier results of Spinka and Winkler (22).

The absolute cross section for the ^{16}O + ^{16}O system at $E_{c.m.}$ = 11.85 MeV, σ = 450± 100 mb, is in good agreement with the previous result of Spinka and Winkler (400± 100 mb,ref.22) as well as the recent result of Wu and Barnes (390 mb,ref.23). The values obtained in the γ-ray work of Kolata et al. (210 mb,ref.24) and Cheng et al. (155 mb,ref.25) have to be considered as lower limits, since no corrections for missing yield nor summing effects (in the close geometries applied) have been taken into account.

The observed yield distribution of the evaporation residues, produced by both reactions at several comparable excitation energies in ^{32}S,are shown in Fig.15. Apart from some minor differences, the distribution is qualitatively similar and suggests that a compound nucleus ^{32}S is formed in both reactions and that direct reaction mechanisms are not apparent.

In the search for entrance channel effects on fusion cross sections, the results for both systems are compared in Fig.16, where the energy scales are matched (for a better comparison) according to the different heights in Coulomb barriers.If dynamic effects of nuclear deformation on fusion cross sections can be neglected, the fusion cross section for the rigid deformed system ^{12}C + ^{20}Ne is expected not to drop as rapidly at energies below the Coulomb barrier as that for the spherical system ^{16}O + ^{16}O. This expectation is however not verified by the observation (Fig.16),where an opposite trend in the energy dependence is indicated. The observation of ^{20}Ne Coulomb excitation, which is a dominant feature in the ^{12}C + ^{20}Ne system at low energies, may indicate,that dynamic deformation effects on heavy ion induced fusion cross sections are important.

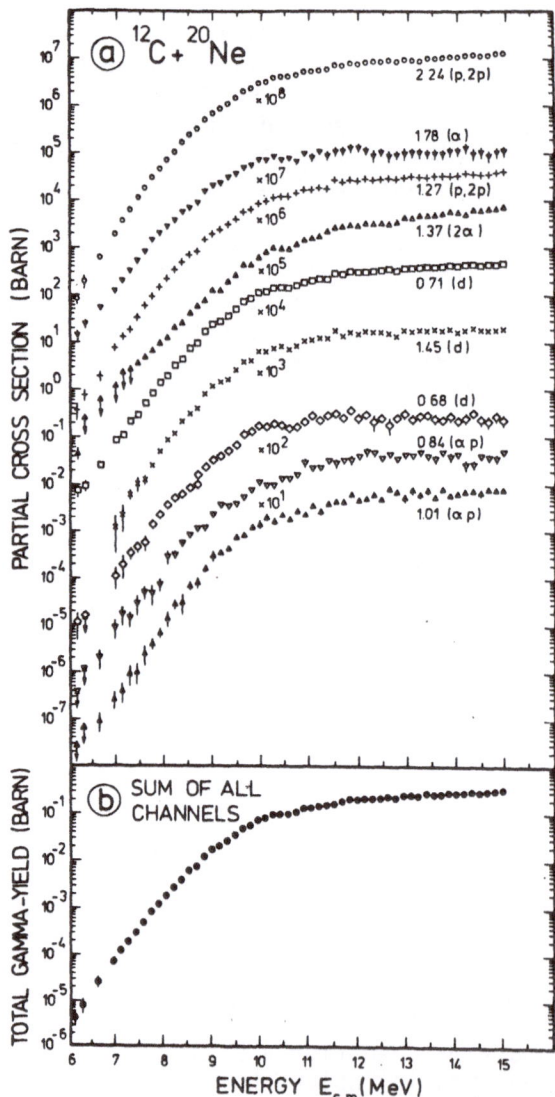

Fig.14: Shown are the excitation functions of the characteristic
        γ-ray transitions from the heavy residual nuclei produced
        in the ^{12}C + ^{20}Ne fusion reaction (Fig.12). The summed
        cross sections are given in the lower part of the figure.

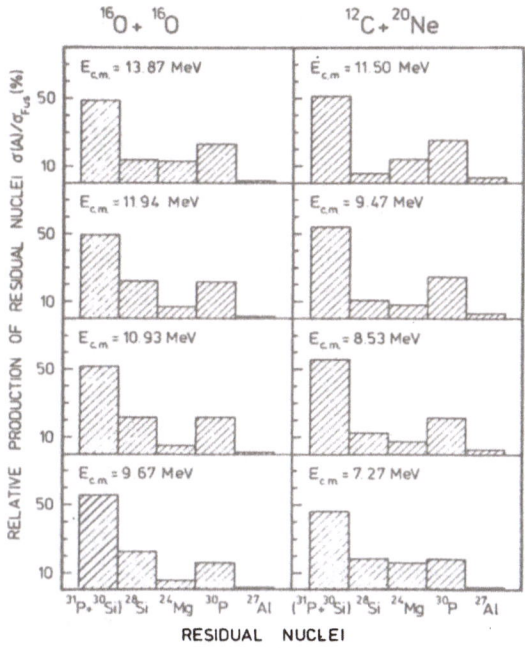

Fig.15: Distribution of evaporation residues in the $^{16}O + ^{16}O$ and $^{12}C + ^{20}Ne$ induced reactions are shown at comparable excitation energies in the compound nucleus ^{32}S.

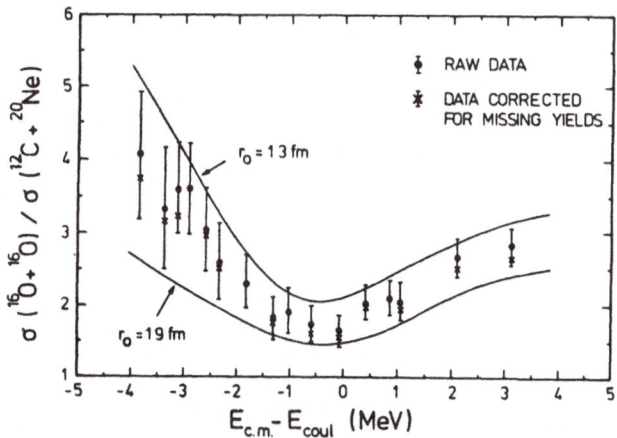

Fig.16: Ratio of fusion cross sections for the reactions $^{12}C + ^{20}Ne$ and $^{16}O + ^{16}O$. The energy scales have been shifted by amounts given by their different heights in Coulomb barriers. The solid lines indicate the results, if these heights are varied according to $r_0 = 1.3 - 1.9$ fm.

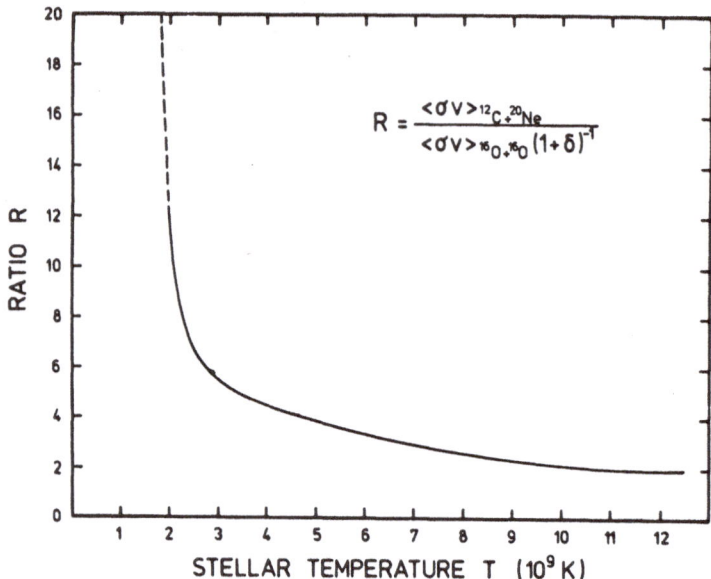

Fig.17:Ratio of stellar reaction rates <σv> for the ^{12}C + ^{20}Ne and
^{16}O + ^{16}O fusion reactions. The dashed line represents an
extrapolation from the measured energy regions.

Finally, the ratio of stellar reaction rates <σv> for both heavy ion systems is illustrated in Fig.17. The results demonstrate
that at low stellar temperatures of $T_9 \leq 2$, the ^{12}C + ^{20}Ne reaction
burns much faster than the ^{16}O + ^{16}O reaction. What role the ^{12}C +
^{20}Ne reaction can play in the late phase of stellar evolution depends however on many factors such as the mass of a star, the
amount and type of nuclei present, the ^{20}Ne photodesintegration
rate etc. Therefore, astrophysically relevant conclusions have to
await the results of such stellar model calculations.

REFERENCES

1. J.N.Bahcall, Space Science Reviews 24(1979)227
2. R.Davis,Jr.,Proc.of the Brookhaven Solar Neutrino Conference,
   BNL 50879,1(1978)1
3. R.W.Kavanagh,T.A.Tombrello,J.M.Mosher and D.R.Goosman,Bull.Am.
   Phys.Soc.14(1969)1209
4. C.Wiezorek,H.Kräwinkel,R.Santo and L.Wallek,Zeitsch.f.Phys.
   A282(1977)121
5. P.D.Parker and R.W.Kavanagh,Phys.Rev.131(1963)2578
6. K.Nagatani,M.R.Dwarakanath and D.Ashery,Nucl.Phys.A128(1969)325
7. C.Rolfs,J.Görres,K.U.Kettner,H.Lorenz-Wirzba,P.Schmalbrock,H.P.
   Trautvetter and W.Verhoeven,Nucl.Instr.a.Meth.157(1978)19

8. G.Hulke,C.Rolfs and H.P.Trautvetter,Zeitsch.f.Phys.(in print)
9. J.Görres,K.U.Kettner,H.Kräwinkel and C.Rolfs,Nucl.Instr.a.Meth.
   (in print)
10.T.Freye,H.Lorenz-Wirzba,B.Cleff,H.P.Trautvetter and C.Rolfs,
   Zeitsch.f.Phys. A281(1977)211
11.T.A.Tombrello and P.D.Parker,Phys.Rev.131(1963)2582
12.H.Kräwinkel,H.W.Becker,L.Buchmann,J.Görres,K.U.Kettner,W.E.Kieser,
   C.Rolfs,R.Santo,P.Schmalbrock,H.P.Trautvetter (all Münster),J.W.
   Hammer (Stuttgart),R.E.Azuma (Toronto) and W.S.Rodney (Washing-
   ton), to be submitted to Zeitsch.f.Phys.
13.E.M.Burbidge,G.R.Burbidge,W.A.Fowler and F.Hoyle,Rev.Mod.Phys.
   29(1957)547
14.W.A.Fowler,G.R.Caughlan and B.A.Zimmerman,Ann.Rev.Astron.Astro-
   phys. 13(1975)69
15.M.G.Mazarakis and W.E.Stephens,Phys.Rev.C7(1973)1280
16.G.Michaud,Astr.J.175(1972)751 and Phys.Rev.C8(1973)525
17.K.U.Kettner,H.Lorenz-Wirzba,C.Rolfs and H.Winkler,Phys.Rev.Lett.
   38(1977)337
18.K.U.Kettner,H.Lorenz-Wirzba and C.Rolfs,Zeitsch.f.Phys.(submitted)
19.H.W.Becker,Diplomarbeit,Universität Münster (1978)
20.H.Feshbach,J.Phys.C5(1976)177
21.G.Michaud and E.Vogt,Phys.Lett.B30(1969)85
22.H.Spinka and H.Winkler,Nucl.Phys.A233(1974)456
23.S.C.Wu and C.A.Barnes,Caltech (to be published)
24.J.J.Kolata,R.M.Freemann,F.Haas,B.Heusch and A.Gallmann,Phys.Rev.
   C19(1979)2237
25.V.K.C.Cheng,A.Little,H.C.Yuen,S.M.Lazarus and S.S.Hanna,Nucl.
   Phys.A322(1979)168

TRITIUM AS A STRATOSPHERIC TRACER

G. Hut

Isotope Physics Laboratory
State University of Groningen
Westersingel 34
9718 CM Groningen   The Netherlands

INTRODUCTION

The radioactivity of tritium was established in 1939 by Alvarez and Cornog[1], who found that the radiation is of very short range and that the half life is long.
Subsequent work has revealed that tritium decays by the emission of a beta particle with a maximum energy of 18 keV and an average energy of 5.7 keV to form ^{3}He and that the half life of tritium is 12.3 years.
Tritium is produced by a number of nuclear reactions, e.g.:

a. ^{2}H + ^{2}H $\rightarrow$ ^{3}H + ^{1}H + 4 MeV
b. ^{6}Li+ ^{1}n $\rightarrow$ ^{4}He + ^{3}H + 4.69 MeV
c. ^{10}B + ^{1}n $\rightarrow$ ^{8}Be + ^{3}H + 0.2 MeV
d. ^{11}B + ^{1}n $\rightarrow$ ^{9}Be + ^{3}H + 9.6 MeV
e. ^{14}N + ^{1}n $\rightarrow$ ^{12}C + ^{3}H $-$ 4.3 MeV
f. ^{14}N + ^{1}n $\rightarrow$ 3^{4}He + ^{3}H $-$ 11.5 MeV
g. ^{2}H + ^{1}n $\rightarrow$ ^{3}H + 6.26 MeV

Tritium is also produced as a fission product at the rate of 1 atom per 1 x 10^4 to 2 x 10^4 fissions for three types of material, natural uranium, enriched uranium and a mixture of transuranium nuclides (Albenesius[2,3]).
Sources of tritium production are:
Light-Water Reactors:
Tritium can be produced in reactors by neutron bombardment of a number of elements and by fission of the reactor fuel. Fast neutrons produce tritium from the ^{10}B of control rods or from the ^{14}N in residual air. These sources, however, are small compared with the thermal-neutron irradiation of the deuterium, ^{3}He or ^{6}Li in the

reactor coolant. Fission produces probably more tritium than all
other sources in light-water cooled reactors. A total typical production rate for a light-water reactor is between 90 and 500 Ci/yr/
1000 MW (Pinford[4]). Part of this reaches the cooling water.

Heavy-Water Reactors:

Heavy water can be used as a moderator, a reflector, and a coolant
in certain types of reactors. Although some fission-produced tritium
can penetrate fuel-element claddings, this addition is negligible
compared with the activity produced from deuterium. External losses
from heavy-water reactors are relatively small. E.g. releases of
tritium to surface water at the Savannah River Plant are due
primarily to releases from heavy-water reactors. The concentration
of tritium in the Savannah River Plant ranges from 10^{-5} to 1.4 x
x $10^{-5}\mu$Ci/ml or about 0.3 to 0.4% of the maximum permissible
concentration (Reinig et al.[5]).

Fuel-Reprocessing Plants:

During fuel reprocessing about 20% of the fission-produced tritium
appears in the dissolver off gas and is released through the stack
in the gaseous phase (Haney et al.[6]). Most of the remaining tritium
follows the aqueous phase and is released to the environment in
low-level waste streams. Therefore all fission-produced tritium can
be considered released to the environment within a few years of its
production. Local contamination problems may result.

Thermonuclear Detonations:

In a deuterium-tritium bomb, the tritium release is composed of a
residual amount of tritium and a production yield due to the neutron
irradiation of nitrogen. The total yield is not very well known.
It has been reported to range from 7 to 50 megacuries per megaton
fusion (MCi/Mt fusion) with a suggested average value of 20 MCi/Mt
(NCRP[7]). (1 g of tritium is equivalent to about 9700 curies).

Natural Tritium Production:

Tritium is produced from atmospheric nitrogen by bombardment with
cosmic ray neutrons. The reaction has an activation cross section
of 11 $\pm$ 2 millibarns for fission neutrons with energies > 4.4 MeV.
Proton bombardment of nitrogen, oxygen and carbon can also yield
tritium. According to Craig and Lal[8] the tritium production in the
atmosphere is approximately 0.25 T-atom/cm^2. sec.

The tritium formed is oxidized to water or is exchanged with atmospheric water vapour; the exchange is readily achieved since the
mass action equilibrium coefficient for the reaction:

$$HT + H_2O \leftrightarrows HTO + H_2$$

is approximately 6 at 25^0C (Jacobs[9]). By condensation and evaporation mainly to and from the oceans, tritium becomes, in the form of
HTO, part of the hydrological cycle. The natural production of
tritium is low and results in a concentration of about 1 Tritium
Unit (Roether[10]), 1 Tritium Unit or 1 TU is equivalent to a T-H ratio
of 10^{-18}. Due to the thermonuclear bombs the tritium concentration
in rain increased strongly with a maximum of about 4000 TU in 1963.

After the Partial Test Ban Treaty (1963) the tritium concentration
decreased steadily almost back to the natural level as can be seen
from figure 1.

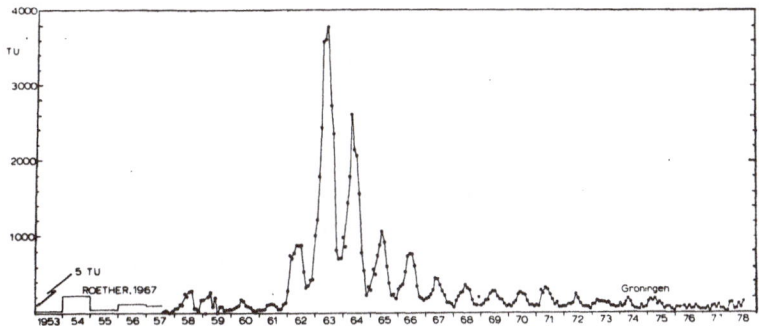

Fig. 1. Variation of tritium activity in precipitation as compiled
        from IAEA data for the Northern Hemisphere (courtesy Van
        der Straaten).

     The study reported in this paper was undertaken to collect
tritiumdata from the troposphere and from the lower stratosphere in
order to obtain information about stratospheric-tropospheric exchange
processes and transport in the stratosphere.
The study was concluded during the authors' stay at the Tritium
Laboratory of the University of Miami. The tritium data were collected by A.S. Mason of the University of Miami. Comparison of the
tritium inventory of the stratosphere with the ^{95}Zr-inventory was
possible through the cooperation of K. Telegadas of the National
Oceanic and Atmospheric Administration at Silver Spring, Maryland.
The data are published elsewhere (Mason, Hut and Telegadas[11]).

PROJECT AIRSTREAM

     The Tritium Laboratory of the University of Miami began programs
of measurement of tritiated water vapour (HTO) and tritium gas (HT
and T_2) collected at ground level in 1968, and from aircraft in 1971.
Previous publications have described the techniques and interpreted
the data (Östlund and Mason[12]; Mason and Östlund[13]; Mason[14]).
Data through 1976 are available in unpublished reports (Östlund,
Mason and Ydfalk[15]; Mason and Östlund[16]).
     Project Airstream (sponsored by the Department of Energy) is
a longterm study of stratospheric radioactivity and chemistry,
carried out by three series of flights annually, covering the
latitude range from the equator to 75^0 north. Four flight levels
above the tropopause are sampled over that span; in addition,
vertical profiles have been flown in the vicinity of Panama,
Republic of Panama; Houston, Texas, and Anchorage, Alaska, for
each flight series since July 1977. The vertical profiles take

samples from 3 to 19 km altitude and the transects are made between
14 and 19 km.

Tritium measurements have been made as part of Project Airstream
since 1975, however, the project is older than that, having begun in
1965 with sampling of particulate and noble gas radioactivity.

Zirconium-95 is a particulate fission product with 65-day half-
life produced by all nuclear tests.

Measurements of Zr-95 have been used extensively to estimate
stratospheric residence times, initial vertical activity distributions
and atmospheric transport from the high yield nuclear tests conducted
by China between 1967 and 1976 (Telegadas[17,18,19]). These data can
now be compared with the tritium (as HTO) data to try to resolve
the question of the significance of particle settling in transport
processes in the upper atmosphere.

EXPERIMENTAL TECHNIQUES

The gas sampler flow diagram is shown in figure 2 (Mason and
Östlund[13]).

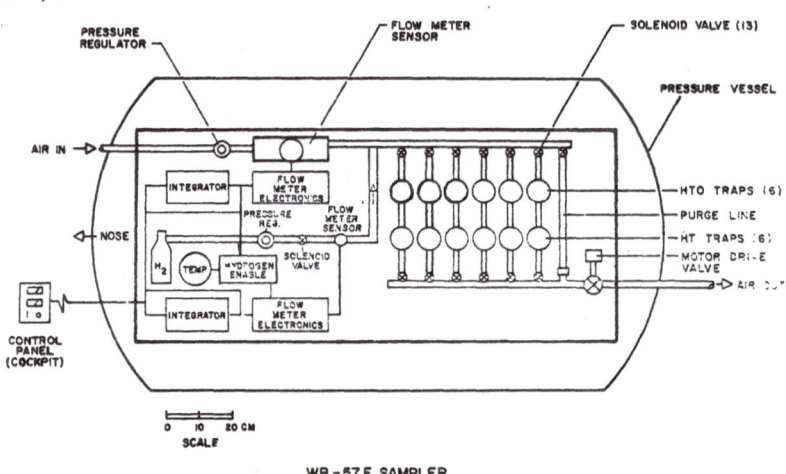

Fig. 2. Plan view of the WB-57F aircraft tritium sampler.

Bleed air from the WB-57F airplane passes through a pressure
regulator and a mass flowmeter into a manifold. Sampling is controlled
from the cockpit by opening one of six pairs of solenoid valves.
The air, to which the sampler adds $1^o/oo$ by volume of tritium-free
hydrogen, passes first through a molecular sieve trap where H_2O and
HTO are adsorbed, and then through a trap of palladiumcoated molecular sieve, where the carrier H_2, plus ambient H_2, HT and T_2 are
oxidized and the resulting water adsorbed.

The mass of air sampled, the location of sampling and the
ambient pressure altitude and temperature are noted by the equipment
operator. The sampling traps are returned to Miami, where the sample

water is extracted by techniques described by Östlund and Mason[12] and the tritium determined by low-level proportional counting (Östlund and Dorsey[20]).

The data take the form of mixing ratios of HTO and HT, i.e., tritium atoms per mg of air. These units may be converted to picocuries per standard cubic meter of air (pCi/SCM) by multiplication by 0.0625 for SCM defined at 1013 mb and 0^0C. Specific activities of atmospheric water vapor and hydrogen cannot be determined accurately due to the very small samples obtained and the consequent use of tritium-free water to flush the samples from the traps. Figure 3 is an example of HTO-data obtained from an Airstream deployment.

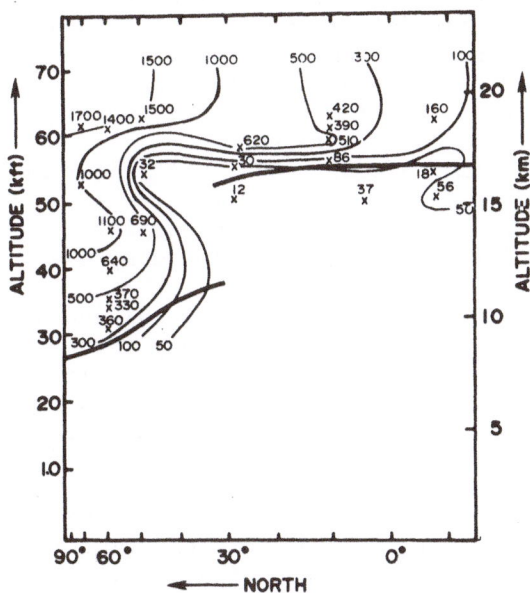

Fig. 3. Distribution of tritium (as HTO) for April 18 - May 6, 1975, approximately 22 months after the Chinese test of June 27, 1973. The numbers represent the observed HTO-concentration (T-atoms/mg air). The heavy lines represent the mean tropopause along the sampling corridor.

THE STRATOSPHERIC TRITIUM AND ^{95}ZR INVENTORY

As was pointed out by Eriksson[21], the source of stratospheric HTO is primarily the testing of thermonuclear (fusion) devices. One would expect that the six reported Chinese thermonuclear tests, all performed at Lop Nor (40^0 N 90^0 E) between 1967 and 1976, would be significant contributors to the stratospheric HTO burden. Changes in the stratospheric HTO burden should provide information about stratospheric-tropospheric exchange processes and transport in the stratosphere.

Telegadas [18],[19] has analyzed the fission product data (primarily ^{95}Zr and ^{144}Ce) following the 27 June 1973 and 17 November 1976 Chinese nuclear tests. Since ^{95}Zr has a relatively short half-life, due to radioactive decay and stratospheric depletion, the stratospheric input from these two events, as those from earlier high-yield Chinese tests, could be followed unequivocally for only about one year. The HTO stratosphere input could be followed for many years due primarily to its much longer radioactive half-life.

Two possible problems exist with using these HTO data for a direct comparison with fission product data: 1. The calculated stratospheric HTO burdens may contain a background from past high-yield tests (attempts will be made to resolve this problem); 2. Although the production of fission products from nuclear tests is fairly well known (Harley et al.[22]), the HTO production from thermonuclear tests has, as mentioned previously, a much larger uncertainty. It is therefore difficult to know with certainty the amount of HTO injected into the atmosphere even if the total yield and fission yield of an event are known.

There were no simultaneous measurements of ^{95}Zr and HTO following the 27 June 1973 high-yield test, whereas there were for three sampling series after the 17 November 1976 test. This test will therefore be discussed first, followed by the analysis of the 27 June 1973 test.

The Northern Hemisphere stratospheric HTO burdens for the seven sampling periods between July 1977 and July 1979 are given in Table 1 together with the ^{95}Zr burdens calculated by Telegadas[19]. The first two columns show the ^{95}Zr burden to about 20 km (based on aircraft sampling) and about 30 km (based on additional balloon sampling from 20 to 30 km). The first line under these two columns shows the ^{95}Zr burden prior to the 17 November 1976 test.

The last significant test prior to this event occured on 17 June 1974 (reported total yield of between 0.2-1 MT) and estimated by Leifer[23] to have a fission yield of 0.4 MT. By the time of 17 November 1976 test, due to stratospheric depletion and radioactive decay, the ^{95}Zr created by the earlier test had decayed below detection limits. The HTO collected between 24 October and 17 November 1976 indicated a background of 3100 kCi of HTO residing in the stratosphere prior to the 17 November 1976 test. This is shown in column A, line 1. Column B shows the observed burden (column A) decay-corrected to the 17 November 1976 test. The burden for the 13 October – 6 November 1978 sampling period listed in column A is questionable due to limited data (Mason et al.[11]).

A line of regression through the decay-corrected HTO burdens given in column B (from July 1977 to July 1979) would indicate a residence half-time of about ten months. It was therefore assumed that the background HTO burden prior to the 17 November 1976 test would be depleted with this same residence time. It should be noted that the (presumably constant) natural background due to cosmic radiation is estimated to be at most 1000 kCi, using the production

TABLE 1. Northern Hemisphere Stratospheric Burden (kilocuries)

Sampling Period	^{95}Zr[1] to ~20km	to ~30km	Tritium (as HTO) to ~20km			
			A	B	C	D
24 Oct.–17 Nov.'76	0^2	0^2	3,100	3,100	3,100	
22 Mar.–10 Apr.'77	50 900	58 100	no data			
6 Jul.–22 Jul.'77	38 400	45 500	18 800	19 500	2 100	17 400
12 Oct.–29 Oct.'77	30 600	36 100	8 700	9 200	1 700	7 500
6 Apr.–21 Apr.'78	21 000	3	5 900	6 300	1 100	5 200
12 Jul.–31 Jul.'78			4 000	4 400	950	3 450
13 Oct.– 6 Nov.'78			5 300[4]	5 900[4]	750	5 150[4]
6 Apr.–24 Apr.'79			3 400	3 900	500	3 400
3 Jul.–26 Jul.'79			2 500	2 900	400	2 500

[1] Decay-corrected to Chinese test of November 1976
[2] Background (last significant test 17 June 1974; total yield 0.2-MT)
[3] No data available above 20 km at this time.
[4] Questionable – see text.

A. Burdens based on observed data.
B. Burdens from column A decay-corrected to 17 November 1976.
C. Background at time of Chinese 17 November 1976 test. Background burdens at later times assumed using a residence half-time of ten months.
D. Residual burden attributed to 17 November 1976 test (B minus C).

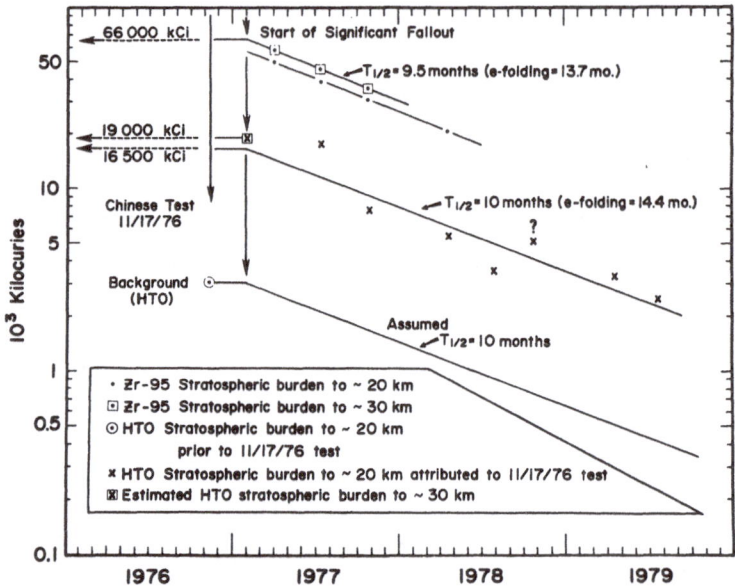

Fig. 4. Stratospheric half-residence-time of the ^{95}Zr and HTO burden
        for the Chinese test of 17 November 1976. All burdens decay
        corrected to time of test.

rate, stratospheric-tropospheric production distribution, and stratospheric residence time of Craig and Lal[8]. The natural background
has been neglected in the calculation of column C due to the uncertainties of the estimate and the absence of asymptotic behaviour
in the data acquired to date. The background burdens at later times
are listed in column C. Column D shows the residual burden attributed
to the 17 November 1976 test, that is, column B minus column C.
The residual burdens are shown in figure 1 together with the assumed
depletion of the background listed in column C.

     The Northern Hemisphere ^{95}Zr stratospheric burdens to 20 and
30 km shown in figure 4 are extrapolated back to 1 February 1977,
when it was estimated that significant fallout started following the
17 November 1976 event (Telegadas[19]). It can be seen from either
figure 4 or table 1 that about 15% of the ^{95}Zr burden resided above
the aircraft altitude of approximately 20 km. ^{95}Zr burden to 30 km
is determined to be 66000 kCi which is equivalent to a fission
yield of 2.7 MT. The reported total yield for this event was 4 MT;
therefore the fusion yield is estimated to have been 1.3 MT.

The line of regression through the tritium burden attributed to the 17 November 1976 test (Fig. 4.) indicates a residence half-time of about ten months with an input into the stratosphere of 16500 kCi of HTO to about 20 km. Increasing this amount by 15% (assuming the same percentage of tritium above the aircraft altitudes as was determined for the ^{95}Zr burden) would indicate an input of 19000 kCi of HTO. Since the fusion yield was estimated to be 1.3 MT, the assumed production of HTO from the 17 November 1976 event would be about 15 MCi/MT (fusion). This seems reasonable, since the range has been reported as 7 to 50 MCi/MT with a suggested average value Of 20 MCi/MT for thermonuclear devices (National Council on Radiation Protection and Measurements[7]).

The first stratospheric sampling following the 27 June 1973 test where a reliable stratospheric inventory could be performed was not until 18 April 1975 (Table 2), nearly two years later.

TABLE 2. Northern Hemisphere Tritium (as HTO) Stratospheric Burden to approximately 20 km (kilocuries)

Sampling Period	Observed[1]	Decay-Corrected[2] (to 27 June '73)	Obser.-Background[3] (Decay-Corrected to 27 June '73)
18 Apr.- 6 May'75	9 400	10 500	7 300
14 Jul.- 5 Aug'75	4 300	4 800	2 200
22 May - 9 Jun.'76	4 900	5 800	4 300
12 Aug.-30 Aug.'76	2 000	2 400	1 300
24 Oct.-17 Nov.'76	3 100	3 800	2 800

[1]Burden based on observed data (not decay-corrected).
[2]Observed burden decay-corrected to 27 June 1973.
[3]Observed burden[1] minus background (decay-corrected to 27 June 1973) where: background = assumed input of 6000 kCi from the 17 June 1974 test had a stratospheric half-residence-time of ten months, decay-corrected to time of measurements in 1975 and 1976.

Shown in Table 2 are the computed Northern Hemispheric strato-spheric HTO burdens to approximately 20 km. Column 1 gives the ob-served inventory. Decay-correcting these inventories to 27 June 1973, the Chinese high-yield test is shown in column 2. The burdens listed in column 2 are shown in figure 5, together with the ^{95}Zr inventories to 30 km attributed to the 27 June 1973 test, reported by Telegadas[18]. A line of regression through the decay-corrected HTO burdens indicates a residence half-time of about 13 months. Extrapolating this regression line back to 15 December 1973, when (as estimated) signi-ficant fallout of ^{95}Zr began (Telegadas[18]), indicates a production of 18400 kCi of HTO. The stratospheric inventory of ^{95}Zr showed about 5% above the sampling altitude of the aircraft. The HTO burden (to

20 km) at time of significant fallout was therefore increased by 5%
for a total input of 19300 kCi of HTO into the stratosphere.

The ^{95}Zr burden to about 30 km indicated a 34000 kCi input which
is equivalent to a fission yield of 1.4 MT. The reported total yield
for the 27 June 1973 event was 2 to 3 MT. Assuming a total yield of
2.5 MT would mean that this event has a fusion yield of about 1.1 MT.
Dividing the HTO stratospheric input of 19.3 MCi by 1.1 MT indicates
a production of 18 MCi/MT (fusion), not too different from that
calculated for the 17 November 1976 test.

The stratospheric residence half-time of the ^{95}Zr burdens,
figure 2, based on only two sampling periods, was five months.
This residence time is shorter than that for the 17 November 1976
test but about the same as Telegadas[17] found for the 17 June 1967
high-yield Chinese test between the sampling periods of Februari 1968
and April 1968 (6 months). The ^{95}Zr burden attributed to the 27 June
1973 test could not be determined beyond April 1974 due to the fact
that the next Chinese test of 17 June 1974 (total yield 0.2-1 MT)
dominated the lower stratosphere of the Northern Hemisphere during
the next sampling period in October 1974.

As mentioned previously, Leifer[23] estimated the fission yield
of the 17 June 1974 test to about 0.4 MT. When this event occured
no mention was made whether it was an all fission or thermonuclear
test. We will assume that this test was thermonuclear and it had a
fission-fusion ratio of 1, that is, 0.4 MT (fusion). Assuming an
HTO production of 15 MCi/MT (fusion) that was calculated for the
17 November 1976 test, this test would have injected 6000 kCi of
HTO into the lower stratosphere of the Northern Hemisphere. It is
further assumed that this HTO input had a stratospheric residence
half time of ten months. This input was then decay-corrected to the
1975 and 1976 sampling periods listed in Table 2 and subtracted from
column 1. Column 3, Table 2, is therefore the residual burden
attributed to the 27 June 1973 high-yield test if the 17 June 1974
test was thermonuclear and had a fission-fusion ratio of 1.

The effect of assuming a fusion yield of 0.4 MT from the 17
June 1974 test upon the production estimate of the 27 June 1973
test is as follows: figure 5 indicates a stratospheric half-residence
time of 16 months for the burden from the 1973 test. Extrapolating
the burden back to the start of significant fallout and adding 5%
to account for burden above the aircraft ceiling indicate an HTO
production from the 1974 test of 9600 kCi, or 9 MCi/MT (fusion).
This is half of the previous estimate which assumed no fusion yield
from the 1974 test.

There are many uncertainties in estimating the stratospheric
HTO inventories following the 27 June 1973 and 17 November 1976
events. One cannot determine unequivocally how much HTO was above
the aircraft sampling altitudes or how much was transported into
the Southern Hemisphere. The HTO inventories attributed to the
27 June 1973 test also had further uncertainties. Sampling did not
start until about two years after the event; there was no overlap
between ^{95}Zr and HTO data, and there is the possibility that the
17 June 1974 test could have contributed substantially to the

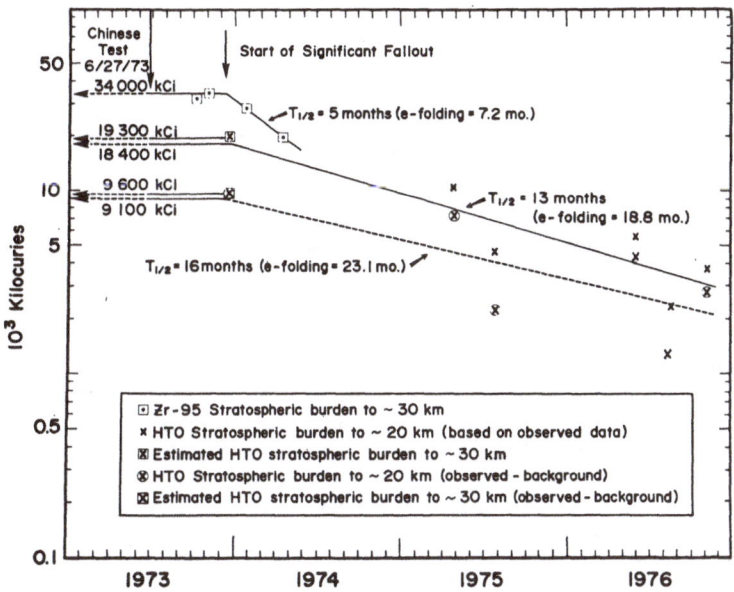

Fig. 5. Stratospheric half-residence-time of the ^{95}Zr and HTO burden for the Chinese test of 27 June 1973. All burdens decay-corrected to time of test.

observed inventories calculated during 1975 and 1976.

There is more confidence in the estimated HTO burdens attributed to the 17 November 1976 test (figure 4) than the estimated HTO burdens from the 27 June 1973 test (figure 5). This is in part due to the fact that there was a partial overlap of the ^{95}Zr and HTO burdens starting eight months after input. Further, the background HTO inventory based on measurements taken shortly before the 17 November 1976 event could be accounted for and subtracted from the observed inventories at later times.

The calculated residence half-time for particulate ^{95}Zr versus gaseous HTO is approximately equal for the 17 November 1976 test indicating that gravitational settling of particles in the lower stratosphere, to at least 20 km, possibly higher, for all partical purposes is negligible. A note of caution is in order – the HTO regression line in figure 4 is weighted toward the first calculated burden based on observations taken between 6-22 July 1977. If these samples were not taken and the questionable burden of October 1978 is not considered, a residence half-time of 15 months would have been calculated. One then would come up with the determination that particle settling is significant in transport processes assuming both the initial distribution of particles and gases were the same. This points out that measurements at early times after a nuclear test are important in determination of residence times.

REFERENCES

1. Alvarez, L.W. and Cornog, R.: Helium and Hydrogen of Mass 3,
     Phys. Rev. |2| 56, 613 (1939)
2. Albenesius, E.L.: Tritium as a product of fission, Phys. Rev.
     Lett. 3 (6), 274-275 (1959)
3. Albenesius, E.L. and Ondrejcin, R.S.: Nuclear fission produces
     tritium, Nucleonics 18 (9), 100 (1960)
4. Pinford, T.H., Keaton, M.J. and Mann, B.J.: Fuel cycles for
     electrical power generation, Teknekron Report no. E E ED 101
5. Reinig, W.C. and Albenius, E.L.: Control of tritium health hazards
     at the Savannah River Plant, USAEC Report DPSPU-62-30-5,
     Savannah River Laboratory (1962)
6. Haney, W.A., Brown, D.J. and Reisenauer, A.E.: Fission product
     tritium in separations wastes and in the ground water,
     USAEC Report HW-74536, Hanford Atomic Products Operation (1962)
7. National Council on Radiation Protection and Measurements: Tritium
     in the environment, NCRP Rpt. No. 62, Washington DC (1979)
8. Craig, H. and Lal, D.: The production rate of natural tritium,
     Tellus, 13, 85-105 (1961)
9. Jacobs, D.G.: Sources of tritium and its behaviour upon release
     to the environment, USAEC (1968)
10. Roether, W.: Estimating the tritium input to groundwater from
     wine samples: Groundwater and direct run-off contribution to
     Central European surface waters, Isotopes in Hydrology, 73,
     IAEA Vienna (1967)
11. Mason, A.S., Hut, G. and Telegadas, K.: Comparison of strato-
     spheric tritium (as HTO) and Zirconium-95 burdens from the
     high yield Chinese nuclear tests of June 27, 1973, and November
     17, 1976, EML-371, Environmental Measurements Laboratory,
     New York NY (1980)
12. Östlund, H.G. and Mason, A.S.: Atmospheric HT and HTO 1. Experi-
     mental procedures and tropospheric data 1968-72, Tellus, 26,
     91-102 (1974)
13. Mason, A.S. and Östlund, H.G.: Atmospheric HT and HTO: Distribu-
     tion and large-scale circulation, in: Behaviour of tritium in
     the environment, (Proc. Sym., San Francisco CA, Oct. 16-20,
     1978), International Atomic Energy Agency, Vienna (1979)
14. Mason, A.S.: Atmospheric HT and HTO 4. Estimation of atmospheric
     hydrogen residence time from interhemispheric tritium gas
     transport, J. Geophys. Res., 82, 5913-5916 (1977)
15. Östlund, H.G., Mason, A.S. and Ydfalk, A.: Atmospheric HT-HTO
     1968-71, Tritium Laboratory Data Report # 2, Rosenstiel School
     of Marine and Atmospheric Science, University of Miami,
     Miami FL (1972)
16. Mason, A.S. and Östlund, H.C.: Atmospheric HT and HTO 1975-1976,
     Tritium Laboratory Data Report # 7, Rosenstiel School of Marine
     and Atmospheric Science, University of Miami, Miami FL (1977)

17. Telegadas, K.: Radioactivity distribution in the stratosphere
    from Chinese and French high yield nuclear tests (1967-1970)
    HASL-281, Health and Safety Laboratory, New York NY (1974)
18. Telegadas, K.: Radioactivity distribution in the stratosphere
    from the Chinese high yield nuclear test of June 27, 1973,
    HASL-298, Health and Safety Laboratory, New York NY (1976)
19. Telegadas, K.: Radioactivity distribution in the stratosphere
    from the Chinese high yield nuclear test of November 17, 1976,
    EML-356, Environmental Measurements Laboratory, New York NY
    (1979)
20. Östlund, H.G. and Dorsey, H.G.: Rapid electrolytic enrichment
    and hydrogen gas proportional counting of tritium, in: Low-
    Radioactivity Measurements and Applications (Proc. Int. Conf.
    High Tatres, 1975), Slovenske Pedagogicke Nakladatelstvo,
    Bratislava (1977)
21. Eriksson, E.: An account of the major pulses of tritium and
    their effects in the atmosphere, Tellus, 17, 118-130 (1965)
22. Harley, N., Fisenne, I., Ong, L.D.Y. and Harley, J.: Fission
    yield and fission product decay, HASL-164, Health and Safety
    Laboratory, New York NY (1965)
23. Leifer, R., Schonberg, M. and Toonkel, L.: Updating stratospheric
    inventories to July 1975, HASL-306, Health and Safety Labora-
    tory, New York NY (1976).

PROTON INDUCED X-RAYS

R.D. Vis

Natuurkundig Laboratorium der Vrije Universiteit
Amsterdam
The Netherlands

I. INTRODUCTION

The creation of inner shell holes in the electron cloud of an
atom under proton bombardment is well known. Nevertheless, from the
physical point of view, still a lot of work is and will be done;
theoretical models are tested with sophisticated experiments in
order to refine wave functions of the electrons in the inner shells.
Every three years an international conference is held entitled:
International Conference on inner shell ionization phenomena, where
progress made is shown.
Starting in 1970, proton induced X-rays are also used in applied
physics for the determination of trace elements in samples from
different origin. The reason for this rather successful application
(called PIXE), is that cross sections for inner shell ionization
are very high compared with the cross sections for nuclear reactions
as for instance the cross section for neutron capture; an
established technique for trace analysis called neutron activation
analysis.
So, in fact there are two different fields to distinguish and they
are called arbitrarely:
i) Fundamental:
    Here, experiments are done to test theoretical descriptions;
    most measurements are done on (differential) cross sections or
    fluorescence yields and also other phenomena like radiative
    electron capture, quasi-molecular X-rays, double ionization
    (satellite X-rays) and Auger-spectroscopy.
ii) Applied:
    Here, experiments are done to measure chemical or biological
    quantities, very often concentrations of trace elements in
    samples to learn about the behaviour of trace elements in the

human body, or in geology, or in technical fields (semi-conductor industry) and so on. The theory is used to optimize the measuring techniques in order to obtain the best results.

In this paper a survey is given with the emphasis on the applied field from the experimental point of view.

II. THE CROSS SECTIONS FOR PROTON INDUCED X-RAYS

A recent review of the theories of charged particle excitation of the inner shells is given by Madison and Merzbacher[1]. Several calculations are given, each of them applicable dependent of the collision involved. Fig. 1 gives a view of the applicability of these descriptions depending on the collision parameters Z_1/Z_2 (atomic numbers of projectile and target) and v/v_K (the collision velocity in units of the velocity of the K-shell electron).

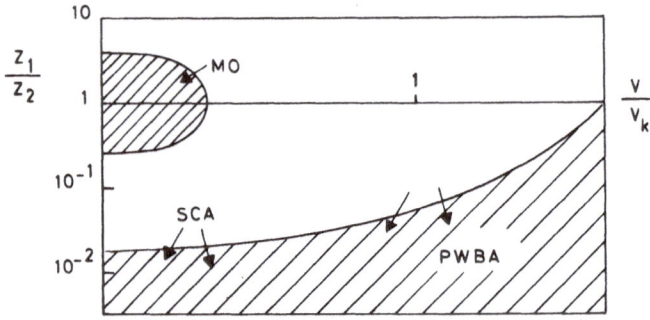

Figure 1.   Regions of validity for theoretical descriptions.

The region in which the molecular orbital method is valid encompasses slow collisions between heavy ions and atoms with Z_1/Z_2 about unity. In this case the impact velocities are small compared with the inner shell electron velocity. Then the electron clouds of the colliding system will overlap and give rise to the so-called Pauli-excitation due to the formation of a short lived quasi molecular state. This kind of collisions is of no interest for analytical chemical applications due to the low cross-sections at low projectile energy and due to the continuous character of the emitted radiation. Limiting ourselves to protons with energies between 1 and 5 MeV (most favourite in the applied field) principally three models are in use for the calculation of cross-sections.

i)   The plane wave Born approximation   (PWBA)
ii)   The binary encounter approximation (BEA)
iii) The semi-classical approximation   (SCA)

     A feature common to these models is that the incoming particle
is treated as a point charge perturbing the target atom. This
requires a projectile to target charge ratio much smaller than unity.
The excitation mechanism considered is the direct Coulomb
interaction between the projectile and the target electron, with the
final state of the electron in the continuum.

i) The PWBA

     The criterion for the validity of the Born approximation is:

$$\frac{Z_1 Z_2 e^2}{\hbar v} \ll 1 \qquad (1)$$

where Z_1 is the projectile charge, Z_2 the nuclear charge of the
target and v the velocity of the projectile relative to the
velocity of the bound electron.
In the first Born approximation, it is assumed that the incident
and inelastically scattered particle can be described by plane
waves. The transition from the initial to the final states of
the atom are described as a transition of the electron's initial
bound state to a state described by a continuum wave function
with the other electrons remaining in their initial states.
Vacancy production due to the excitation of the electron to
unoccupied orbitals is neglected. The PWBA formula for the
differential cross-section for the ejection of a target electron
with final energy E_f is given by:

$$\frac{d\sigma}{dE_f} = \frac{4\pi}{\hbar^2} Z_1^2 e^4 \frac{M_1}{E_1} \int_{q_0}^{\infty} \frac{dq}{q^3} J \qquad (2)$$

with

$$J = \sum_f \left| \int e^{i\vec{q}\vec{r}} \psi_f^* (\vec{r}) \psi_i (\vec{r}) d\vec{r} \right|^2 \qquad (3)$$

ħq denotes the momentum transfer with minimum value $ħq_0$. Most of the existing estimates have further approximated the initial state with hydrogenic wave functions with an effective charge Z_s, to account for screening. In these calculations the parameters

$$\theta_s = \frac{n^2 u_s}{Z_s^2} \qquad \text{and} \qquad \eta_s = \frac{1}{Z_s^2} \left(\frac{ħv}{e^2}\right)^2$$

where n is the principal quantum number of the sth shell, and u_s is the binding energy, are often used. In terms of these parameters, the total cross-section for vacancy production is given by:

$$\sigma_s = (8\pi Z_1^2 \, a_0^2 / Z_s^4 \eta_s) \, f_s \, (\theta_s, \eta_s) \tag{4}$$

where a_0 is the Bohr radius and f_s is the result of the integration of eq. (3).

ii) The BEA

The universal validity of the theory for Rutherford scattering is used to describe the ionizing collision between the projectile and the atom as a classical binary encounter between the projectile and a particular atomic electron. The rest of the target atom is regarded as a spectator in the collision. Theoretical models which use this physical picture as a starting point for approximations are generally referred to as binary encounter models. Their applicability to inner shell ionization calculations has been stressed by Garcia[2,3,4]. The result is given by

$$\sigma_i(v_1) = N_i \int_0^\infty \sigma_i(v_1, v_2) \, f(v_2) \, dv_2 \tag{5}$$

where $f(v_2)$ is the velocity distribution of the bound electron, N_i the number of electrons having binding energy u and

$$\sigma_i = \int_u^{E_1} \frac{d\sigma}{dE'} \, dE'$$

E' is the energy exchange between projectile and electron
E_1 is the projectile energy

If hydrogenic velocity distributions are used in eq. (5) these results obey a scaling law for a given subshell, which states that the product of the binding energy squared and the cross-section is an universal function of the incident energy, expressed in units of the binding energy:

$$u^2 \sigma_I = Z_1^2 \; f\left[\frac{E_1}{uM_1} \; , \; M_1 \right] \qquad (6)$$

iii) <u>The SCA</u>

In this model the motion of the impinging particle in the field of the target nucleus is treated classically, whereas the transition of the inner shell electron to the continuum is studied from the point of view of quantum mechanics. The treatment of the atomic Coulomb excitation in the SCA formulation has been based on first order time dependent perturbation theory in impact parameter form and the use of unperturbed single electron wave functions. The Coulomb interaction between the bound inner shell electron and the projectile is used as the perturbing potential

$$V = \frac{Z_1 e^2}{|\vec{r} - \vec{R}(t)|} \qquad (7)$$

$\vec{R}$ and $\vec{r}$ are the position vectors of the projectile and the electron. By the introduction of hyperbolic paths in eq.(7) it may be shown that the deflection of the projectile plays an important role in cross-section calculations in the low energy region. However, the straight line SCA-calculations are in frequent use[5],[6].

<u>Results of the various calculations compared with experimental data</u>

As an illustration in fig. 2 cross sections are given for protons on Al calculated with the three models given and compared with experimental results. At low energies the PWBA values are too high due to the fact that the inequality in expression (1) is no longer fulfilled. From the fundamental point of view there is still interest in these cross sections, especially the 3 fold, 2 fold and single differential cross sections because the less integrations are performed the more sensitive the test on the wave functions. Fig. 3 and fig. 4 show the measurement of single differential cross sections with the results. Note the deviation at small impact parameter due to nuclear effects.

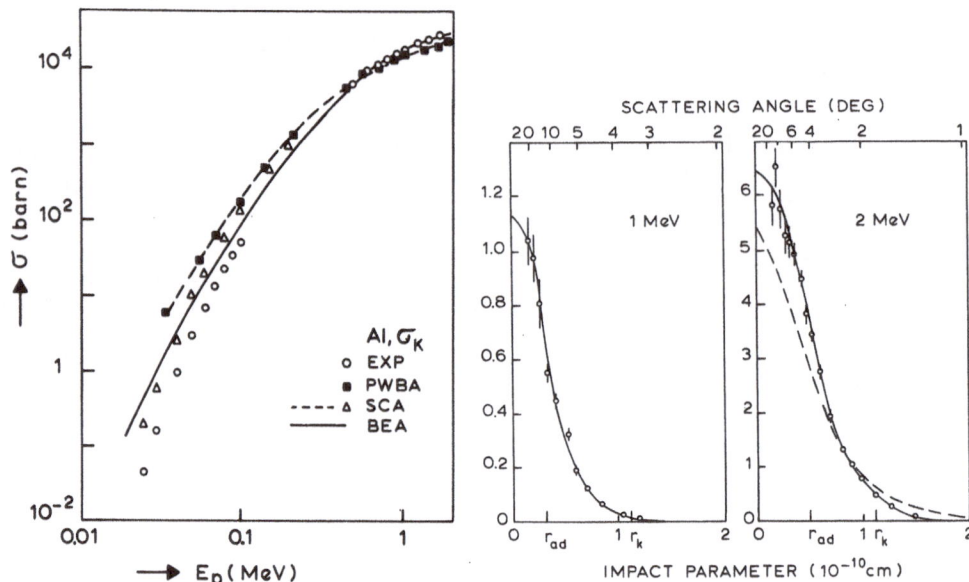

Fig. 2. Total cross sections
for protons on Al

Fig. 3. K-shell ionization
(absolute probability,1p(10^{-5})
versus angle and impact parameter
for 1 and 2 MeV protons on Ag(...
...) Prediction of SCA model

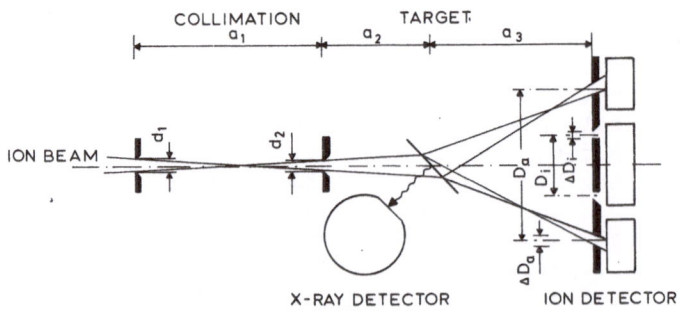

Fig. 4. Experimental set-up for measuring impact-parameter
dependence of X-ray emission (ref. 7)

III. THE TECHNIQUES USED FOR APPLICATIONS

For special applications it is sometimes advantageous to measure other radiation than X-rays; moreover, it is worthwhile to make comparisons with related techniques and we can frame the following matrix:

in	out	$h\nu$ (X-rays)	$h\nu$ (γ-rays)	P	$\bar{e}$
P		PIXE	PRA	NBS	
$h\nu$		XRF	–	–	ESCA
$\bar{e}$		SEM	–	–	AES

Fig. 5 shows the cross sections for the different processes. Incoming electrons are only used in combination with imaging, using an electron microscope adjusted such that the electrons are

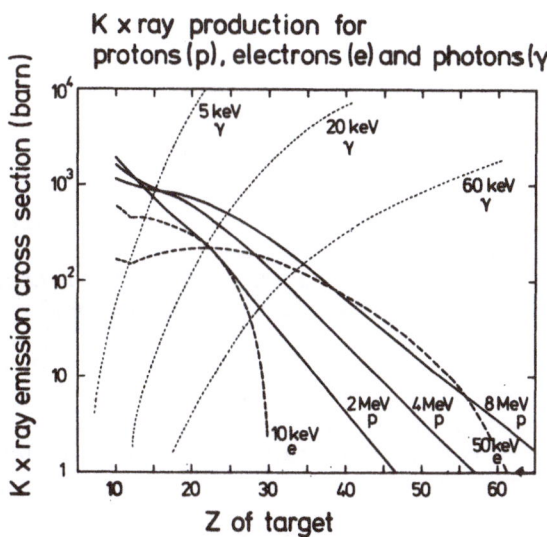

Fig. 5   The X-ray production cross section for different
         incoming radiation.

at sufficient high energy to excite X-rays or Auger electrons (AES).
Detection power is poor due to the high background caused by
bremsstrahlung of the electron beam (SEM = scanning electron micro-
probe).
For real trace analysis (concentration $\sim$ <10^{-4}) excitation with
X-rays is often used and called X-ray fluorescence (XRF) and, if
especially for surface layer analysis electrons are detected,
electron spectroscopy for chemical analysis (ESCA). Frequently
X-ray tubes are in use for the excitation; sometimes radio-isotopes
(and for ESCA lasers). A promising source for the excitation is
synchrotron radiation from an electron storage ring due to the
high photon flux and especially due to the high degree of
polarization of the synchrotronradiation decreasing the Compton-
background in the spectrum.

According to the title of this paper in more detail the use of
protons will be described.
Calculation of the cross section for (p,X) reactions show that σ
is maximal when the velocities of the incoming projectile equals the
electron orbit velocity; the so-called velocity matching peak.
This means that for the ionization of elements with $Z \sim 30$, the
highest cross-sections are obtained with projectile energies of
about 20 MeV.
On the other hand, experimentally one finds that the continuous
background of electromagnetic radiation is very strong at low
energies (bremsstrahlung of secondary electrons) and is rapidly
decreasing when the energy becomes larger than the value T_m:

$$T_m = \frac{4m_e M_p}{(M_p + m_e)^2} E_p \approx \frac{4m_e E_p}{M_p} \tag{8}$$

T_m represents the maximal energy that can be transferred from
a projectile to a free electron of mass m_e.
When T_m is larger than the energy E_x of the characteristic X-rays
from the trace element the background will strongly increase the
detection limits.
Hence, it is the energy region $E_x > T_m$ that is of interest for
trace analysis. Folkmann et al.[9] calculated optimal proton energies
using a determination limit defined as a concentration giving a
peak to background ratio of 1 (fig. 6), using the background
calculations of ref. 9.

The calculated limits are of the order of 10^{-7} g/g, in good
agreement with experimental values [10,11]. A major advantage of

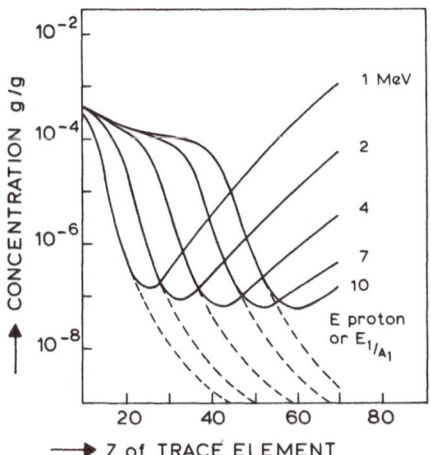

Fig. 6.   Calculated detection limits for different proton energies.

PIXE is that very small samples are sufficient (flux density of
proton beams are very high) setting absolute detection limits as
low as 10^{-10} grams. In consequence of this advantage a lot of work
in the PIXE field is done on biological specimen[12] and as an
illustration the next paragraph shows a few examples of such
applications. It is noteworthy that complementary to the PIXE
technique detection of elastically scattered protons may give
information about the very light elements and/or about the target
thickness, important for concentration assignment. (NBS = nuclear
backscattering). For special elements prompt emitted γ-radiation
during the proton bombardment is valuable, for instance for Na and
Al (PRA = prompt radiation analysis) using the $(p,p'\gamma)$ nuclear
reaction.[13]

IV. PIXE FOR TRACE ANALYSIS IN BLOODSERUM

PIXE is often applied for the determination of trace elements
in serum trying to correlate some kind of disease with deviations
in the concentration of a particular element.
Normally, only 10 µl of serum is used, placing this droplet on a
backing foil, irradiating the dried specimen with protons of a few
MeV energy and recording the X-ray spectrum with a Si(Li) or Ge-
detector (fig. 7) during several minutes.

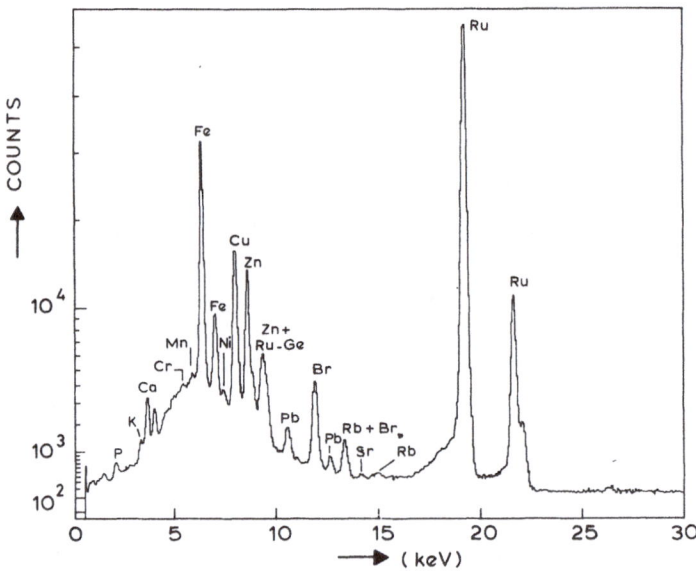

Fig. 7 X-ray spectrum of bloodserum
$(E_p = 2.5$ MeV$)$

Concentration assignment is performed either with the aid of an internal standard or using the bremsstrahlungsbackground for normalization.[14] In both cases one normalizes the number of protons hitting the sample and the sensitivity (G) of the technique is then expressed as:

$$G = \frac{\sigma \bar{N}_p \; N_0 \; \Delta \Omega \; \varepsilon \; C}{A} \qquad (9)$$

$\bar{N}_p$ = normalized number of protons ε = detector efficiency
N_0 = Avogadro's number C = absorption from X-rays
$\Delta\Omega$ = solid angle A = atomic weight of the
σ = X-ray production cross section   element involved

commonly expressed in counts/μg.μC.

Using theoretical cross sections and an appropriate efficiency curve of the detector concentration calculation can be done. Alternatively, one can use standard targets from various elements determining the sensitivity as a function of Z.[15] Correlations are found between the Se-concentrations in serum and various kinds of cancer[16], between Cu and the progress of pregnancy and the behaviour of trace elements during hemodialysis is studied[17]. Due to the fact that several trace element deviations appear more pronounced in tissue or hair a lot of work is performed nowadays on these kinds of samples. In order to measure trace element distributions in for instance these tissues or across and along the hair a lot of effort is put in the construction of proton microprobes, focussing and/or collimating the proton beam down to several μm[18],[19]. Especially in case of hair analysis the possibility is then created to distinghuish between deviations in the metabolism and influences from the environment.

V. CONCLUSIONS

Low energy accelerators are frequently used for (p,X) reactions both for fundamental and for applied purposes. Especially the latter field is rapidly growing for instance in medicin due to a growing awareness that deviations from normal trace element concentrations can lead to pathological effects. Experiments on processes in micro-biological systems indicate an important role of trace elements. The speed and the possibilities for automation make the techniques very suitable for environmental studies (aerosol work) and in technology. A further increase of the use of PIXE is to be expected if cheaper and easy to operate small accelerators will become available.

REFERENCES

1. B. Craseman, Atomic Innershell Processes, Academic Press (1975)
2. J.D. Garcia, Phys. Rev. A1 (1970) 280
3. J.D. Garcia, R.J. Fortner, T.M. Kavenagh, Rev. Mod. Phys. 45 (1973) 111
4. J.D. Garcia, Phys. Rev. A1 (1970) 1402
5. J.M. Hansteen, O.M. Johnsen, L. Koebach, J. Phys. B7 (1974) 271
6. J.M. Hansteen, O.M. Johnsen, L. Koebach, At. Data, Nucl. Data Tables 15 (1975) 305
7. N.J. Stern, H.O. Lutz, P.H. Mokler, P. Armbruster, Phys. Rev. A5 (1972) 2126
8. E. Laegsgaard, J.V. Andersen, L.C. Feldman, Phys. Rev. Lett. 29 (1972) 1206
9. F. Folkmann, C. Gaarde, T. Huus, K. Kemp Nucl. Instr. Meth. 116 (1974) 487
10. R.D. Vis, H. Verheul, J. Radioanal. Chem. 27 (1975) 447
11. R.C. Bearse, D.A. Close, J.J. Malanify, C.J. Umbarger, Anal. Chem. 46 (1974) 499
12. Proceedings of the Second International Conference on PIXE and its Analytical Applications (Lund, Sweden 1980)
13. R.D. Vis, K.J. Wiederspahn, H. Verheul, J. Radioanal. Chem. 45 (1978) 407

14. Y.J. Uemura, Y. Kumo, Y. Koyama, T. Yamazaki, P. Kienle,
    Nucl. Instr. Meth. 153 (1978) 573
15. R.D. Vis, thesis, Vrije Universiteit, Amsterdam (1977)
16. H.P.M. Kivits, thesis Technische Hogeschool, Eindhoven (1980)
17. R.D. Vis, P.L. Oe, A.J.J. Bos, H. Verheul,
    Radiochem. Radioanal. Letters 41 (1979) 245
18. J.A. Cookson, A.T.G. Ferguson, F.D. Pilling,
    J. Radioanal. Chem. 12 (1972) 39
19. R. Nobiling, Y. Civelekoglu, B. Povh, D. Schwalm, K. Traxel,
    Nucl. Instr. Meth. 130 (1975) 325

PARTICIPANTS

AARTS, E.H.L.   K.V.I., Zernikelaan 25, Groningen,
                          The Netherlands

ABRAHAMS, K.   E.C.N., Westerduinweg 3, Petten,
                          The Netherlands

ACKERMANN, K.   Strahlenzentrum JLU-Giessen,
                          Leihgesternerweg 217, Giessen,
                          W.Germany

ALLAART, K.   Nuclear Physics Department,
                          Free University, Amsterdam.
                          The Netherlands

ANTHONY, I.   Department of Physics, University
                          of Edinburgh, West Mains Road,
                          Edingburgh, U.K.

ARAVANTINOS, A.E.   N.R.C. Demokritos, Agia Paraskevi,
                          Athens, Greece

ARCISZEWSKI, H.F.R.   R. v.d. Graafflab., Pb. 80000,
                          Utrecht, The Netherlands

ASLANOGLOU, X.A.   N.R.C. Demokritos, Agia Paraskevi,
                          Athens, Greece

BAKKUM, E.L.   R. v.d. Graafflab., Princetonplein 5,
                          Utrecht, The Netherlands

BARRETO VIVAS, E.L.   Institut Physique Nucléaire Orsay,
                          B.P. 01, Orsay, France

BECKER, A.   R.J. v.d. Graafflab., Sorbonnelaan 4,
                          Utrecht, The Netherlands

BECKER, H.   G.S.I., 6100 Darmstadt 11,
                                    W. Germany

BHOWMIK, R.K.   K.V.I., Paddepoel, Zernikelaan 25,
                                    Groningen, The Netherlands

BIJKER, R.   K.V.I., Zernikelaan 25, Groningen,
                                    The Netherlands

BIJL, L.T. van der   Vrije Universiteit, De Boelelaan,
                                    Amsterdam, The Netherlands

BIRENBAUM, Y.   Physics Dept. N.R.C.N., P.O.B. 9001,
                                    Beer Sheva, Israel

BLAND, L.C.   Tandem Accel. Lab., Univ. of Pennsylvania,
                                    209 S. 33 St., Philadelphia,
                                    Penna. 19104, U.S.A.

BLASI, N., Miss   K.V.I., Univ. of Groningen,
                                    Zernikelaan 25, Groningen,
                                    The Netherlands

BLOK, H.   Nat. Lab. Vrije Universiteit, Pb. 7161,
                                    1007 MC Amsterdam, The Netherlands

BLOMMESTIJN, G.J.F.   I.K.O., Pb. 4395, 1009 AJ Amsterdam,
                                    The Netherlands

BLUNDEN, P.G.   Physics Dept., Queen's Univ.,
                                    Kingston, Ontario, Canada

BOS, A.J.J.   Nat. Lab. Vrije Universiteit,
                                    De Boelelaan, Amsterdam, The Netherlands

BROWN, J.D.   Dept. of Physics, Birmingham Univ.,
                                    Birmingham, U.K.

CHUNG, C.   Phys. Dept., BNL, Upton,
                                    N.Y. 11973, U.S.A.

COWERN, N.E.B.   A.E.R.E. Harwell, Didcot, Oxon, U.K.

CUNNINGHAM, M.A.   Yale University, New Haven, U.S.A.

CURRAN, A.R.   Kelvin Lab., Nel, East Kilbride, U.K.

DELFINI, M.G., Miss   E.C.N. Westerduinweg 3, 1755 ZG Petten,
                                    The Netherlands

DIEPERINK, A.E.L.   K.V.I., Zernikelaan 25, Groningen,
                             The Netherlands

DUVAL, P.D.   University of Arizona,
                             Tucson, Arizona, U.S.A.

ELLIOTT, J.P.   Univ. of Sussex, Physics Division,
                             Brighton BN1 9QH, U.K.

ENGELBERTINK, G.A.P.   Fysisch Lab., Princetonplein 5,
                             Postbus 80000, 3508 TA Utrecht,
                             The Netherlands

EIJNDE, J.P.H.W. van den   Technische Hogeschool, Pb. 513,
                             5600 MB Eindhoven, The Netherlands

FRATAMICO, G.   Instituto di Fisica, V. Irnerio 46,
                             Bologna, Italy

GELDER, P.C.M. de   Lab. voor Kernfysica, Proeftuinstraat 86,
                             B-9000 Gent, België

GLAUDEMANS, P.W.M.   Physics Dept., Univ. of Utrecht,
                             Sorbonnelaan 4, Utrecht,
                             The Netherlands

GÖKTÜRK, E.H., Miss   Chemistry Dept., M.E.T.U.,
                             Ankara, Turkey

GONGGRYP, S.   K.V.I., Zernikelaan 25, Groningen,
                             The Netherlands

HAMILL, J.J.   Nuclear Phys. Lab., Univ. of Colo.,
                             Boulder, Colo, 80309, U.S.A.

HANSEN, G.,   Institute of Physics, Univ. of Aarhus,
                             DK 8000 Aarhus, Denmark

HANSEN, P.G.   Aarhus University, Dept. of Physics,
                             Langelandsgade, DK 8000 Aarhus-C,
                             Denmark

HARO, R. de   I.K.P. - K.F.A. - Jülich,
                             D-5170 Jülich, W. Germany

HEES, A.G.M. van   Fys. Lab. Rijksuniversiteit Utrecht,
                             Princetonplein 5, 3508 TA Utrecht

HENRIQUEZ, A.   Fysisk Institutt, Pb. 1048,
                             Blindern, Norway

HEUSI, P.   c/o SIN, CH-5234 Villigen,
                                   Switzerland

HILDINGSSON, L.   Research Institute of Physics, Fack,
                                   S-10405 Stockholm, Sweden

HOOFT, G. 't   Rijksuniversiteit Utrecht,
                                   Princetonplein 5, 3508 TA Utrecht

HOLZMANN, R.   I.P.C., Chemin du Cyclotron, 2
                                   1348 Louvain-la-Neuve, Belgium

HOVE, M.A. van, Miss   Institut de Physique Corpusculaire,
                                   2, Chemin du Cyclotron,
                                   1348 Louvain-la-Neuve, Belgium

HOYLER, F.   Phys. Institut der Univ. Tübingen,
                                   D-7400 Tübingen, W.Germany

HÜRLIMANN, W   K.F.A. - I.K.P., P.O.B. 1913,
                                   D-5170 Jülich, W. Germany

HUT, G.   Lab. voor Alg. Natuurkunde,
                                   Westersingel 34, Groningen, The Netherl.

IACHELLO, F.   K.V.I., "Paddepoel", Zernikelaan 25,
                                   Groningen, The Netherlands

JONG, J. de   Westersingel 34, 9718 CM Groningen,
                                   The Netherlands

KAUP, U.   Institut für Kernphysik, Univ. Köln,
                                   Köln, W. Germany

KEIZER, P.H.M.   I.K.O., Pb. 4395, Amsterdam,
                                   The Netherlands

KERN, Th.   Fak. f. Physik, Hermann Herderstrasse 3,
                                   78 Freiburg, W.Germany

KHAZAIE, F., Miss   Physics Dept., Univ. of Birmingham,
                                   B15 2TT Birmingham, U.K.

KOHLER, M.M.   Physics Institute, University of
                                   Zürich, Schönberggasse 9,
                                   CH-8001 Zürich, Switzerland

KONIJN, J.   Institute of Nuclear Research, I.K.O.,
                                   Oosterringdijk 18, Amsterdam,
                                   The Netherlands

KÜHNER, G.   Institut für Kernphysik, THD,
                               Schlossgartenstrasse 9,
                               6100 Darmstadt, W. Germany

LEVENSON, S.M.   Argonne Nat. Lab., Argonne,
                               Illinois 60439, U.S.A.

LEWIS, P.M. miss   Phys. Dept., Birmingham Univ.,
                               P.O.B. 363, Birmingham, U.K.

LIFSHITZ, M.   Dept. of Physics, Technion, Haifa,
                               Israel

LIGTHART, H.J.   Westersingel 34, 9718 CM Groningen,
                               The Netherlands

LIMA, A.P. de   Phys. Dept., Univ. of Coimbra,
                               3000 Coimbra, Portugal

LÖH, H.   Physikalisches Institut, Univ. of
                               Erlangen-Nürnberg, Erwin Rommel-
                               strasse 1, D 8520 Erlangen,
                               W. Germany

LOZANO, M.   Dept. de Física Atómica y Nuclear,
                               Universidad de Sevilla, Spain

LÜCKE, K.   Freie Universität Berlin,
                               Gottschedstr. 20, 1 Berlin 65,
                               W. Germany

LUNARDI, S.   Instituto di Fisica, Via Marzolo 8,
                               35100 Padova, Italy

MAC GREGOR, I.J.D.   Kelvin Lab., N.E.L., East Kilbride,
                               U.K.

MAGNUSSON, S.M.   Niels Bohr Institute, Blegdamsvej 17,
                               Copenhagen, Denmark

MATULEWICZ, T.   University of Warsaw, Hoża 69, Warsaw,
                               Poland

MCGRORY, J.B.   Oak Ridge National Lab., P.O.B. X,
                               Tennessee 37830, U.S.A.

MITSUNARI, T.   Univ. of London, Reactor Centre,
                               Silwood Park, Ascot, Berks., U.K.

MOLNÁR, G.   Institute of Isotopes, H-1525 Budapest,
                               Hungary

MOOY, R.   R. v.d. Graaff Lab., Princetonplein 5,
                                  Utrecht, The Netherlands

MUKHOPADHYAY, D.   Fachbereich Physik der Univ. Marburg,
                                  Renthof 6, Marburg, W. Germany

NIFENECKER, H.   CEN Grenoble, B.P. 85X, F-38041,
                                  Grenoble-Cedex, France

NISHIOKA, H.   Dept. of Physics, Univ. of Surrey,
                                  Guildford, Surrey, U.K.

OLANDERS, P.   P.O. Box 725, S-220 07 Lund 7,
                                  Sweden

PALTEMAA, R.H.   Univ. of Helsinki, Accelator Lab.,
                                  Siltav. Penger 20M, SF-00170 Helsinki,
                                  Finland

PAPAIOANNOU, S.   Tandem Accelerator Lab.,
                                  NRC "Demokritos", Aghia Paraskevi,
                                  Attiki, Greece

POEL, C.J. van der   R.J. v.d. Graaflab., Princetonplein 4,
                                  De Uithof, Utrecht, The Netherlands

POLIKANOV, S.M.   EP division CERN, CH 1211 Geneve 23,
                                  Switzerland

PRINGLE, D.M.   Nuclear Physics Lab., Keble Road,
                                  Oxford, U.K.

RICHTER, A.   Inst. f. Kernphysik, T.H. Darmstadt,
                                  Schlossgartenstr. 9, W. Germany

ROBERTSON, J.L.   Kelvin Lab., N.E.L., East Kilbride,
                                  U.K.

ROLFS, C.   Institut für Kernphysik, Münster,
                                  W. Germany.

ROODEN, C.D. van   Nat. Lab. Vrije Univ., De Boelelaan
                                  1081, Amsterdam, The Netherlands

RUYL, J.F.A.G.   E.C.N., Westerduinweg 3, Petten,
                                  The Netherlands

RUYVEN, J.J. van   Nat. Lab. V.U., De Boelelaan 1081,
                                  1081 HV Amsterdam, The Netherlands

RYDSTRÖM, L.O.   A.F.I., S-104 05 Stockholm, Sweden

SAMBATARO, M.   K.V.I. "De Paddepoel", Zernikelaan 25,
                                  9247 AA Groningen, The Netherlands

SAMI, T., miss   PNT-CRN, B.P. 20 CR 67037 Strasbourg
                                  Cedex, France

SCHERPENZEEL, D.E.C.   R.J. v.d. Graaff Lab., P.O. Box 80000,
                                  Utrecht, The Netherlands

SCHIFFER, J.P.   Argonne Nat. Lab., Building 203,
                                  Argonne, Illinois 60439, U.S.A.

SCHWALM, D.   G.S.I., 6100 Darmstadt 11, W. Germany

SEGATO, G.F.   Instituto di Fisica, Via Marzolo, 8
                                  Padova, Italy

SELIG, A.M.   I.K.O., Oosterringdijk 18, Amsterdam,
                                  The Netherlands

SENE, M.R.   Dept. Physics, Univ. of Edinburgh, U.K.

SHEPPARD, H.M., miss   Univ. of Liverpool, P.O.B. 147,
                                  Liverpool L68 3BX, U.K.

SOBOLEWSKI, J.   Inst. of Nucl. Res., Dept. IA,
                                  05-400 SWIERK, Poland

STEFFEN, W.   Institut f. Kernphysik, THD, Schloss-
                                  gartenstr. 9, 6100 Darmstadt, W. Germany

STORM, M.H., miss   Dept. of Natural Philosophy, Glasgow,
                                  U.K.

TANG, X.   Max-Planck-Inst. f. Kernphys.,
                                  Postfach 103980, 69 Heidelberg,
                                  W. Germany

TIELENS, T.A.A.   ECN, Westerduinweg 3, Petten,
                                  The Netherlands

TIMM, W.   Inst. f. Theor. Physik I, Münster,
                                  Correnstr., W. Germany

VENEMA, W.   K.V.I., Zernikelaan 25, Groningen,
                                  The Netherlands

VIS, R.D.   Afd. Kernfysica, Vrije Univ., De
                                   Boelelaan 1081, Amsterdam, The Nether-
                                   lands

WARD, N.J., miss   Univ. of Liverpool, Oliver Lodge Lab.,
                                   P.O. Box 147, Liverpool, U.K.

WEISE, W.   Inst. f. Theor. Phys., Universität
                                   Regensburg, 8400 Regensburg,
                                   W. Germany

WENES, G.C.M.   I.N.W., Proeftuinstraat 86, 9000 Gent,
                                   Belgium

WIENKE, H.   Nat. Lab., Vrije Univ., De Boelelaan
                                   1081, Amsterdam, The Netherlands

WIERZBICKI, J.   Łodz, Narutowicza 68, Inst. Fizyki,
                                   Univ. of Łodz, 90 - 136 Łodz, Poland

WOODS, P.W.   Schuster Lab., Brunswick Str. The Uni-
                                   versity, Manchester, M13 9PL, U.K.

WOUDE, A. van der   KVI, "Paddepoel", Zernikelaan 25,
                                   Groningen, The Netherlands

YAZICI, F.   Phys. Dept. (83), Haceteppe Univ.
                                   Ankara, Turkey

ZALMSTRA, J.J.A.   Vrije Univ., De Boelelaan 1081,
                                   Amsterdam, The Netherlands

ZAMFIR, N.V.   Inst. of Physics and Nuclear Engi-
                                   neering, P.O.B. 5206, Bucharest,
                                   Romania

ZEMEL, A.   Dept. of Nuclear Phys., Weizmann Inst.,
                                   Rehovot, Israel

ZHAN, X.   Niels Bohr Institutet, København,
                                   Denmark

ZWARTS, D.   R.J. v.d. Graafflab., Princetonplein 5,
                                   Utrecht, The Netherlands

ŻWARTS, F.   K.V.I., Zernikelaan 25, Groningen,
                                   The Netherlands

SECRETARIAT

RAVENSWAAY, R.O. van	Bureau Congresses, Ministry of Education and Science, Nieuwe Uitleg 1, 2514 BP The Hague, The Netherlands
DRUNEN, J.J.M. van	Bureau Congresses, Ministry of Education and Science, Nieuwe Uitleg 1, 2514 BP The Hague, The Netherlands
KLERK, miss B.A.S. de	Bureau Congresses, Ministry of Education and Science, Nieuwe Uitleg 1, 2514 BP The Hague, The Netherlands
SALVERDA, A.G.	Bureau Congresses, Ministry of Education and Science, Nieuwe Uitleg 1, 2514 BP The Hague, The Netherlands

SCIENTIFIC ORGANIZING COMMITTEE

GLAUDEMANS, P.W.M., chairman

ENGELBERTINK, G.A.P., secretary

KONIJN, J., treasurer

ABRAHAMS, K.,   )
ALLAART, K.,   ) editorial committee
DIEPERINK, A.E.L.,)

INDEX